INFRARED AND MILLIMETER WAVES

VOLUME 3 SUBMILLIMETER TECHNIQUES

CONTRIBUTORS

T. G. BLANEY

A. F. GIBSON

A. HADNI

W. M. KELLY

M. F. KIMMITT

F. K. KNEUBÜHL

EIZO OTSUKA

CH. STURZENEGGER

MICHAEL VON ORTENBERG

G. T. WRIXON

INFRARED AND MILLIMETER WAVES

VOLUME 3 SUBMILLIMETER TECHNIQUES

Edited by **KENNETH J. BUTTON**

NATIONAL MAGNET LABORATORY
MASSACHUSETTS INSTITUTE OF TECHNOLOGY
CAMBRIDGE, MASSACHUSETTS

1980

ACADEMIC PRESS
A Subsidiary of Harcourt Brace Jovanovich, Publishers
New York   London   Toronto   Sydney   San Francisco

ACADEMIC PRESS, INC.
111 Fifth Avenue, New York, New York 10003

United Kingdom Edition published by
ACADEMIC PRESS, INC. (LONDON) LTD.
24/28 Oval Road, London NW1 7DX

Library of Congress Cataloging in Publication Data
Main entry under title:

Infrared and millimeter waves.

Includes bibliographies and indexes.
CONTENTS: v. 1. Sources of radiation.——
v. 2. Instrumentation.——v. 3. Submillimeter
techniques.
1. Infrared apparatus and appliances.
2. Millimeter wave devices. I. Button, Kenneth J.
TA1570.I52   621.36'2   79–6949
ISBN 0–12–147703–7 (v. 3)

PRINTED IN THE UNITED STATES OF AMERICA

80 81 82 83   9 8 7 6 5 4 3 2 1

CONTENTS

v

LIST OF CONTRIBUTORS

Numbers in parentheses indicate the pages on which the authors' contributions begin.

T. G. BLANEY (1), *Division of Electrical Science, National Physical Laboratory, Teddington TW11 0LW, England*

A. F. GIBSON (181), *Rutherford Laboratory, Chilton, Didcot, Oxfordshire, England*

A. HADNI (111), *Laboratoire Optique IR et du Solide, University of Nancy I, 54037 Nancy—Cedex, France*

W. M. KELLY (77), *Microelectronics Research Center, University College, Cork, Ireland*

M. F. KIMMITT (181), *University of Essex, Colchester, Essex, England*

F. K. KNEUBÜHL (219), *Physics Department, Eidgenössische Technische Hochschule, CH-8093 Zurich, Switzerland*

EIZO OTSUKA (347), *Department of Physics, College of General Education, Osaka University, Toyonaka, Osaka 560, Japan*

CH. STURZENEGGER* (219), *Physics Department, Eidgenössische Technische Hochschule, CH-8093 Zurich, Switzerland*

MICHAEL VON ORTENBERG (275), *Physical Institute of the University of Würzburg, 8700 Würzburg, Federal Republic of Germany*

G. T. WRIXON (77), *Microelectronics Research Center, University College, Cork, Ireland*

* Present address: Sprecher & Schuh AG, CH-5036 Oberentfelden, Switzerland.

PREFACE

The keynote of this book is T. G. Blaney's opening chapter, which reviews submillimeter and short millimeter wave detection techniques. Blaney's overview of detection techniques is backed up by three separate chapters on the most interesting detectors. The Schottky barrier diodes certainly have an interesting immediate future. We were very fortunate to get long-time leaders in that field, Kelly and Wrixon, to write the first, but not the last, analysis of this important subject. We shall watch them and their noted colleagues in order to obtain a "Schottky Revisited" chapter as soon as possible. Suggestions from our readers are solicited herewith.

Professor Armand Hadni has written the chapter on pyroelectricity; pyroelectric detectors are of most value whenever a good broadband, room temperature detector is needed. The photon drag detector (Chapter 4) is not as widely applied, but we publish here a full treatment of its principles and performance, by experts indeed, Maurice Kimmitt and Alan Gibson.

This treatise could never claim distinction without a chapter on "Electrically Excited Submillimeter Wave Lasers." These HCN and H_2O lasers not only started many of us in the practice of submillimeter spectroscopy but also represent today the least expensive and most practical way for new investigators to enter submillimeter research. Only very well-endowed laboratories can afford the alternative sources of radiation that were described in Volume 1. This chapter did not appear in Volume 1 (Sources of Radiation) because of schedules of Professor Fritz K. Kneubühl and Dr. Charles Sturzenegger.

Two outstanding examples of submillimeter spectroscopic techniques have been selected to demonstrate the practical importance of submillimeter wave developments up to the present time. These examples described by Professors Michael von Ortenberg and Eizo Otsuka are most important to those who wish to enter the specialty of submillimeter wave spectroscopy by using lasers. The two chapters together represent the most advanced submillimeter wave techniques. They are totally adequate to use as a guide for any sort of research of this kind. This subject was introduced by B. L. Bean and S. Perkowitz in Volume 2. The other important spectroscopic technique, Dispersive Fourier

Transform Spectroscopy, was described by J. R. Birch and T. J. Parker in Volume 2.

Finally, we call attention to the contents for the forthcoming book on Millimeter Systems. We hope that we can do justice to it by using the only formula known to us, namely, the selection of the best authors, followed by the application of appropriate patience.

CONTENTS OF OTHER VOLUMES

CHAPTER 1

Detection Techniques at Short Millimeter and Submillimeter Wavelengths: An Overview

T. G. Blaney

Division of Electrical Science
National Physical Laboratory
Teddington, England

ISBN 0-12-147703-7

I. Introduction

Short millimeter- and submillimeter-wavelength techniques are still largely the domain of the research scientist. However, the research applications cover a wide field of activities, and the variety of requirements made on measurement systems, including detectors and receivers, is equally wide. This variety of techniques, especially in detectors and receivers, has been perpetuated by the lack, relative to much of the rest of the electromagnetic spectrum, of standard lines of commercial equipment. This has resulted in many types of device being used, these often being designed for quite specific research applications.

This chapter attempts to give a broad picture of radiation detection problems and techniques in the short millimeter- and submillimeter-wavelength range. The wavelength range to be covered is taken approximately from 3 mm to 20 μm, i.e., 100 GHz to 15 THz in frequency. For brevity, this range will be called the SMSMR.

The wavelength limits are to some extent arbitrary, but do approximately enclose a region where (i) there are, as yet, no large-scale commercial applications, (ii) there is little standardization of components or calibration procedures, and (iii) the atmospheric attenuation is generally high. These three factors, which are of course very much interrelated, have considerably influenced the way in which SMSMR techniques have developed. However, particularly as regards detection, the techniques inevitably change as one moves through the more than two decades of frequency considered here. One area of commonality is that the performance of detectors and receivers is often considerably less than ideal throughout the region and extensive development is required in some techniques, such as heterodyne systems.

Excellent reviews of SMSMR detection techniques have been given, particularly by Putley and Martin (1967), Warner (1969), Kimmitt (1970), Putley (1973), Arams (1973) and Robinson (1973). This chapter follows to some extent the account given by the author in Blaney (1978a). However, this material puts considerably more emphasis on general problems and gives little precise detail on presently available practical devices. These more detailed discussions are left to other authors in this and the companion volumes of the treatise.

The present review begins with a discussion of the peculiarities of the SMSMR vis-à-vis detection, and goes on to discuss the general properties of detection devices and receivers that are of interest for the region. An overall view of practical devices and their performance is then given, including a discussion on the increasingly pressing problem of calibration techniques. In the Conclusion some indication is given of future trends in both requirements and actual devices.

II. The Nature of SMSMR Techniques and the Demands Made on Detectors and Receivers

A. INTRODUCTION

Before discussing how the radiation is to be sensed, it is necessary to consider the properties of the radiation itself and how this can affect the choice of the receiving system. The demands made on the receiver can be considered as arising from three main interrelated factors:

(i) The source of radiation to be observed.

(ii) The way in which the radiation is propagated from the source to the receiver.

(iii) The general features of the application for which the system is being used and the sort of information required.

These factors are discussed in the following paragraphs, particularly as they influence the SMSMR.

B. SOME GENERAL COMMENTS ON RADIATION SOURCES AND PROPAGATION TECHNIQUES

Table I lists the radiation properties that are most important when choosing the most appropriate sensor. Most of these would apply to any region of the spectrum, and many are sufficiently simple to require no further discussion. However, an important point worthy of further discussion is that of the spatial and angular distribution of power in the beam of radiation and the nature of étendue.

In an optical system in which the limiting apertures are large compared with wavelength and through which spatially incoherent radiation is propagating, the étendue or throughput of the system is readily defined. For example, consider a spatially incoherent source (such as a hot body) of area A_s that illuminates the input to the optical system (which ultimately terminates with a receiver) of input aperture A_0 (see Fig. 1). If the source subtends a solid angle of Ω_s at the position of the aperture A_0, then the input étendue is defined as the product $A_0 \Omega_s$. This is equal to the product $A_s \Omega_0$ (see Fig. 1).

The étendue is the geometrical factor that determines the amount of energy propagating through the system. If the focusing properties and apertures of the various optical components through the system are properly chosen, the étendue can be maintained through the system and, with the

TABLE I

PROPERTIES OF THE SOURCE RADIATION OF IMPORTANCE IN THE CHOICE OF A
DETECTOR/RECEIVER

Property	Definition/example/question
(1) Wavelength range	e.g., Continuum, narrow line, or monochromatic (tunable or fixed in frequency)
(2) Spatial and angular distribution of power in beam	(a) Free space, guided wave, or waveguide
	(b) Single mode, multimode, or spatially incoherent
	(c) Étendue? (see Section II)
(3) Power level	Is it large enough to be observed?
	Is absolute power of radiation required?
(4) Polarization	Linear, circular, elliptical or unpolarized
(5) Amplitude or frequency modulation	e.g., Pulsed, cw, chopped
(6) Radiation at unwanted wavelengths	Power outside range of interest, but which may affect the detector
(7) Background radiation	Is the source being viewed against a background of other radiation sources or through a medium that is itself emitting?
(8) Spatial variation of intensity across source	Is this variation of interest?
	Is some sort of imaging system required?

limit of geometrical optics applying, no energy is lost by purely geometrical means. However, the étendue of the receiver, $A_R \Omega_R$ (see Fig. 1), and the individual values of A_R and Ω_R (receiver area and angle of view) must be matched to the final stage of the propagation system to optimize the received energy.

This simple idea of étendue is valid only for a spatially incoherent source, such as a thermal emitter radiating directly into free space. [For a general discussion on coherence, the reader is referred to Troup and Turner (1974) and Kastler (1964).] The situation is somewhat different for a spatially coherent source. Two obvious examples of such sources in the SMSMR are

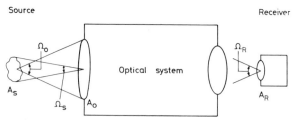

FIG. 1 Illustrating the concept of étendue for a system in which radiation from a multimode source propagates through an optical system to a receiver.

(i) radiation from a laser cavity oscillating in a single cavity mode, or (ii) a microwave-type oscillator (e.g., klystron, IMPATT diode) coupled into free space by a fundamental-mode hollow waveguide and a suitably shaped horn. The important peculiarity of such coherent radiation is the phase correlation between electromagnetic disturbances at points spatially separated on the wavefronts. Thus, radiation emitted from any point on the wavefront can not only interfere with itself (if notionally separated into two beams and recombined) but also with radiation from other points on the wavefront. The nature of the phase correlation across the wavefront, combined with Huygens' principle, determines the way in which the beam converges or diverges into space as it progresses.

Simple diffraction calculations show that if there exists a plane coherent wave of wavelength λ, emanating from an area A, then the wave will spread into a solid angle Ω given by $A\Omega \sim \lambda^2$. In particular, a system used to propagate this radiation can, in principle, do so effectively if the étendue is $\sim \lambda^2$. If such a system also restricts the radiation to a single polarization, then it is said to be propagating in a single radiation mode.

As discussed earlier, there is no such restriction on étendue for a thermal source emitting into free space. Even if a thermal source is provided with a bandpass filter of very narrow bandwidth, there is no spatial coherence across a surface perpendicular to the propagation direction to produce extra directionality if the effective aperture is increased. Thus étendue is not restricted to λ^2. In the example of Fig. 1, the maximum input étendue of the system can be increased by increasing A_0, although there is an obvious useful upper limit to the étendue of $2\pi A_s$. The radiation from such a source is thus said to be propagating in many modes, the number at a wavelength λ being given by the étendue divided by $\frac{1}{2}\lambda^2$. The "$\frac{1}{2}$" factor arises from the single polarization of each mode, while thermal radiation is unpolarized.

It is possible to select one spatial mode from thermal radiation. For example, if the radiation from a thermal source is launched into free space via a waveguide that will support only the fundamental propagating mode at the wavelength of interest, then the resulting radiation will have spatial coherence at that particular wavelength. However, the selection of only one mode does, of course, drastically reduce the power available from a thermal source (as discussed in Section C.2). If the waveguide coupling the source to free space can support higher modes of propagation, then the resultant radiation will be multimode and, in the limit, will again become totally spatially incoherent. Of course, it is in principle possible to destroy spatial coherence by introducing a diffuser (i.e., a device introducing random phase shifts across the beam).

The considerations above result in various conclusions of importance in choosing a detector or a receiver.

(i)   For a spatially incoherent source of radiation, the étendue of a receiver, defined as the product of the effective receiving area and the angle of view, should be at least as great as that of the optical system that delivers the radiation to the receiver, if the power detected is to be optimized. Of course, the receiving area and angle of view must themselves be suitably matched to the system by suitable optical or other components. However, the étendue should not be larger than is necessary as this may result in detection of unwanted radiation, which could decrease the sensitivity of the receiver (Section IV.C.2).

(ii)   If the receiver is sensitive to a single mode only (heterodyne and some video systems are restricted in this way), this severely reduces the power available from a multimode or spatially incoherent source.

(iii)   For observation of radiation from a single-mode source, or from a source that has been filtered by a single-mode selector, the étendue required of the receiver is $\sim \lambda^2$. In principle, a receiver that can receive only a single mode is most appropriate in these circumstances.

(iv)   It is obvious from the definition of the étendue for a single mode that if the physical area of the detector/receiver is $\gtrsim \lambda^2$, then not even a single mode can effectively be received. This is obvious in that λ^2 is the order of the diffraction-limited area of a focused spot of radiation. However, it is important in the SMSMR where, on occasion, devices of an area suitable for infrared work are used at wavelengths where physical size becomes an important limitation.

C.   RADIATION SOURCES IN THE SMSMR

1.   Introduction

Compared with many other parts of the spectrum, the SMSMR is poorly provided with sources both for fundamental physical reasons and because of the relatively undeveloped state of the technology in the region. This often puts particular pressure on obtaining the very best detector performance. The sources fall into two main groups: the continuum sources, usually thermal in character, and monochromatic and coherent sources.

2.   Continuum Sources

By the nature of emission from thermal sources at readily achievable temperatures, black or grey body emission is weak at the long wavelength end of the SMSMR. Data on blackbody emission in this region is given in Figs. 2 and 3.

At low temperatures thermal sources are poor emitters over the whole spectrum. However, as temperature rises, the overall power emitted increases

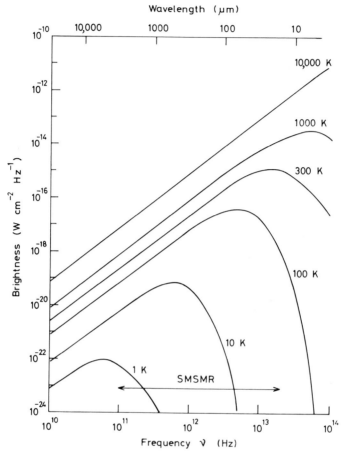

Fig. 2   Spectral brightness (power emitted per unit area per unit frequency interval into a hemisphere) for black bodies obeying the Planck radiation law at various temperatures.

as T^4, but the emission moves increasingly to shorter wavelengths. In the long-wavelength "tail" of the distribution (i.e., where $hv < kT$, where v is the radiation frequency), the emission increases only approximately linearly with temperature.

In the 100–1000 GHz range (3–0.3 mm), readily available laboratory sources, such as mercury-discharge lamps, have quite modest temperatures of up to 3000 K (Robinson, 1973). Some discharge-tube microwave noise sources that operate to beyond 100 GHz have effective temperatures of up to around 10,000 K, but may only be usable in single-mode or weakly multi-moded systems (Costley *et al.*, 1978). To obtain the much higher effective

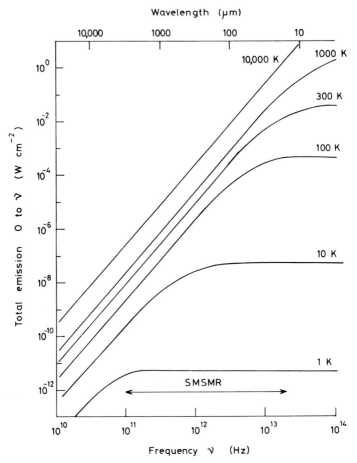

FIG. 3   Planck blackbody emission integrated from zero frequency to a frequency v (power per unit area into a hemisphere) plotted as a function of v for various source temperatures.

temperatures necessary for powerful emission, much less readily available sources are required and are usually pulsed (e.g., magnetically contained high-temperature plasmas emitting cyclotron radiation). Thus, from Fig. 3 we see that the total emission from a thermal source that is likely to be available in the laboratory is of the order of 1 μW or less at frequencies below a few hundred gigahertz. The power increases at higher frequencies, and at >1000 GHz the measurement of thermal radiation poses much less of a detection problem, at least for sources at $\lesssim 100$ K.

The data in Figs. 2 and 3 assume that a multimode propagation system and receiver is used to handle the power emitted. If only a single mode of the thermal emission is detected, then the power, $P(v)\,dv$, available in a frequency range dv, is given by

$$P(v)\,dv = \frac{hv\,dv}{\exp(hv/kT) - 1},\qquad (1)$$

$P(v)$ is plotted for various source temperatures in Fig. 4. For $hv \ll kT$ (the thermal limit), $P(v)\,dv$ is just $kT\,dv$, as with Johnson noise from a resistor. For $hv \gg kT$ (the quantum limit), $P(v)$ falls with increasing v. In most conditions the thermal limit will probably be of most interest in the SMSMR, but it is clear that at temperatures below 300 K, quantum effects will be significant at the shorter wavelengths of the range.

3. Monochromatic Sources

Up to around 300 GHz, various "microwave tube" and solid-state sources are available, or at least have been demonstrated, while cw backward-wave oscillators have been used up to around 1000 GHz [see, e.g., Kantorowicz

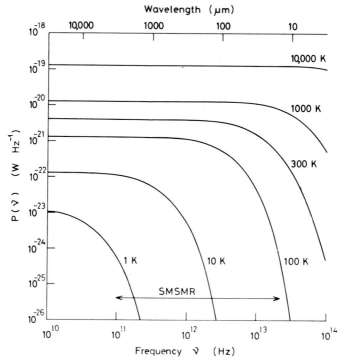

FIG. 4 Spectral power density, $P(v)$ (power per unit frequency interval), emitted into a single mode of the radiation field from a Planck blackbody for various source temperatures. See Eq. (1).

et al. (1979)]. However, power levels decrease rather rapidly as the frequency increases. Typically, cw power levels of several hundred milliwatts are readily (if quite expensively) available at 100 GHz, dropping to a few tens at 300–400 GHz, and dropping rapidly beyond that point. However, it is worth noting that gyrotron sources of cw power 1.5 kW have been reported at about 300 GHz (Andronov *et al.*, 1978). Harmonic generation from lower-frequency microwave sources is often used, but power levels are usually below a milliwatt, and often below a microwatt, depending on frequency. Above 500 GHz, tunable sources are presently available only at considerable expense. Above 1000 GHz (0.3 mm), and as far as the highest frequencies considered in this review ($\sim$15 THz), there are, as yet, virtually no practical tunable sources.

Well over a thousand cw and pulsed discrete-frequency gas laser sources are available in the SMSMR [see, e.g., Hodges (1978) and Knight (1979)]. Of the cw sources, most are weak and powers of more than 10 mW are probably only available at a few tens of wavelengths in readily attainable systems. Pulsed powers of up to a few tens of kilowatts peak have been produced at some wavelengths, usually with pulse lengths of $\gtrsim$ 100 ns, with some amplified pulses rising to over 1 MW (Hodges, 1978).

Thus the overall position as regards monochromatic sources is still rather poor and is likely to remain so for some time; progress is likely to be most rapid at the longest wavelengths (Martin and Mizuno, 1976). This situation has significantly affected the history of the development of detectors and receivers for the SMSMR. First, even with monochromatic sources, power levels are still often sufficiently low so that good detector sensitivity is required. Second, the shortage of tunable sources has resulted in forms of spectroscopy, such as Fourier transform spectroscopy, being developed that do not require such sources and which demand a different performance from receivers. Third, insufficient power and/or tunability in sources suitable for local oscillators has severely hindered the development of heterodyne systems. Fourth, it is worth remarking that while the sources mentioned here are nominally monochromatic, some often produce radiation in more than one spatial mode, sometimes complicating both propagation and reception.

D. PROPAGATION TECHNIQUES IN THE SMSMR

At the longer wavelengths of the region, wavelengths are often comparable to the size of the components, and guided-wave propagation techniques are used. At much shorter wavelengths, the "directed-wave" methods of classical optics are more common. However, as discussed by Martin and Lesurf (1978), an approach which combines elements of both these methodologies may be appropriate in many cases.

Up to about 300 GHz (1 mm), hollow waveguides, usually of metal, are commonly used, particularly in conjunction with microwave-type monochromatic sources. Single-mode propagation is usually preferred, but to reduce attenuation, oversized guides are often used with the possibility of overmoding, particularly if the beam must negotiate bends. Above 300 GHz, relatively little work has been done in low-order mode waveguides [but see, e.g., Crenn (1979) and Harris *et al.* (1978)]. A considerably oversized waveguide, usually in the form of a cylindrical metal tube, is often used for transmission of laser and incoherent radiation. Unless care is taken in the way in which this type of waveguide is excited by a single-mode source, very considerable overmoding occurs, and it is then often called a *light pipe*. In these conditions there is a generally unpredictable structure to the variation of intensity across any cross section of the tube and to the divergence of the partially coherent beam when it emerges from the guide. Once propagated in this way, such radiation cannot generally be received effectively by a single-mode receiver. Once launched into free space again, such radiation cannot always be efficiently collected for reception by a multimode receiver.

Between 100 and 300 GHz, free-space optical techniques are common, and are used at higher frequencies in the vast majority of cases. Single-mode laser beams can be propagated in this way; however, with spatially incoherent sources of dimensions considerably greater than a wavelength, geometrical optical principles apply. To most effectively propagate and receive power from the latter type of source, the system is usually designed to image the source on the sensitive area of the receiver, care being taken to see that étendue is not unnecessarily restricted at any point.

E. APPLICATIONS OF THE SMSMR

Most of the work carried out so far in the SMSMR has been in the laboratory, and most of the field work (say in astronomy or atmospheric studies) has been carried out by researchers familiar with the design of the apparatus. The lack of substantial commercial applications has meant that ruggedness, convenience, and long-term reliability have not usually been foremost in receiver design. The low power levels from many of the sources of interest has resulted in receiver sensitivity often being given priority.

Historically the main applications have been in spectroscopy. The best developed broadband spectroscopic techniques in the SMSMR are those based on the Fourier transform technique (Chantry, 1971). With such instruments an extended continuum source is usually employed, and it is desirable in most circumstances (a major exception being that when exceptionally high spectral resolution is required) to have sources, optics, and receiver with as high an étendue as possible. Such systems can also be used over large instantaneous spectral bandwidths. With only a few exceptions (Costley, 1979), they are used with nonpulsed sources, and unless the instrument is quickly scanned, high detector speed is not usually a prerequisite. Thus the demand here has been for receivers of fairly high étendue, good sensitivity, large instantaneous spectral range, but not necessarily high speed. Multimode video detectors have been commonly employed with single-mode sources such as lasers. Single-mode video detectors (such as rectifiers, see Section V.C) are usually employed only for their often high response speed, which is particularly useful for pulsed sources.

Unlike the microwave and longer-wavelength millimeter-wave regions where tunable active systems for such applications as communications or radar demand the frequency discrimination and sensitivity of heterodyne receivers, there has until recently been little demand for heterodyne systems in the SMSMR. Of course, demand is dependent to some extent on the expectation of what can reasonably be supplied, and there is no doubt that there have been, and still remain, technological problems in devising heterodyne systems of sufficiently useful performance. The requirements for astronomy are now motivating the development of increasingly sensitive and convenient heterodyne receivers up to 300 GHz and beyond. Another pressing requirement for heterodyne systems exists at frequencies up to about 1000 GHz for plasma diagnostics using laser scattering (Evans, 1976). The further development of tunable monochromatic sources is sure to motivate the development of and demand for heterodyne systems, particularly at the longer SMSMR wavelengths, but also throughout the whole region. It is worth noting that heterodyne techniques are increasingly being used at wavelengths of 10 μm and shorter, so that these techniques will not spread into the SMSMR exclusively from longer wavelengths.

Another requirement that has not yet been satisfactorily met is that for simple and effective imaging devices. These are particularly useful for setting up optical systems, assessing the intensity distribution or mode structure in laser or similar beams, or for displaying fringe patterns in interferometer systems.

While there is likely to be increasing commercial activity related to both civil and military applications at the longer wavelengths of the range, the demands for receivers at present are usually research oriented and often highly specialized. It is still common for individual experimenters to develop their own devices for particularly demanding applications. This, and the relatively low commercial activity, has not encouraged the development of proper standards and recognized calibration techniques for many aspects of measurement in the SMSMR, in particular the measurement of receiver performance. Reliable quantitative comparisons between the performances of different types of receivers built in different laboratories is thus sometimes difficult. The reader is warned that in assessing performance data given in the literature, it is important to take into account how the measurements were performed.

III. Radiation Detection Devices: General Properties

A radiation-sensitive device is a transducer that converts radiation into a form more readily accessible to and measurable by the human senses. For the vast majority of cases currently of interest in the SMSMR, we require from the detection system an output that ultimately can be observed on an electrical meter or in an analogous way. We are thus primarily interested in devices that produce an electrical output, although not all the devices mentioned here will necessarily be of this type.

Up to now we have used the terms detector, detection device, and receiver almost synonymously, but it is now necessary to be more rigorous in the use of these terms. We shall understand a detector or detection device to be the primary transducer in the receiver and that which responds in some way to the incident radiation. A radiation receiver is a system of which a detector forms a part, and which gives an output related to the character of the incident radiation. This differentiation is necessary because a single detector can be used in different types of receiver (e.g., video and heterodyne), each with quite different performance characteristics.

A wide range of physical effects can provide radiation detection devices with an electrical output, but most fall into three main groups:

(i) Rectifiers: In these, currents are induced in the device at the radiation frequency, and by a nonlinear voltage–current relationship in the device, a dc component, related to the amplitude, or more commonly to the power, of the ac input is obtained.

(ii) Thermal devices: In this type the radiation is absorbed to give heat, and a resulting rise in temperature is then sensed by a thermometer, usually with an electrical response.

(iii) Photon (or quantum) devices: In these, individual quanta of the input radiation release charge carriers, usually by excitation across a forbidden energy band between bound and "free" current-carrying states (e.g., as in photoconductors).

Table II lists the main parameters by which performance is specified for a detection device. In practice, many of the performance characteristics of a video receiver (see Section IV.C) are virtually identical to those of the detection devices used. The properties of practical devices, usually when used in video receivers, are discussed in Sections V–VII.

The parameters in Table II are most appropriate to devices providing an electrical output, but some would obviously not apply, for example, to imaging devices providing a direct visual display (see Section VII). Two provisos are necessary when applying this particular list to actual devices. First, even for some of the most commonly used detection devices in the SMSMR,

TABLE II

SPECIFYING PARAMETERS FOR DETECTION DEVICES

Property	Definition/comment
(1) Spectral range	The range of wavelengths over which a useful response can be obtained.
(2) Spectral response	Responsivity as a function of wavelength. The "instantaneous spectral response" may often be varied (within the spectral range) for example, by changing the input coupling system.
(3) Effective sensitive area	Usually related to physical area or dimensions of device. In rectifiers is related to antenna characteristics of radiation pickup structure and is generally polarization dependent.
(4) Angle of view/sensitivity versus angle of incidence	Determined by simple geometrical considerations, or other characteristics of the device.
(5) Etendue	Effective area × solid angle of view (useful étendue may be different when incorporated into a receiver).
(6) Input-coupling efficiency	Fraction of incident radiation effectively absorbed by the device. In photon-sensitive devices, quantum efficiency is fraction of quanta usefully absorbed.
(7) Responsivity	Voltage output per watt of incident power. May be referred to a specific electrical-circuit arrangement.
(8) Electrical output impedance	Important in matching to postdetector amplifiers and choice of biasing circuits.
(9) Response time	Expresses speed at which device responds to change in incident power. May depend on output impedance and circuitry.
(10) Noise characteristics	Level and spectrum of device noise, and dependence on other parameters. May limit the NEP of the receiver.
(11) Polarization sensitivity	All radiation properties may be sensitive, particularly (3), (5), (6), and (7).
(12) Miscellaneous	Dynamic range, linearity. Operating temperature, refrigeration power required. Bias (dc or ac) and bias stability required. Sensitivity to environmental factors (e.g., temperature, vibration, magnetic and electric fields). Reproducibility of performance. Repeatability (e.g., if matched devices are required). Damage thresholds (radiation or electrical). Age-dependent effects.

information is not readily available on some of the listed properties (particularly those under "miscellaneous"). Even for the most obviously important parameters, such as étendue and responsivity, measured values are sometimes insufficiently precise for good system design. Second, different examples of the same detection device may have significantly different properties and the range of values of a given parameter may be quite large.

IV. Receiver Systems: General Properties

A. INTRODUCTION

Four types of receiver will be mentioned in this review: video, heterodyne down-converting, imaging, and heterodyne up-converting. Most attention will be given to the first two, the others having been relatively little investigated in the SMSMR. Much more detailed discussions of receiver principles can be found in accounts by Arams (1973), Jones (1953a), Ross (1966), and Tiuri (1964, 1966). In principle, with all types of receivers the signal radiation can be amplified before being applied to the detector system. Section B discusses this possibility and subsequent sections discuss the general properties of the receiver types mentioned.

B. SIGNAL AMPLIFICATION

To be useful in a receiver, an amplifier must be of sufficiently low noise so that the signal-to-noise ratio in the whole system is improved by its use. Useful low-noise amplifiers will almost certainly be single-mode devices.

To the author's knowledge, no low-noise amplifiers have yet been developed in the SMSMR. However, amplifiers would be useful in the range of, say, 100–1000 GHz, particularly as a first stage in a heterodyne receiver. However, the noise temperature of such an amplifier would have to be of the order of 1000 K or considerably less to be competitive with the best existing mixers (see Table VIII). Solid-state masers have not yet been developed for use above 100 GHz, although the cyclotron resonance maser (already developed in electron-beam structures such as the gyrotron) may be a possibility in semiconductors (Andronov et al., 1978). Many of the processes which may be or are used for generation of SMSMR radiation (Martin and Mizuno, 1976) could be usable for amplification. Some preliminary work has been carried out on "optically pumped" lasers as low-noise amplifiers (Galantowicz, 1977; Merat et al., 1976) but no quantitative results on performance appear to have been published.

Parametric amplifiers based on semiconductor varactors are not presently available above 100 GHz. The nonlinear reactance of Josephson junctions may be used in several ways as the basis of parametric amplifiers (Richards,

1977a; Wahlsten *et al.*, 1978), and although so far tested only at the longer millimeter wavelengths (Chiao, 1979), there is a hope of low-noise performance at low signal levels to several hundred gigahertz.

C. VIDEO RECEIVERS

1. *General*

Sometimes called "direct" or "base-band" receivers, video receivers are presently the most used type in the SMSMR. A schematic diagram of such a receiver is shown in Fig. 5a. Apart from the possibility of an amplifier, which we presently neglect, the "fore components" in front of the detection device prepare the radiation so that the detector can use it most effectively.

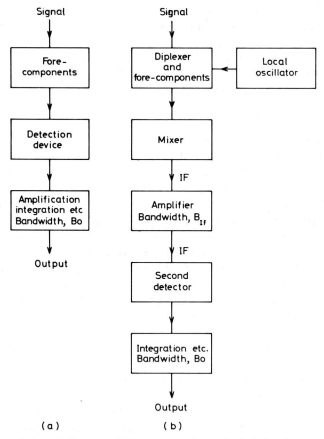

FIG. 5   Schematic diagrams of (a) a video receiver and (b) a heterodyne receiver.

Typically, the fore components would be in three parts:

(i) A spectral filter to reject radiation of unwanted wavelengths. If necessary, the filter may be cooled to reduce thermal emission toward the detector.

(ii) A spatial filter, typically an aperture or series of apertures, to restrict the field of view to that required. This also may be cooled.

(iii) A radiation coupler to focus or locate the radiation energy for most effective absorption by the detector. Examples of components used are lenses, mirrors, waveguides or resonant cavities fed by horns, or integrating cavities. Some parts of the coupling structure are often considered to be intrinsic to the detection device (e.g., the antenna wire of a rectifying device, see Section V.C.2).

The examples just given are most relevant to radiation propagating in free space, but would have their analogues in guided-wave systems. The electrical output from the detector is amplified, and often time-integrated in order to improve the signal-to-noise ratio in the final output. Steady or quasi-steady radiation signals are usually modulated in amplitude, often by chopping.

Depending on where the modulation is carried out, chopping or similar modulation may avoid problems due to drift in the background level of radiation (Dicke, 1946). It also allows the convenience and often the signal-to-noise advantage of an ac signal, which can be amplified by a phase-sensitive system (Smith, 1951). For some types of detector, such as Golay cells or pyroelectrics, intensity modulation is essential.

In most cases it is desirable to have the electrical output voltage proportional to the intensity (i.e., power) of the incident radiation. Most of the detection devices used are assumed to have an intrinsically linear relationship between output voltage and input power at sufficiently low power levels.

Table III lists the main specifying parameters for video receivers and compares them with heterodyne receivers. The performance is largely determined by that of the detection device, modified by the fore components. Most video receivers used in the SMSMR are multimode receivers. An important exception to this is the rectifying type of detector, which can only except a single mode via its coupling system (Section V.C).

The smallest change that can be observed in the radiation signal depends on the fluctuations in the output voltage. For most of the noise sources of interest (see below), the rms noise voltage observed is proportional to $(B_0)^{1/2}$, where B_0 is the final output bandwidth of the system. The change in radiation power to give a change in the output voltage equal to the rms fluctuation level is usually called the noise equivalent power (NEP) of the system, and this is also proportional to $(B_0)^{1/2}$. Noise equivalent power is usually given for $B_0 = 1$ Hz and is expressed in $W\,Hz^{-1/2}$. The video NEP, NEP_v, is usually a

TABLE III

SPECIFYING PARAMETERS AND CHARACTERISTICS OF RECEIVERS

Parameter	Direct (video) receiver	Heterodyne receiver
Spectral range		Limited by detector/mixer range and LO availability
Spectral response	Determined by properties of the detection device, as modified by fore optics	Instantaneous spectral response determined by LO frequency and IF bandwidth (see text)
Étendue, E		$\gtrsim \lambda^2$ (see text)
Polarization sensitivity		Single polarization only
Response time	Limited by detector and/or electronics	Ultimately limited by mixer and/or electronics, $\gtrsim (B_{IF})^{-1}$
Expression of NEP	NEP_V: usually given for $B_o = 1$ Hz (units: W Hz$^{-1/2}$)	Heterodyne or predetection NEP (NEP_H): usually given for $B_{IF} = 1$ Hz (units: W Hz^{-1}) Postdetection NEP (NEP_o) after second detector: usually given for $B_o = 1$ Hz (units: W Hz$^{-1/2}$) $NEP_o \approx NEP_H \cdot B_{IF}^{1/2}$ [a,f]
Specific detectivity, D^*	$\dfrac{Area^{1/2}}{NEP_V}$ [b]	Not usually defined
Receiver noise temperature, T_R [c]	$T_R \approx \dfrac{NEP_V}{k(B_{in})^{1/2}}$	$(T_R)_{DSB} \approx \dfrac{NEP_H}{2k}$ $(T_R)_{SSB} \approx \dfrac{NEP_H}{k}$
ΔT_S [d]	$\Delta T_S \approx \dfrac{NEP_V}{2kB_{in}} \dfrac{\lambda^2}{Étendue}$ [e]	$\Delta T_S \approx \dfrac{NEP_H}{2k(B_{IF})^{1/2}}$ [f,g]

The bracket on the left spans Spectral range, Spectral response, Étendue, Polarization sensitivity.

[a] Assumes signal-to-noise ratio at the IF is less than unity [see Eq. (20) and Smith (1951)].

[b] Strictly speaking, D^* is valid only for detectors for which the NEP_V varies as the square root of area, as is the case for ideal devices limited by background fluctuations [e.g., Eq. (4)]. For real devices in the SMSMR this is often not true and the use of D^* will be largely avoided in this article.

[c] T_R is defined only for single-mode receivers, where $hv_s \ll kT_S$ and $hv_s \ll kT_R$. It is not usually defined for video receivers, but the expression given (which assumes single-mode behavior) may be useful in comparisons with other receivers. B_{in} is the input spectral bandwidth. For the heterodyne case the formulas are for double- and single-sideband receivers.

[d] When used as a radiometer, ΔT_S is the change in source temperature to give a change in output equal to the rms noise fluctuations with $B_o = 1$ Hz (units: K Hz$^{-1/2}$).

[e] Only valid for $hv_s \ll kT_S$. B_{in} is the input spectral bandwidth.

[f] These formulas are correct to within a factor of order 2. Exact relationships depend on details of the receiver construction (Smith, 1951; Tiuri, 1966).

[g] Given for a double-sideband receiver.

function of many variables, and it is desirable to quote at least the wavelength, the chopping frequency, and the value of B_0 at which it was measured.

The noise fluctuations at the output of the receiver can be considered as arising from three sources:

(i) Noise in the detection device and the postdetector electronics. In most cases this noise is independent of the radiation falling on the detector and could arise from, for example, Johnson noise and shot noise in the detection device and electronics, electrical noise at contacts to the detector element, fluctuations in the temperature of the detector (e.g., in a bolometer), microphonics (e.g., in pyroelectrics), or $1/f$ and other noise in the postdetector amplifier. The level and relative importance of such sources of noise obviously depend on the type of detector being used. To a great extent, sources of noise can be reduced by good design, although there are limits below which it is impossible to go (e.g., Johnson and shot noise).

(ii) Noise arising from fluctuations in the radiation falling on the receiver from sources other than the source of interest (i.e., background radiation). In the fundamental limit this noise arises from fluctuations intrinsic to the nature of thermal radiation and will have contributions from all thermally emitting bodies within the field of view of the detection device, including windows, filters, aperture stops, etc. The detector will also be subject to fluctuations from the radiation which it, itself, emits.

(iii) Fluctuations in the radiation from the signal source, which might be thermal (i.e., incoherent) or coherent in nature. Once again, these fluctuations cannot be reduced below certain fundamental limits.

The noise listed in (i) is very much a function of the detector device and associated electronics, and more comment will be made in the discussions on practical devices (Sections V–VIII). The noise sources in (ii) and (iii) above can be discussed much more generally, as in the immediately following sections.

2. *Limitations to Receiver Sensitivity Arising from Fluctuations in Background Thermal Radiation*

Thermal radiation can be considered as a superposition of an infinite number of sources, the properties of which are treated in a statistical way. Thus, for example, the power in a stream of thermal radiation is subject to fluctuations, the relative size of which are determined by the length of time over which the observations are averaged (Lewis, 1947). These fluctuations will be observed on a receiver, as also will the fluctuations in the thermal radiation emitted by the receiver itself, and will be one of the fundamental limits

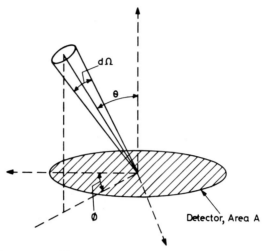

FIG. 6 Geometry used in the discussion of the effects of background radiation.

to the smallest signal that can be observed with the receiver. These background limits have been discussed in some detail by McLean and Putley (1965) and Putley (1964), but we shall review the main points.

First consider the detector, of sensitive area A, in a video receiver (see Fig. 5a), and for the moment assume the detector responds to energy rather than to the number of photons (i.e., photon detectors are presently excluded). The mean square fluctuations in power falling on the detector can be found by considering contributions from thermal radiation incident from an element of solid angle $d\Omega$ about a direction described by the angular coordinates θ and ϕ (Fig. 6). The angular response of the detector to energy is assumed proportional to a factor $f(\theta, \phi)$ such that $f(\theta, \phi)$ would be unity for a perfectly black detector at normal incidence. In general, and certainly for single-mode receivers, $f(\theta, \phi)$ is also a function of radiation frequency, but for many thermal detectors used in the video mode, $f(\theta, \phi) \sim \cos \theta$ throughout the wavelength range in which they are good absorbers.

If thermal radiation, characterized by temperature T, is incident from the direction θ, ϕ, then the mean square fluctuation in power $\langle \Delta P(\theta, \phi, v)^2 \rangle_{\text{av}}$ falling on the detector in $d\Omega$ and in a frequency range dv is given by (Lewis, 1947)

$$\langle \Delta P(\theta, \phi, v)^2 \rangle_{\text{av}} = \frac{4h^2 v^4}{c^2} \frac{e^x}{(e^x - 1)^2} AB_o f(\theta, \phi) \, dv \, d\Omega, \qquad (2)$$

where B_o is the bandwidth in which the fluctuations are observed (i.e., the output bandwidth of the receiver) and $x = hv/kT$. In the general case, T will be a function of θ and ϕ and possibly also of frequency if nonblack thermal sources are considered. However, we shall consider only rather more simple cases. The NEP of the detector, $(NEP_B)_{ext}$, when limited by this external background radiation, will be given by

$$(NEP_B)_{ext} = \left[\int_\Omega \int_{v_1}^{v_2} \langle \Delta P(\theta, \phi, v)^2 \rangle_{av} \right]^{1/2}, \tag{3}$$

where the integrals are taken over the radiation frequency limits of the detector's response and the total solid angle over which the detector views the source of the background radiation. For the simple case of a detector viewing background radiation in a cone of half-angle α about its normal

$$(NEP_B)_{ext} = 2\sqrt{\pi} \sin \alpha \, \frac{(kT)^{5/2}}{ch^{3/2}} A^{1/2} B_0^{1/2} [J_4(x_1) - J_4(x_2)]^{1/2}, \tag{4}$$

where x_1 and x_2 correspond to v_1 and v_2 and the integrals $J_n(x)$ are given by

$$J_n(x) = \int_0^x \frac{x^n e^x}{(e^x - 1)^2} \, dx, \tag{5}$$

and are tabulated by Rogers and Powell (1958).

For blackbody radiation received at all frequencies and over 2π radians, Eq. (4) becomes [see Putley (1964)]

$$(NEP_B)_{ext} = 4\left[\frac{(\pi kT)^5}{15c^2 h^3} AB_0 \right]^{1/2} \tag{6}$$

or

$$2.50 \times 10^{-15} T^{5/2} A^{1/2} B_0^{1/2} \quad W,$$

where A is in squared meters and B_0 is in hertz.

The background fluctuations are usually taken to include also those arising from emission from the detector. This emission NEP contribution, which we can call the "internal" background limited NEP, $(NEP_B)_{int}$, contributes to the total background NEP $(NEP_B)_{tot}$ such that

$$(NEP_B)^2 = (NEP_B)_{int}^2 + (NEP_B)_{ext}^2. \tag{7}$$

$(NEP_B)_{int}$ is given by similar expressions to those in Eqs. (3), (4), and (6), with the temperature replaced by the effective temperature of the device itself.

Table IV gives numerical examples [some drawn from Putley (1964)] of background-limited NEPs. For multimode video detectors they lie in range 10^{-10}–10^{-16} W for a 1 Hz output bandwidth for most conditions of interest in the SMSMR.

TABLE IV

Some Examples of Video NEP Limited by Fluctuations in Background
Thermal Radiation

Example[a]	$(\text{NEP})_B - \text{W Hz}^{-1/2}$[b]
1. Detector at 1.5 K, background at 1.5 K.	9.8×10^{-17}
2. Detector at 1.5 K, background at 300 K.	3.9×10^{-11}
3. Detector at 300 K, background at 300 K.	5.5×10^{-11}
4. Detector at 1.5 K, background at 300 K.	3.6×10^{-13}
Only sensitive from $\lambda = 100$ μm to $\lambda = \infty$. Field of view 0.2 sterads about normal.	
5. As 4, but photoconductive detector, detecting signal at $\lambda = 100$ μm.	6.4×10^{-13}
6. As 5, but signal at $\lambda = 1$ mm.	6.4×10^{-14}
7. As 4, but only sensitive to 10% bandpass at $\lambda = 500$ μm.	1.8×10^{-14}
8. Single-mode detector at 1.5 K.	3.2×10^{-15}
Background at 300 K. Sensitive from $\lambda = 1$ mm to $\lambda = \infty$ (0 to 300 GHz)	

[a] Unless otherwise stated, the data are calculated for an ideal energy-detecting device of area A of 100 mm^2, obeying Lambert's law, having an angle of view of 2π steradians, and absorbing and emitting as a blackbody at all wavelengths. Several of the examples are drawn from Putley (1964).

[b] This is the background-limited NEP for an output bandwidth B_0 of 1 Hz, and includes fluctuations in both energy received and emitted by the detector [Eq. (7)].

The expressions given above are for energy detectors. For photon detectors the fluctuations in the rate of arrival of photons is of primary importance. Thus for constant external background conditions (in particular fixed v_1 and v_2), the value for $(\text{NEP}_B)_{\text{ext}}$ [and also $(\text{NEP}_B)_{\text{int}}$] will depend on the frequency v' of the radiation signal of interest such that, for example,

$$(\text{NEP}_B)_{\text{ext}} = hv' \left[\int_\Omega \int_{v_1}^{v_2} \langle \Delta n(\theta, \phi, v)^2 \rangle_{\text{av}} \right]^{1/2}, \qquad (8)$$

where

$$\langle \Delta n(\theta, \phi, v)^2 \rangle_{\text{av}} = \frac{4v^2}{c^2} \frac{e^x}{(e^x - 1)^2} AB_0 f(\theta, \phi) \, dv \, d\Omega. \qquad (9)$$

Thus, for photon detectors, the equation analogous to Eq. (4) is

$$(\text{NEP}_B)_{\text{ext}} = 2\sqrt{\pi} \sin \alpha \frac{(kT)^{3/2}}{ch^{1/2}} v' A^{1/2} B_0^{1/2} [J_2(x_1) - J_2(x_2)]^{1/2}, \qquad (10)$$

where the J_ns are defined as before. Examples are given in Table IV. For small fractional bandwidths, $(v_2 - v_1) \ll v'$, the background-limited NEPs for both energy and photon detectors converge to the same value.

The above arguments are for multimode detectors and receivers receiving radiation of all polarizations. For a single-mode receiver (e.g., a rectifier used in a video system) there are restrictions in étendue and polarization. In particular

$$A \int_{\Omega} f(\theta, \phi, v) \, d\Omega \simeq \lambda^2 = v^2/c^2 \tag{11}$$

and only one polarization is received. Thus, for example, the $(NEP_B)_{ext}$ for such a detector, if energy detecting rather than photon detecting, is given by

$$(NEP_B)_{ext} = \left\{ \frac{2k^3 T^3}{h} B_o [J_2(x_1) - J_2(x_2)] \right\}^{1/2}. \tag{12}$$

A similar expression gives $(NEP_B)_{int}$, and by use of Eqs. (8)–(10) the expressions for single-mode photon detectors can be calculated. [Single-mode video photon detectors have not been much investigated in the SMSMR, although one possible form of them has recently been discussed by Schwarz and Ulrich (1977).]

Single-mode detectors are often used with rather small input bandwidths, and one case of interest is when $hv \ll kT$ and $\Delta v \ll v$. In that case

$$(NEP_B)_{ext} = kT(2B_o \, \Delta v)^{1/2}. \tag{13}$$

If $hv, \gg kT$ and $\Delta v \ll v$

$$(NEP_B)_{ext} = hv \exp\left(-\frac{hv}{2kT} \right)(2B_o \, \Delta v)^{1/2}. \tag{14}$$

Although the heterodyne receiver is a single-mode system, the expressions for fundamental limits to NEP are derived in a slightly different way because of the different way in which the receiver operates (see Section IV.D).

As discussed in later sections, only a few SMSMR video receivers, mainly based on cooled bolometers and photoconductors are background limited in practical application.

3.   Limitations to Receiver Sensitivity Arising from Signal Fluctuations

Even in the absence of background radiation, discussed in the previous section, and even if we assume that a receiver of negligible internal noise could be constructed, there will still be a lower limit to the smallest power that could be observed. As discussed by McLean and Putley (1965), the signal should be strong enough so that at least one photon arrives during virtually every observing period T_d, where $T_d \sim B_o^{-1}$. This means that the mean signal should be high enough so that the fluctuations in the signal do not produce a large probability of no photons arriving during T_d.

The signal radiation can usually be considered as coming from a thermal-type incoherent source, or from coherent sources such as lasers or microwave-type oscillators. For a thermal type of source, these "photon-limited" sensitivities will be of consequence (i.e., comparable with or larger than the background-type fluctuations discussed in 2) only if $hv \gg kT$. With this latter condition, the photons from the source arrive at the detector such that the fluctuations from the mean rate of arrival follow a Poisson distribution, and this is the same as occurs in a beam of photons from a coherent source (McLean and Putley, 1965; Mandel 1964). On this basis, the criterion that there should be at least a 50% probability of a photon arriving during each period T_d is satisfied if the mean rate of arrival is one photon per T_d (McLean and Putley, 1965). This defines the ultimate minimum power which can be observed as

$$(P_s)_{min} = hv/T_d \qquad (15)$$

or

$$(P_s)_{min} \sim hvB_o. \qquad (16)$$

As discussed by Gelinas and Genoud (1959), a factor of between 1 and 10 appears in the exact form of expression (16), depending on exact definition of $(P_s)_{min}$ and B_o.

If the detector/receiver absorbs only a fraction η of the incident photons, $(P_s)_{min}$ will be increased by a factor η^{-1}. Even in principle, this photon limit is unlikely to be met in SMSMR video receivers. For example, at 1 mm wavelength $(P_s)_{min}$ is $\sim 2 \times 10^{-23}$ W for $B_o = 1$ Hz and will only be achievable if the background temperature is considerably less than 1 K. At 10 μm wavelength, these figures rise to 2×10^{-21} W and 100 K, respectively.

D. HETERODYNE DOWN-CONVERSION RECEIVERS

1. General Properties

The scheme of a frequency down-conversion heterodyne receiver is shown in Fig. 5b. This type of receiver is also called a coherent detection system. The detection device now acts as a frequency mixer, and is subjected to radiation not only from the signal source but also from a much more powerful local oscillator (LO), which is usually a single-mode monochromatic source. We assume that the detector gives an output voltage proportional to the input radiation power such that

$$V_{out} = K\langle p_{in} \rangle, \qquad (17)$$

where K is a constant and $\langle p_{in} \rangle$ is the input power, time averaged for a time that is long compared with the period of the radiation cycle but shorter than, say, the response time τ_R of the detector. If the difference between the signal and LO frequencies, the "intermediate frequency" (IF or v_{IF}), is less than $1/\tau_R$ then

$$V_{out} = K[p_L + p_s + 2(p_L p_s)^{1/2} \cos(2\pi v_{IF} t - \phi)], \qquad (18)$$

where p_L and p_s are the LO and signal powers, and ϕ is the phase difference (at $t = 0$) between the signal and LO radiations. It is worth noting that if we had considered a detector that was fast enough to respond to changes at the radiation frequency (e.g., the classical square-law device) then its output would also contain sums and harmonics of the frequencies applied (Torrey and Whitmer, 1948). However, we shall neglect this possibility here.

Eq. (18) shows that if $p_L > p_s$, then the rms IF voltage is larger than that which would have resulted from the signal alone, without the LO. In this simple case the IF voltage can be increased without limit by increasing p_L. However, in practice, the *power* gain at the IF is limited both for practical and sometimes fundamental reasons. A parameter often quoted, particularly for rectifier-type mixers is the conversion efficiency, E_c, defined by

$$E_c = \frac{\text{power available at the IF from the mixer}}{\text{power available from the signal source}}. \qquad (19)$$

For many mixers, even if losses in the input coupling of the signal are allowed for, E_c is less than unity. Even so, power has been converted to a frequency, usually at low microwave or radio frequencies, where amplifiers of relatively low noise are available. To obtain the final output of the receiver, the amplifier IF signal is rectified in a second detector, which operates at the IF and has a final output bandwidth B_o.

Some of the properties of the heterodyne receiver are summarized in Table III. The main points are the following:

(i)   The receiver responds only to signals which down convert in frequency to lie inside the IF bandwidth B_{IF}. Signals at $\pm v_{IF}$ from the LO frequency are received. B_{IF} is limited at high frequencies by the availability of good high-frequency IF amplifiers and by the response time of the mixer. If the spectrum of the LO source is sufficiently narrow, B_{IF} can be made very small to provide very high spectral resolution.

(ii)   The IF power is proportional to signal power. IF noise will usually be proportional to B_{IF}. It is thus quite common to quote the heterodyne or "predetection" NEP, NEP_H, given in $W\ Hz^{-1}$, which is the radiation signal power that gives a power signal-to-noise ratio of unity at the IF with $B_{IF} = 1\ Hz$. It is only after the second detector that the performance can be directly compared with that of a video receiver. The (voltage) signal-to-noise ratio at the output $[SN]_o$ is given by

$$[SN]_o \approx [SN]_{IF}\left(\frac{B_{IF}}{B_o}\right)^{1/2} \tag{20}$$

when the (power) signal-to-noise ratio at the IF $[SN]_{IF}$ is less than unity. (This is discussed in detail by Smith, 1951 and Tiuri, 1964; 1966). Thus the final output NEP, NEP_o, for $B_o = 1$ Hz is given by

$$NEP_o \approx NEP_H(B_{IF})^{1/2} \tag{21}$$

and is expressed in W $Hz^{-1/2}$ where NEP_H is for $B_{IF} = 1$ Hz. The NEPs of heterodyne receivers are discussed more fully in Section 2.

(iii)   Signal phase information is retained at the IF [see Eq. (18)]. The IF outputs from several spatially distributed mixers can, in principle, be interfered to give spatial information about the source. This is usually called aperture synthesis in astronomy.

(iv)   Arising from the phase sensitivity, the receiver will respond properly only to signal radiation which, in effect, is spatially coherent with the LO over the sensitive area of the mixer. In practice, this means that the receiver has antenna properties and will respond only to a single mode of the signal radiation field, even if the mixer is a multimode detection device (e.g., Siegman, 1966).

Thus the heterodyne receiver offers quite different properties than a video system. It is particularly useful where a high-sensitivity, single-mode receiver giving high spectral resolution is required. For some purposes, particularly those not requiring the resolution or single-mode reception, video receivers may certainly be preferable (Blaney, 1975; Richards, 1977b).

2.   The Sensitivity of Heterodyne Receivers

In a properly designed heterodyne system, noise arising at stages up to and including the IF amplifier should firmly dominate noise from the later stages. In practice, the major noise sources will usually be noise intrinsic to the mixer device (Johnson noise, current shot noise) and IF amplifier noise, as well as miscellaneous sources such as amplitude or frequency modulation in the LO. In addition, there will be fundamental lower limits to the noise set by fluctuations in the background and signal radiations, and it is worth considering a receiver's sensitivity in this ideal limit.

As indicated in the previous section, it is usual to describe the sensitivity of the system in terms of the heterodyne NEP, NEP_H, with the final receiver NEP, NEP_o, being obtained from an expression of the form given in Eq. (21). If we consider a detector/mixer which usefully absorbs a fraction C_{in} of the incident signal (equal to quantum efficiency for a photon detector) and which views a signal source against a background of temperature T_s, then the ideal minimum power $(P_s)_{min}$ that can be observed at the IF stage is given by

$$(P_s)_{min} \approx \frac{h\nu B_{IF}}{C_{in}} + \frac{h\nu B_{IF}}{\exp(h\nu/kT_s) - 1}. \tag{22}$$

This expression can vary by a small factor (of order 2) depending on the mechanism of the mixer (McLean and Putley, 1965; Arams, 1973; Blaney, 1975).

NEP_H is given by Eq. (22) with $B_{IF} = 1$ Hz, so that for the case $C_{in} = 1$,

$$(NEP_H)_{min} \approx h\nu + \frac{h\nu}{\exp(h\nu/kT_s) - 1}. \tag{23}$$

This expression is plotted in Fig. 7.

The second term on the right-hand side of Eq. (23) arises from fluctuations in the thermal radiation received, similar to that derived for video receivers in Section IV.C.2. (We have neglected thermal emission from the detector, but this will also have a contribution to make, depending on the temperature of the detector.) The first term in Eq. (23) is associated with photon noise in the signal, as discussed in Section IV.C.3.

When $h\nu \ll kT_s$, the thermal background noise dominates in Eq. (23), and $(NEP_H)_{min} \sim kT_s$. (Note that for a receiver that receives on both IF sidebands,

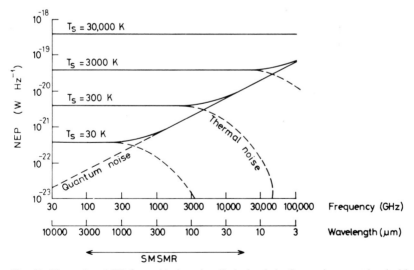

FIG. 7 Heterodyne NEP for an ideal receiver limited only by fluctuations associated with a source or background of temperature T_s filling the field of view. See Eq. (23).

$(NEP_H)_{min}$ would rise to $\sim 2kT_s$.) The photon-noise term is only of consequence when $hv \gtrsim kT_s$, as shown in Fig. 7. For most of the background temperatures of interest, this transition from thermal noise to quantum noise as the limitation takes place in the SMSMR. It is also worth noting from Eq. (22) that if $C_{in} < 1$, the quantum noise term could begin to dominate even when $hv < kT_s$.

These fundamental limits can be approached only if the mixer conversion efficiency is sufficiently large so that noise associated with the fundamental fluctuations can be "amplified" sufficiently to dominate the mixer's intrinsic noise and the IF amplifier noise. In principle, for, say, a photoconductive or thermal detector/mixer, this heterodyne amplification might be increased to a sufficiently high level by increasing the LO power [e.g., see Eq. (18)]. In practice, there are limits to the LO power that can usefully be employed; e.g., saturation or heating of the detector, LO-induced change of the detector's output impedance leading to a poor match to the IF amplifier, excess mixer noise at high LO power. For these and other reasons discussed in Section VI, the fundamental limits to heterodyne reception have not yet been approached in the SMSMR.

At the present time most of the interest in SMSMR heterodyne receivers lies at the longer wavelengths of the range, 0.3 mm and longer. The receivers are currently far from being quantum noise limited, and so the formalism of the thermal noise limit is employed. Under these conditions the concept of receiver noise temperature and noise factor are used (see Table III), and it is useful to consider generally the factors that contribute to receiver sensitivity in these circumstances. The following analysis is approximate, but good enough for present purposes. More precise formalisms have been given by Mumford and Scheibe (1967), Messenger and McCoy (1957), and Meredith et al. (1964). For a system with input-coupling efficiency C_{in} and a mixer conversion efficiency of E_c [see Eq. (19)], the power P'_{IF} delivered to the IF amplifier as the result of a signal P_s is given by

$$P'_{IF} = C_{in} E_c C_{IF} P_s, \tag{24}$$

where C_{IF} is the efficiency with which the mixer's IF output is coupled to the IF amplifier. If the signal is from a thermal source of temperature T_s then, assuming $hv \ll kT_s$ in Eq. (1),

$$P_s = kT_s \Delta v. \tag{25}$$

For a heterodyne system that receives on both IF sidebands, $\Delta v = 2B_{IF}$. Thus the signal power in this double-sideband (DSB) case, $(P_s)_{DSB}$, is given by

$$(P_s)_{DSB} = 2kT_s B_{IF}. \tag{26}$$

For the single-sideband receiver

$$(P_s)_{SSB} = kT_s B_{IF}. \tag{27}$$

It is worth pointing out that this thermal power is just noise, so that at the IF stage in an ideal receiver, the signal and noise powers are just equal: it is only by time integration further on in the receiver that the signal-to-noise ratio can become greater than unity [Eq. (20)].

In addition to the "signal" noise, there will be the noise produced in the receiver itself. For convenience, we shall define the receiver noise in three parts:

(i) Noise intrinsic to the mixing element and assumed to be independent of signal level. Assuming this noise has a flat frequency spectrum within B_{IF}, we can define a temperature T_D such that the effective noise power input to the IF amplifier is $kT_D B_{IF} C_{IF}$.

(ii) IF amplifier noise, defined by a noise temperature T_A so that the effective noise input power to the amplifier from this source is $kT_A B_{IF}$.

(iii) Miscellaneous noise sources (e.g., caused by amplitude or frequency modulation of the LO), giving rise to an effective input at the IF amplifier of $kT_{misc} C_{IF} B_{IF}$.

Thus the receiver noise P_R referred to the input of the IF amplifier is

$$P_R = kB_{IF}(T_D C_{IF} + T_A + C_{IF} T_{misc}). \tag{28}$$

Using Eqs. (24), (26), and (28), we see that this noise power is equivalent to that which would be produced from a thermal signal source of temperature $(T_R)_{DSB}$ given by

$$(T_R)_{DSB} = \frac{T_D}{2C_{in} E_c} + \frac{T_A}{2C_{in} C_{IF} E_c} + \frac{T_{misc}}{2C_{in} E_c}. \tag{29}$$

We define $(T_R)_{DSB}$ as the double-sideband noise temperature of the receiver. (As indicated earlier, we have assumed that noise sources at later stages in the receiver are negligible.) $(T_R)_{DSB}$ is the noise temperature most appropriate to the situation in which the receiver is used as a radiometer and in which signal power occurs in both IF sidebands of the LO. For narrowband signals, falling inside only one IF sideband of the LO, we can define a single sideband (SSB) noise temperature $(T_R)_{SSB}$, which is, in our approximation [Eqs. (24), (27), and (28)], just twice $(T_R)_{DSB}$. The receiver noise temperature is related to NEP_H in a simple way, as given in Table III. Eq. (29) makes it obvious that a high conversion efficiency reduces the effect of noise sources, such as the amplifier. A term often used is the (single sideband) mixer noise temperature T_M, given by

$$T_M = T_D/E_c. \tag{30}$$

E_c and T_M are two of the most important parameters of a mixer, as also is the LO power to optimize the values of E_c and T_M. As already made clear, the above formalism is valid only in the thermal limit, which means that $T_R \gg hv/k$. In a pragmatic way it is thus possible to define an ultimate lower limit for a receiver noise temperature of hv/k. For $\lambda = 1$ mm, this temperature would be ~ 14 K, and ~ 46 K at 0.3 mm.

E. FREQUENCY UP-CONVERSION RECEIVERS

In a frequency up-conversion receiver, a long-wavelength photon is combined with one or more photons from a shorter wavelength source to produce a photon in or near the visible range, where quantum-noise-limited detectors (e.g., photomultipliers) are available. Such a receiver might, in principle, be useful in the SMSMR under conditions in which quantum (rather than thermal) noise is the ultimate limit to sensitivity, i.e., at the shorter wavelengths of the region. The methods which have been discussed at least for infrared up-conversion, fall into two main classes:

(i)   The coherent interaction of the signal and a higher frequency pump in a medium with nonlinear properties (such as certain crystals and vapors) leading to sum and difference frequency generation (Warner, 1971; Boissel, 1975; Abbas et al., 1976).

(ii)   The incoherent "addition" of photons in solid or gaseous materials with suitable energy levels, leading to the so-called quantum counter (Bloembergen, 1959; Lengfellner and Renk, 1977; Gelbwachs et al., 1978).

Systems falling into both these classes have been investigated at wavelengths of 10 μm and less, where their potential for creating visible images from infrared radiation is also of interest (Boyd, 1977; Krishnan et al., 1978). However, even at these wavelengths, fully developed competitive systems are not yet available.

In the SMSMR, systems of the first type do not appear to have been investigated experimentally for purposes of low-level detection and are not particularly promising. The two major problems for the SMSMR are the choice of sufficiently low-loss and nonlinear materials and filtering out the "pump" radiation from the required up-converted signal. The quantum-counter approach has been the subject of preliminary investigations in ruby crystals at 0.337 mm wavelength (Lengfellner and Renk, 1977) and in semi-conductors at 28 μm wavelength (Gundersen, 1974). However, both of these are some way from being competitive as working detectors, and will not be discussed further here.

F. IMAGING RECEIVERS

The general properties of these receivers will be discussed briefly in conjunction with an account of practical devices in Section VII.

V. Properties of Practical Detectors in Video Systems

A. INTRODUCTION

Most of the present use of detectors in the SMSMR is in video receivers, and the data available is often for receivers rather than for basic devices. However, in many cases the receiver and detection device properties are virtually identical (see Table III). The following is only a brief discussion of the principles and performance of practical devices, together with some of the key references. Much more detailed accounts of many of the topics are given in other reviews in this series, and further references are given by Blaney (1978a).

Most of the data are drawn from the literature. In most cases it is not possible to make reliable estimates of the uncertainties in the quantitative data. In some cases there is a basic lack of suitable standards; in others, the original authors may not have given sufficient information on which to make an assessment. Thus, in most cases, no attempt is made to attach uncertainties. In some cases the data given in the literature are for the "best" performance, sometimes omitting a balanced discussion of the practical difficulties and what can be expected on a routine basis. The reader should assess the published information in some detail before attempting rigorous quantitative comparisons between devices.

B. THERMAL DETECTORS

1. General Properties

A thermal detector consists of a radiation absorber to which is attached a thermometer. The absorber/thermometer combination, which is usually designed to be of low thermal mass, is by some means connected to a thermal reservoir via a thermal conductance. The spectral response is determined by the spectrum of the absorber. The time response, which should be separated into rise and decay times, depends on the distribution of the radiative absorption, the effective conduction properties and thermal mass of the absorber/thermometer, and the effective thermal conductance to the reservoir. The internal noise of the device is often determined by the noise properties of the thermometer, while at a usually more fundamental level are temperature fluctuations arising from fluctuations across the thermal conductance. Many of the sources of noise decrease as the operating temperature is decreased. An excellent short review of thermal detectors has been given by Putley (1977b). Table V compares some of the properties of the most used types.

The discussion below is not meant to be exhaustive. For example, Putley (1973, 1977b) mentions several other types of thermal device that have been tried in the infrared region. In principle, at least some of these might be applicable to the SMSMR.

The use of thermal detectors in power standards is discussed in Section VIII.

2. Room-Temperature Thermopiles and Thermistors

At the longer wavelengths of the submillimeter region, the thermopile is considerably slower in response and less sensitive than the Golay cell and the best pyroelectric devices. At shorter wavelengths, say < 100 μm, where black-paint absorbers are effective and where devices of relatively small area (and hence quite fast response time) can be used, thermopiles can still be competitive, particularly where a rugged and nonmicrophonic device is required (Astheimer and Weiner, 1964; Stevens, 1970). Generally similar remarks can be made about bolometer (i.e., thermistor) devices at room temperature (Allen et al., 1969).

Thin metal film bolometers with short response times have been used at 10 μm wavelength (Block and Gaddy, 1973), and might reasonably be adapted to measuring intense pulses from submillimeter lasers. One approach to improving the sensitivity of these thin-film bolometers is to make them of very small area. Hwang et al. (1979b) have described bismuth bolometers with dimensions of a few microns. These are coupled to the radiation field by an evaporated film antenna, and thus the system is single-mode. Preliminary experiments at 119 μm wavelength indicate a video NEP of $\sim 2 \times 10^{-10}$ W Hz$^{-1/2}$ at modulation frequencies up to 100 Hz, remaining within a factor of ten of this up to 25 MHz modulation.

3. Pneumatic Detectors

The pneumatic detector senses pressure changes in a gas cell heated in some way by the incident radiation. The most popular form is the Golay cell (Golay, 1947, 1949; Hickey and Daniels, 1969; Chatanier and Gauffre, 1972) which can have an NEP approaching (within an order of magnitude) that of an ideal room-temperature device. Despite rather slow speed (see Table V) and quite severe microphony, it is a well-established commercially available device for laboratory work. Radiation absorption in this type of cell is by a thin metal film within the cell, heat being transferred to the gas by conduction. When their electrical resistance per square is a suitable fraction of the impedance of free space, such metal films can absorb up to half the electromagnetic energy per pass with virtually flat spectral response (Silberg, 1957). In the Golay cell the pressure cell is connected to a larger gas reservoir via a small orifice, so that the response is transient.

An "optoacoustic" cell device designed to monitor the output from far-infrared lasers has been described by Busse (1979), although this general type of device has received little attention in the SMSMR so far.

4. Pyroelectric Detectors

The pyroelectric effect exists in noncentrosymmetrical crystals in which the spontaneous electric polarization is temperature dependent and produces a varying surface charge in a particular direction as a result of temperature changes. Thermal radiation detectors based on the effect are usually operated at room temperature, but cooled devices may offer some sensitivity advantage (Hadni et al., 1978). The theory and characteristics have been reviewed by several authors, notably Putley (1970, 1977a), Kremenchugskii and Roitsina (1976), and Liu and Long (1978).

The signal produced by the devices is usually proportional to the rate of change of temperature, so the detection effect is transient. Although responsivity falls with increasing modulation frequency, typically at least as quickly as the reciprocal of frequency above 10–100 Hz, suitable examples of pyroelectric devices can be used to 100 MHz or more and for the observation of fast laser pulses (Roundy and Byer, 1972; Baker, 1975). The pyroelectric effect occurs in many materials, including ceramics and polymer films, so that devices can be tailored in size and shape. The most sensitive devices use triglycine sulphate and its derivatives (Putley, 1977a; Keve, 1975), although they tend to be more fragile than those made from other materials. Of the more rugged materials, lithium tantalate can provide good sensitivity, and lead zirconate titanate and similar materials may be preferred when high powers may be encountered (Beerman, 1975).

Most of the development of pyroelectric devices has been aimed at detection around 20 μm and shorter wavelengths, and most commercial devices are of a size (say a few squared millimeters or less) and have windows (where required) suitable for such wavelengths. It is usual to employ the pyroelectric material itself as the radiation absorber, although thin metal films and absorbing paints can be used (Annis and Simpson, 1974; Geist et al., 1976). Many of the materials used maintain high absorption throughout much of the SMSMR, although in some cases, the efficiency of absorption would drop significantly at the longer wavelengths.

The noise sources that limit the video NEP are Johnson noise in load resistors and from dielectric loss in the pyroelectric material, amplifier noise, and thermal fluctuations in the pyroelectric material. The relative

TABLE V

THERMAL DETECTORS

Type	Spectral range	Area A mm^2	Responsivity R V W^{-1}	Response time τ Range of chopping frequency	NEP W Hz$^{-1/2}$
Thermopiles (room temp.)	Depends on absorber. Usually $\lambda < 50~\mu m$ with black paint absorber.	$10^{-2} - > 10^2$	$\sim 10^{-1} - > 10^2$ (depending on design)	$\sim 10^{-2}$ s $- > 10$ s	Can be $\sim 10^{-9}$ with $\tau \sim 10^{-1}$ s and A of a few mm^2
Golay cell (room temp.)	Can cover from λ of a few mm through SMSMR. Details depend on window.	7 and 20 (commercial)	$10^5 - 10^6$ (typical commercial packages)	Usual working range 5–20 Hz	$\sim 10^{-10}$ (typical commercial)
Pyroelectric (room temp.)	In principle, can cover all of SMSMR. In practice, depends on material, absorber, and windows.	1–20 (typical commercial) Could be up to several hundred.	$10^2 - 10^6$ (typical commercial packages at 10 Hz chopping).	With much reduced responsivity, have been constructed with $\tau \sim 10^{-9}$ s	$\sim 5 \times 10^{-10}$ (typical commercial at $\lambda \sim 10~\mu m$, A of a few mm^2, and 10 Hz chopping)
^{4}He-cooled semiconductor bolometers ($\gtrsim 1.2$ K)	Virtually all of SMSMR. Non-composite types may fall in responsitivity at both ends of range, depending on material and coating. Composite types (thin metal film absorber) probably fairly flat, extending to milli-meter wavelengths. Also a dependence on window material.	Element typically a few mm^2, and 0.1–0.5-mm thick. In composite types, absorber up to ~ 20 mm^2. Effective area depends on optical input arrangement to bolometer chamber.	10^3 up to $\sim 10^6$	Usually 5 to 100 ms.	Carbon types, $\sim 10^{-12}$ (Corsi et al., 1973). Germanium and silicon types typically 10^{-14} to 10^{-13}. Best Ge type (Nishioka et al., 1978): electrical NEP 3×10^{-15} (for $T = 1.2$ K, $\tau = 25$ ms).

Type					
³He-cooled germanium bolometers (usually ~ 0.3 K)	As for ⁴He types.	Similar to ⁴He types.	Up to ~ 10^7.	Similar to ⁴He types in practice.	Examples: Nolt et al. (1977): 2.5×10^{-14} for $T = 0.31$ K, $\tau = 4$ ms. Nishioka et al. (1978): 6×10^{-16} (electrical) for $T = 0.35$ K, $\tau = 6$ ms, radiation NEP within a factor of 2 of this.
Superconducting transition edge (⁴He-cooled)	Similar to semiconductor types. Composite construction now preferred.	Generally similar to semiconductor types, but see references in text.	Up to ~ 10^5.	Typically 50–100 ms.	Examples: Composite Al bolometer (Clarke et al. 1977): 2×10^{-15} (electrical) for $\tau = 83$ ms, $T = 1.3$ K. Composite Sn bolometer (Green, 1978; Green et al. 1979): 5×10^{-15} (predicted electrical) for $\tau = 100$ ms, $T = 3.7$ K.

importance of these sources in specific circumstances has been discussed by
Putley (1970, 1971, 1977a, 1978a), Stokowski (1976), and Liu and Long (1978).
Pyroelectric devices are also piezoelectric, so that they are quite microphonic,
while performance can be significantly dependent on ambient temperature
(but see Gabriel, 1974; Byer *et al.*, 1975). At modulation frequencies of
around 10 Hz, the NEP can be similar to that of the Golay cell (see Table V)
for devices of a few square millimeters in area (Firth and Davies, 1971;
Putley, 1977b). At higher modulation frequencies, NEP increases at least as
quickly as the square root of frequency between 10^2 and 10^5 Hz (Baker *et al.*
1972; Hadni, 1969).

Thus, pyroelectric devices compete well with other room-temperature
detectors for some applications, but further optimization of performance
for the SMSMR is desirable and possible.

Effects similar to the pyroelectric effect have been suggested for detection
purposes. Examples are the pyromagnetic effect (Walser *et al.*, 1971) and the
pyrionic effect (Sher *et al.*, 1976), although neither appear to have been
developed for working detectors.

5. *Cooled Thermal Detectors*

Most of the cooled thermal detectors in the SMSMR operate at around 4 K
and below, and virtually all are of the bolometric type (i.e., depending on a
variation of electrical resistance with temperature). Cooling to cryogenic
temperatures not only reduces noise associated with temperature fluctuations, Johnson noise, and emission from the detector, but also allows access
to bolometric effects of rather high responsivity (see Table V).

In a typical bolometer the bolometer element is suspended by fine electrical
leads in an evacuated cavity that has a suitable window for the signal radiation. The electrical leads also serve as the thermal conductance from the
bolometer element to the heat reservoir. The absorption losses in the cavity
should be dominated by absorption in the bolometer element (see below).
For the best bolometers it is important to maximize the collection area for
radiation and to concentrate the radiation efficiently onto the bolometer
while restricting the angle of view, as far as is as appropriate, to reduce
unwanted background flux. While this can be done by cooled baffles, the
so-called "heat-trap field optics" (Winston, 1970; Harper *et al.*, 1976;
Keene *et al.*, 1978), which involves the use of specially shaped radiation
horns, now appears to be favored for this (Nishioka *et al.*, 1978).

The theory of bolometers has been discussed by Jones (1953b), Low (1961),
and Zwerdling *et al.* (1968). Aside from fluctuations associated with background radiation, the NEP is ultimately limited by Johnson and thermal
fluctuation noise, which decrease with both the operating temperature and
the thermal conductance G between the element and the reservoir (Low,

1961). On the other hand, the thermal time constant increases with decreasing G (as well as being proportional to the thermal capacity of the element), so that this may be traded off against NEP. In practice, thermal time constants in the range 10–100 ms are usually chosen (Table V). In practice the best bolometers approach these limits, or are limited by background fluctuations. Typical bolometers are limited by other miscellaneous noise sources, including the post-bolometer amplifier. Cooled amplifiers are sometimes used (Knotek and Gush, 1977). As the NEP is possibly a strong function of background radiation level, the "electrical" NEP of the bolometer, i.e., the video NEP determined by electrical rather than radiative heating, is sometimes quoted. The electrical NEP is usually smaller than the radiation NEP, although not always (Nolt et al., 1977).

Most of the cooled bolometers in use are based on semiconducting materials operating in the liquid-^{4}He range (1–4 K). Germanium elements with various dopants have been used by Low (1961, 1965, 1966), Zwerdling et al. (1968), Chanin (1972), Richards (1970), Nokagawa and Yoshinaga (1970), Coron et al. (1972), Hauser and Notarys (1975), and Nishioka et al. (1978). Doped silicon elements have been discussed by Bachmann et al. (1970), Chanin (1972), Kinch (1971), Pankratov et al. (1978), and Chanin et al. (1978). Relatively simple devices based on carbon resistors, but with usually poorer NEPs, have also been employed for some years (Boyle and Rogers, 1959; Richards and Tinkham, 1960; Ginsberg and Tinkham, 1960; Chanin, 1972; Corsi et al., 1973; Dall'Oglio et al., 1974). Thallium selenide is also suggested as giving a promising performance (Nayar and Hamilton, 1977), while stannic oxide could be useful where the detector may be subjected to magnetic fields (Von Ortenberg et al., 1977).

In practice, the NEP of good germanium and silicon-based bolometers appear similar [see Table V and Chanin (1972), Kinch (1971), Kleinmann (1976), and Pankratov et al., (1978)]. The most sensitive bolometers tend to be of germanium, although in principle, silicon with its lower heat capacity and other possible advantages (Summers and Zwerdling, 1974; Pankratov et al., 1978) should be at least as good. In any case, the modes of construction and use of the bolometers described in the literature are rather disparate, thus making detailed comparisons of NEP unreliable. There have been few direct comparisons of radiation NEP of different types of bolometer (but see Dall 'Oglio et al., 1974).

Further reduction of NEP is possible by operating at lower temperatures, in the liquid-^{3}He range, and this can now be done with relatively little extra inconvenience in operation. Not only is the fundamental limit to bolometer NEP reduced by operation at a lower temperature (Low, 1961), but in practice the semiconductor crystal lattice contribution to the heat capacity is reduced and the variation of resistance with temperature is increased. Germanium bolometers have been operated at 0.3 K by several workers, including Drew and Sievers (1969), Radostitz et al. (1978), and Nishioka et al. (1978), and operation at 0.1 K has been described by Draine and Sievers (1976). The 0.35 K germanium "composite" bolometer (see below) described by Nishioka et al. (1978) and Woody et al. (1978), with an electrical NEP of 6×10^{-16} W Hz$^{-1/2}$ and a radiation absorption efficiency of 50 % is probably the most sensitive, working far-infrared bolometer yet reported.

As noted above, the "electrical" NEP is often quoted for bolometers. The radiation NEP depends on the efficiency of absorption in the bolometric element. At the shorter wavelengths, say 0.1 mm or less, the absorption of the semiconductor elements is usually sufficiently high, although a coat of a suitably absorbing paint is sometimes added (Low, 1966). At longer wavelengths, the paints are less effective and the absorption of the doped semiconductors also usually decreases, particularly at near-millimeter and millimeter wavelengths (Dall'Oglio et al., 1974; El-Atawy, 1979). To reduce this problem, "composite" bolometers have been developed (Clarke et al., 1974, 1977; Hauser and Notarys, 1975; Nishioka et al., 1978). These employ bismuth films of suitable resistance on a sapphire substrate as the radiation absorber (Silberg, 1957), the bolometer element being mounted on the opposite side of this high-thermal-conductivity substrate. Such absorbers have a rather uniform absorption throughout the SMSMR. A feature that may give rise to rather nonuniform spectral response, particularly in noncomposite types, is that the semiconductor may show photoconductive as well as bolometric response [see Nokagawa and Yoshinaga (1970) and Section D].

The highly temperature-dependent resistance of materials going through a normal-to-superconducting transition can also provide sensitive ^{4}He bolometers that are particularly attractive for applications where operation at the lowest ^{4}He temperatures or in ^{3}He is not possible. The theory and practice of these have been described by Martin and Bloor (1961), Bertin and Rose (1968, 1971), Maul and Strandberg (1969), Gallinaro and Varone (1975), Clarke et al. (1977), Green (1978), and Green et al. (1979). Examples of performance are given in Table V. Another ^{4}He bolometer, based on the temperature dependence of the supercurrent in a Josephson type of junction (see Section C) was described by Clarke et al. (1974, 1977). Although offering a good NEP, it had some disadvantages in comparison with more conventional superconducting bolometers.

At the present time cooled bolometers offer the lowest video NEPs of any broadband detectors in the SMSMR. However, at wavelengths of ~ 0.1 mm and less, photoconductors offer similar NEPs at often higher working temperatures and with shorter response times (see Section D).

C. LUMPED-CIRCUIT RECTIFYING DETECTORS

1. *General Properties*

Some detectors respond to the electrical currents induced in them at the radiation frequency, as would a nonlinear component such as a rectifier in an electrical circuit at lower frequencies. This type of device, when used in the SMSMR, is usually much smaller than the wavelength, and usually can be considered as a lumped-circuit component. Power from the radiation field is transferred to the detection device via some form of antenna. An antenna has spatially coherent characteristics and can essentially only collect energy from a single mode of the radiation field (Jasik, 1961) so that the étendue is limited to $\sim \lambda^2$ (see Section II.B). Thus, a single rectifying detector also has this limitation and is best suited to single-mode applications.

Rectifying devices, usually in the form of diodes are, of course, common at millimeter and longer wavelengths. In the SMSMR, the problems of fabricating effective rectifier-type devices are more difficult than at the longer wavelengths, as is the problem of coupling radiation to the devices. Three general types of rectifier have been used in the SMSMR: semiconductor–metal diodes, metal–insulator–metal diodes, and Josephson junctions. The properties of these are briefly reviewed in Sections C.3–C.5 and their video properties are compared in Table VI.

By definition, a rectifying device must respond at the radiation frequency, i.e., it is intrinsically a fast detector. This particular feature is the one that is most attractive for SMSMR work in such applications as

(i) video detection of short or structured pulses from lasers or similar sources;
(ii) frequency mixing for heterodyne receivers;
(iii) use of the electrical nonlinear characteristics of the devices to generate harmonics of a monochromatic source for use in heterodyne receivers or frequency metrology (Knight and Woods, 1976).

In this and the immediately following sections, the discussion will be directed largely at the video applications of these detectors, but much of the information is also relevant to the heterodyne detection application discussed more fully in Section VI.

For effective performance in the SMSMR, the criteria in designing a rectifying detector are much the same as at lower frequencies (Burrus, 1966) but the high frequency considerably increases the problems of fabrication and use. Thus, for example, the reactive impedances of the device, such as the shunt capacitance, must not too severely reduce the high-frequency currents in the useful nonlinear detecting part of the device. This problem is obviously

TABLE VI

Rectifying Detectors

Type	Material	Wavelength	Examples of performance[a]			Reference
			Type of input coupling	Responsivity V W^{-1}	Video NEP W Hz$^{-1/2}$	
Metal–semiconductor point contact (300 K)	Metal–Ge	1 mm 0.3 mm	3-mm diam. circular wave-guide (overmoded)	— —	$\sim 3 \times 10^{-10}$ $\sim 2 \times 10^{-8}$ (increasing 10 dB per octave)	Becklake et al. (1970)
	Be bronze–InSb	0.337 mm	Whisker antenna	20–25	$\sim 4 \times 10^{-10}$	Dyubko and Efimenko (1971)
	W–GaAs	0.337 mm	Whisker antenna	Up to 10^3	$\sim 2 \times 10^{-9}$	Frayne et al. (1978)
Lithographically prepared Schottky diode (300 K)	N-type GaAs, 0.5 µm diameter	0.496 mm 0.119 mm 42 µm	Whisker antenna	10 0.4 $\gtrsim 10^{-2}$	$\sim 10^{-8}$ $\sim 10^{-7}$ $\sim 10^{-5}$	McColl et al. (1977)
Metal–metal point contact (300 K)	W–Co	0.337 mm	Whisker antenna	~ 1	—	Kobzev et al. (1975)
	W–Ni	0.337 mm–3.89 µm	Whisker antenna	Falling by ~ 10 times over this range	—	Sakuma and Evenson (1974)
Point-contact Josephson junction ($\lesssim 4$ K)	Nb–Nb	2.5 (± 0.8) mm	Whisker antenna	5×10^5	3×10^{-15}	Tolner et al. (1976)
	Nb–Nb	1.5 mm	Fundamental-mode wave-guide mount 20-mm diam. light guide	3×10^5	5×10^{-15} ($\Delta T_s \sim 10^{-2}$ K Hz$^{-1/2}$)	Poorter and Tolner (1979)
"Broadband" video mode	Nb–Nb	0.7 mm		10^3	5×10^{-12}	Divin and Nad (1973)
	Nb–Nb	0.337 mm	Whisker antenna	Up to 10^3	$\sim 10^{-12}$	Blaney (unpublished)

[a] As far as is obvious from the literature, all the data are for the overall performance of the detectors, i.e., not correcting for input-coupling losses.

increasingly difficult as the frequency increases. In practice, this usually means that SMSMR rectifier-type devices are small in area, say a few microns square or even less. "Point-contact" or "cat's whisker" devices are often necessary. If energy is to be efficiently transferred from the radiation field to the detector, the spatial sensitivity pattern of the antenna must be appropriate to the geometry of the radiation beam, while the impedance of the antenna and detector should be appropriately matched (Jasik, 1961). This particularly critical question of radiation coupling to the device is discussed in the next section.

In the discussions on the performance of actual devices in Sections C.3 to C.5, it is assumed that the response is rectifierlike. However, it is possible that thermal, and possibly even photon effects, could occur in some devices. Detection effects of the latter sort would not have the intrinsic very high-speed or harmonic-generating ability of the rectifier.

2. The Coupling of Radiation to Rectifier-Type Detectors

The system that couples energy from the radiation field, in either a waveguide or free space, can take many forms. As already mentioned, the two main problems are to provide (i) some form of antenna that effectively matches the geometry of the incoming radiation beam (this is more appropriate to free-space radiation), and (ii) a means of matching the impedance of the detector to that of the radiation-coupling system. These problems become particularly difficult as the frequency increases, and much more work is still required in the SMSMR.

At microwave and longer millimeter wavelengths, diode detectors are often mounted in fundamental-mode hollow waveguide cavities. The diode is supported and electrically contacted by two wires mounted across the guide, which may be of reduced height to reduce characteristic impedance (see Fig. 8a). In principle, two adjustments are required to produce an impedance match to the detector; these are usually provided by a reflecting plunger behind the detector, and an adjustable tuning stub at a suitable distance in front of it in the guide (Torrey and Whitmer, 1948). At millimeter wavelengths the stub is often omitted, since its effectiveness is offset by losses associated with it in practical systems. The tunable cavity so formed acts as a transformer between the waveguide and detector impedances. If coupling to free-space optics is required, the waveguide can be fed from a horn, the antenna characteristics of which can usually be tailored appropriately (Jakes, 1961). These techniques have been extended to short millimeter and submillimeter wavelengths, most often for heterodyne receivers (Warner, 1969; Kerr et al., 1977a; Wrixon, 1974; Blaney, 1978b), and fundamental-mode mounts have been used at least down to 0.36 mm wavelength (Dryagin and Fedoseev, 1969; Fedoseyev and Shabanov, 1973). However, at submillimeter wavelengths the problems of small physical dimensions, the increasing effect

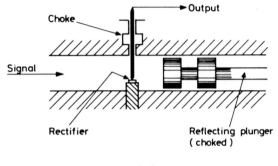

(a)

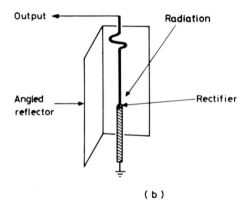

(b)

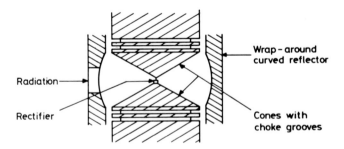

(c)

FIG. 8   Sketches of three methods used for coupling energy to rectifying detectors. (a) Waveguide cavity; (b) free-space antenna, with reflector; (c) biconical cavity, of the general type originated by Gustincic (1977).

of surface resistance losses, and the provision of effective radiation-frequency chokes at rectifier lead-throughs become severe, and other methods are used and are being developed.

One approach is to couple energy via an antenna of the general type used at radio frequencies, but scaled down in size. In its crudest form, often used with point-contact detectors, the antenna is just wire or wires which form or contact the detector, the coupling being pragmatically optimized by variation of the focusing conditions and the angular position of the device. The spatial antenna patterns of fairly simple detector configurations have been shown to agree quite well with the predictions of long-wire antenna theory right through the SMSMR (Matarrese and Evenson, 1970; Twu and Schwarz, 1975). The angular antenna patterns of such structures are not very favourable for efficient coupling from the radiation field, but the addition of reflectors around the wire can significantly simplify and improve the directivity of the antenna pattern (Krautle et al., 1977, 1978; Fetterman et al., 1978; see also Fig. 8b). The fabrication of small antennas by planar techniques will obviously be important in the future and promising results have been reported by Hwang et al. (1979a) at 0.12 mm wavelength. It is also worth pointing out that the difficulties of characterizing the properties of small antennas at the present time means that modeling the systems at longer wavelengths will be valuable (Rutledge et al., 1978; Daiku et al., 1978).

One difficulty with antenna structures of the type just discussed is that impedance matching between the antenna and the detector must be built-in to the design at the fabrication stage, as there is little opportunity for optimization in use. The resonant cavity structures do, however, provide some degree of impedance adjustment. Some structures based essentially on oversize cavities can be described as hybrids between the cavity and antenna approaches. The use of oversize cavities also reduces the problems of small-size and resistive losses associated with the fundamental-mode-only systems. One such system is the biconical cavity system described by Gustincic (1977), which has been used with some success up to nearly 700 GHz in heterodyne work (Gustincic et al., 1977). The system is sketched in Fig. 8c.

One feature of all the coupling systems used with rectifier-type detectors is that they are resonant, and this leads to an instantaneous spectral response that is generally very structured, with peak response occurring over rather narrow frequency ranges (Blaney, 1978a).

The development of all these techniques is at an early stage in the SMSMR. There are, as yet, no reliable direct quantitative comparisons between the performance of the various coupling systems. The performances given in the literature for rectifying SMSMR detectors have been obtained in widely varying and often ill-defined circumstances. Thus the figures given in Table VI can be regarded only as very approximate in absolute terms.

3. *Point-Contact Metal–Semiconductor Diodes and Schottky Diodes*

Small-area contacts between sharply pointed whiskers of metals such as tungsten and flat pieces of semiconductor have been used for many years at millimeter wavelengths as video detectors and heterodyne mixers (Burrus, 1966), and can operate also in the SMSMR. Although various detection mechanisms might occur in such devices (Burrus, 1966; Kerecman, 1973), the devices of most interest in the SMSMR are generally considered as nonlinear resistors of the Schottky-barrier diode type (Irvin and Vander-wal, 1969). In recent years Schottky-barrier devices produced by planar techniques have become much more important in the SMSMR and will be discussed below. However, the point contact device is still of interest, if only because it is simple to fabricate.

Video detection with metal–semiconductor point contacts has been reported with various semiconductors and in various coupling structures. [For a list of references, see Blaney (1978a).] Results differ considerably, and there is no consensus on the materials or doping levels which give the best NEPs or responsivities. Responsivity and NEP are sometimes enhanced by a dc bias current (Frayne *et al.*, 1978). Both responsivity and NEP deteriorate significantly with decreasing wavelength, although quite significant response, possibly not from the Schottky-barrier nonlinearity, has been observed at 10 μm wavelength in W–Si and W–Ge devices (Tsang and Schwarz, 1977; Champlin and Eisenstein, 1977). It is anticipated that Schottky diode performance is likely to be significantly modified at the highest frequencies of the SMSMR due to semiconductor plasma resonances and other effects (van der Ziel, 1976; Champlin and Eisenstein, 1978).

Some typical results are given in Table VI. As video detectors, the NEPs are not competitive with other devices at wavelengths shorter than 1 mm, except where a very fast detector is required. However, they are simple and quite inexpensive to fabricate, although usually mechanically unstable. It is usually assumed that the metal–oxide–metal diode (see Section C.4) may be superior at the shortest SMSMR wavelengths due to its lower spreading resistance, but good comparative data is presently lacking.

Small-area Schottky-barrier diodes are increasingly being fabricated by microelectronic techniques, i.e., epitaxial materials, evaporation, etc., with component features being delineated by means of photolithography or electron-beam lithography (McColl *et al.*, 1972; Wrixon, 1974; Clifton, 1977). Diodes as small as 0.1 μm in diameter have been reported (McColl *et al.*, 1977). Epitaxial gallium arsenide has received most attention, with gold, platinum, or a suitable alloy as the metal side of the barrier. In most cases so far the electrical connections to the small-area metal contact of the diode have been made with a pointed metal whisker, although the electrical characteristics of the devices are essentially determined by the plated contact (Clifton, 1977; Carlson et al., 1978). Little work has been done so far in "all-planar" electrical connections [but see Murphy et al. (1977)], although this technique would obviously be desirable with, for example, strip-line coupling, planar antennas, or arrays of devices.

Microfabricated Schottky diodes are presently of most interest in hetero-dyne mixing, and this is discussed in Section VI. The video performance has been relatively little studied, but results have been reported by Becklake et al. (1970), Fetterman et al. (1974a, b), Mizuno et al. (1975), McColl et al. (1977), and Murphy et al. (1977). Most of the results have been obtained with "open-structure" wire antenna coupling, and the deterioration of perfor-mance with increasing frequency (Table VI) almost certainly reflects poorer coupling efficiencies, as well as intrinsic device falloff at higher frequencies. The NEPs obtained appear similar to those achieved with point-contact Schottky diodes.

The intrinsic noise of Schottky diodes can be reduced, in some circum-stances, by reduction of the working temperature (Weinreb and Kerr, 1973), but video results on cooled diodes do not appear to have been reported. At temperatures below 4 K, a modification of the Schottky diode, the super-Schottky, in which the normal metal is replaced by a superconductor, has been studied (Silver et al., 1978; McColl et al., 1979). Results up to 30 GHz have indicated an excellent video performance (as well as a low-noise mixing performance), but extension to above 100 GHz is not presently certain.

4. Metal–Oxide–Metal Diodes

Small-area contacts between metals with a thin insulating barrier between (usually oxide) show detection properties as far as visible wavelengths [see Heiblum et al. (1978) for references]. There is some disagreement regarding the mechanisms involved, and these probably vary with wavelength. In the SMSMR it is usually assumed that an electron-tunneling effect gives rise to the nonlinear resistance. The best electrical performance has usually been obtained with point-contact devices, although these are generally mechani-cally unstable. However, there is increasing interest in devices made by litho-graphic techniques (Wiesendanger and Kneubühl, 1977; Heiblum et al., 1978).

In the SMSMR, and at higher frequencies, the main use for these devices so far has been in frequency mixing and harmonic mixing for frequency metrology (Knight and Woods, 1976; Evenson et al., 1977b). The small amount of data on SMSMR video performance (Green et al., 1971; Kobzev et al., 1975; Sakuma and Evenson, 1974; Gehre, 1974) show significantly poorer responsivity and NEP than metal–semiconductor devices at wave-lengths longer than 0.3 mm, but possibly a slower deterioration at shorter wavelengths. However, for the present, the metal–semiconductor devices appear preferable for SMSMR detection and heterodyne receivers.

Discontinuous metal films produced by evaporation have shown detection effects at 10 μm wavelength (Okamura and Ijichi, 1976). The mechanism may be similar to that of the metal–oxide–metal diode, and the authors suggest that a useful detector may be possible.

5. *Josephson Junctions*

Josephson effects occur when a weak electrical contact is made between two pieces of superconducting material. With presently known superconductors, this necessitates operation at < 20 K, and in practice most usually in the liquid ^{4}He range.

The mechanisms of Josephson "junctions" as detectors or mixers are distinctly different from those of the rectifying devices already discussed, in that Josephson devices are active and generate internal oscillating supercurrents. It is the interaction of these currents with externally applied currents that provides the nonlinear properties which, in turn, give rise to detection and frequency mixing. The effects have been reviewed by Waldram (1976), while Richards (1977a) has discussed the various ways in which they may be applied to high-frequency receivers. For fundamental reasons related to the nature of superconductivity, the performance of Josephson devices made from the presently known superconductors will fall rapidly once frequencies around 1 THz are exceeded, although the effects will persist to at least several times this frequency (Weitz *et al.*, 1978a, b).

For operation above 100 GHz, the "point-contact" form of device is usually employed. This is a sharpened superconducting wire touching against the flat and usually lightly oxidized surface of a second piece of superconductor. In most cases, to achieve the best electrical performance, devices that can be continuously adjusted in use have been employed. These generally have the instability problems common to most devices of this form, compounded to some extent by the necessity of cryogenic operation. However, some considerable progress has been made with mechanically stable "preset" devices (Cross and Blaney, 1977; Taur and Kerr, 1978) and also encapsulated devices (Blaney and Cross, 1980). Small-area [i.e., < 1 $(\mu m)^2$] devices made by planar techniques are now being investigated (Daalmans and Zwier, 1978; Hu *et al.*, 1979; Voss *et al.*, 1979). These should approach the electrical performance of the point contact and also have stability and repeatability. However, quantitative results above 100 GHz on such devices have not yet appeared.

As video detectors, the performance of point contacts has been demonstrated in several modes of operation (Nad, 1975; Richards, 1977a). The instantaneous spectral response is usually very structured, predominantly because of the spectral characteristics of the coupling structures used, but also possibly for fundamental reasons (Tolner, 1977). Under some conditions a spectrally selective response, tuned by the bias voltage, can be obtained (Blaney, 1971; Tolner, 1977; Divin and Nad, 1979). In the SMSMR, theory predicts that the intrinsic responsivity of the devices in typical video modes would fall as the square of wavelength.

In the long-wavelength part of the SMSMR, the video NEPs obtained from experimental Josephson detectors are outstanding for rectifying types of device, and could even compete with good bolometers (see Table VI). The combination of such good sensitivity and high speed is unique, although of course the device is restricted to single-mode detection. However, the dependence on point contacts and other experimental difficulties have resulted in the device being virtually unused as a practical detector. With the current improvements in device fabrication under way, there is an increasing probability that practical Josephson video detectors will emerge (Hartfuss *et al.*, 1979; Kadlec *et al.*, 1979).

D. PHOTOCONDUCTIVE DETECTORS

1. *General Properties*

Semiconductor photoconductive detectors depend on the change of electrical conductivity produced by external radiation and by processes other than heating of the material. Photoconductive processes of relevance to detection can usually be arranged in three groups:

(i) Intrinsic photoconductivity, which involves the radiative excitation of electrons from the valence to the conduction band of the semiconductor.

(ii) Impurity (or extrinsic) photoconductivity, resulting from radiative excitation either of electrons bound to a donor impurity into the conduction band (*n*-type material) or holes from an acceptor impurity into the valence band (*p*-type material).

(iii) "Hot-electron" photoconductivity, where already existing "free" carriers (which are only weakly coupled to the crystal lattice) are "heated" by the radiation to produce a change in carrier mobility and hence conductivity.

The first two types are quantum detectors in that individual photons excite the current carriers. For a given energy difference between the bound and conducting states, there is obviously a minimum photon energy required, and hence a long-wavelength cutoff for detection. At wavelengths much shorter than the cutoff wavelength, the detector responsivity will again fall because the number of photons for a given power level decreases. At present, intrinsic photoconductors only marginally extend from the infrared into the shortest wavelengths of the SMSMR (see Table VII). Impurity photoconductivity provides practical detectors to about 0.3 mm, with prospects for

TABLE VII

Photoconductive Detectors

Type	Approx. wavelength range[a] (with peak response) (μm)	Reference	Response speed	Examples of video performance[b]
Extrinsic germanium				
Ge–Cd	2–23 (18)	Levinstein (1965) Hughes–SBRC info.)	Typically 100 ns (amplifier circuit allowing).	Ge–Cu (Shivanandan et al., 1975): NEP 2.5×10^{-13} W Hz$^{-1/2}$ (267 Hz modulation).
Ge–Cu	2–30 (22)	As Ge–Cd	~1 ns possible with more heavily compensated materials.	
Ge–Zn	2–40 (35)	As Ge–Cd		
Ge–In	20–120 (100)	Oka et al. (1968)		
Ge–B	30–135 (100)	Levinstein (1965)		Ge–Ga stressed crystal (Haller et al., 1979): NEP 6×10^{-17} W Hz$^{-1/2}$ (150 Hz modulation, at 2 K).
Ge–Sb	40–140 (120)	Oka et al. (1968)		
Ge–Ga	30–120 (100) (and possibly to 160)	Moore and Shenker (1965) Jeffers and Johnson (1968) Moore (1976)		
Ge–Ga (stressed crystal)	~70–200 (150)	Haller et al. (1979)		
Extrinsic silicon				
Si–As (<20 K)	1–23 (21)	Hughes–SBRC info.	Probably similar to Ge photoconductors.	Si–As: D* $\geq 2.5 \times 10^{10}$ cm Hz$^{1/2}$ W^{-1} (0.25–5 mm diam.) (Hughes–SBRC info.).
Si–P (<20 K)	2–30 (28)	Bratt (1977)		

Material (temperature)	Wavelength range[a]	Reference	Response time	NEP[b]
Si–As, Si–P, Si–Sb, D⁻ state photo-conductivity (1.6 K)	100–500 (200)	Norton (1976a, b) Norton et al. (1977)	Probably ~1 ns.	D^- photoconductivity: NEP probably $\sim 10^{-11}$ W Hz$^{-1/2}$ (Norton 1976a).
Gallium arsenide (from ~4 K to ~6 K)	100–350 (280)	Stillman et al. (1977)	~10 ns	NEP down to 4×10^{-14} W Hz$^{-1/2}$ (Shivanandan et al. 1975).
Indium antimonide Transformer coupling (4.2 K)	~300 to >3000	Kinch and Rollin (1963)	Typically $>10^{-4}$ s (circuit-limited).	NEP 6×10^{-13} W Hz$^{-1/2}$ in range 0.5–8 mm (Kinch and Rollin, 1963).
Transistor amplifier type	Approx. as above	Nakajima et al. (1978)	Down to $\sim 10^{-7}$ s.	NEP 4×10^{-13} W Hz$^{-1/2}$ at 1 mm (Nakajima et al., 1978).
With magnetic field ($\sim 0.6\,T$, $\lesssim 2$ K)	200 to >3000	Putley (1965)	~200 ns	NEP $\sim 5 \times 10^{-12}$ W Hz$^{-1/2}$ (Putley and Martin, 1967).
With magnetic field ($>0.7\,T$)	30–200 (tunable, except 50–60)	Brown and Kimmitt (1965)	100–1000 ns	NEP 10^{-11} W Hz$^{-1/2}$.

[a] These wavelengths should only be taken as an approximate guide. More details will be found in the references given. Spectral response may vary from sample to sample in some cases.

[b] The figures usually apply to the peak in the spectral response. The references provide more detail on the measurement conditions, which may be considerably different for the various examples given.

possible further extension to longer wavelengths. In the hot-electron type, which presently provides useful performance at about 0.3 mm and longer wavelengths, the spectral response is largely determined by the "classical" absorption of the radiation by the free-electron gas. This is constant at long wavelengths, but falls as λ^2 at the short wavelength end. Liquid helium temperatures are required for most of the SMSMR photoconductive devices, although devices for the shortest wavelengths may tolerate somewhat higher temperatures.

In video applications the detector element is usually mounted in an integrating cavity, and its change in resistance to a bias current is monitored either directly across the element or across a series load resistor. The relaxation times associated with photoconductivity are typically in the microsecond to nanosecond range so that the detectors can have a fast response. However, the quantum types can have a very high resistance in low-background conditions, so that maintaining a high bandwidth in the postdetector circuit and amplifier can be difficult. The use of cooled amplifiers adjacent to the photoconductive element is now common (Feldman and McNutt, 1969; Kunz and Madey, 1974; Wyatt et al., 1974; Haller et al., 1979). NEP is often limited by amplifier noise, but the best devices (see Table VII) may be background limited even under low-background conditions. The best photoconductive devices at the shorter wavelengths of the SMSMR are superior to the best bolometers as regards NEP, and are also much faster in response. At wavelengths of about 0.2 mm and longer, the best bolometer NEPs become superior.

Another approach to alleviating the impedance and bandwidth problems is to bias and monitor the photoconductor's response at microwave rather then dc or quasi-dc frequencies (Crouch, 1978). However, in the SMSMR such systems do not presently appear to be competitive with the more conventional arrangement (Crowley et al., 1976).

Detailed discussions of photoconductive detectors are given in articles such as those by Bratt (1977), Stillman et al. (1977), Putley (1977c), and Shivanandan (this volume). However, for completeness, we give a brief discussion of the main devices in the succeeding sections, together with some performance details in Table VII.

2. Intrinsic and Impurity Photoconductors

Little effort has been put into developing materials that would be intrinsic photoconductors in the SMSMR. A result by Saur (1968) on a mercury cadmium telluride device, the response of which extended to about 33 μm from shorter wavelengths, give a performance considerably poorer than impurity devices working at the same wavelength.

The best-established impurity photoconductors in the SMSMR are those based on germanium, with a range of different dopants being used (Table VII). Materials with cutoff wavelengths extending to about 160 μm are used, and this can be extended to about 200 μm by applying stress to Ge–Ga in order to further reduce the impurity binding energy (Haller et al., 1979). The NEP figures given as examples in Table VII are for devices operating at relatively low chopping frequencies. It is possible, with some degradation of responsivity and NEP, to reduce the response time to around 1 ns by increasing the level of compensating impurities that capture current carriers and hence reduce relaxation time. Thus, when speed and/or a good NEP are required at shorter wavelengths in the SMSMR, the germanium photoconductors are presently the most obvious choice.

Other photoconductive mechanisms operate in germanium and extend to longer wavelengths than those indicated in Table VII. Very shallow impurity states, now termed as D^- and A^+ centers, can exist in semiconductors, and a brief discussion of their possible use in detection has been given by Kimmitt (1977). The long-wavelength limit may stretch to beyond 1 mm with the type of photoconductivity, but low temperatures ($\gtrsim 1.5$ K) are required, and the mechanism requires some irradiation with shorter wavelength infrared radiation (e.g., 300 K background radiation) to populate the centers involved. There are presently no published data on detector performance of germanium materials in this mode, although some work has been reported using a similar effect in silicon (see below).

When silicon is used as the basis for conventional extrinsic photoconductors, the spectral response does not extend so far into the SMSMR as it does with germanium, and the longest cutoff wavelength discussed by Bratt (1977) is about 30 μm. In this short-wavelength edge of the SMSMR, the silicon detectors offer no distinct advantage when compared with germanium types. Long-wavelength photoconductivity in extrinsic silicon, arising from the D^- states mentioned above for germanium, has been investigated and shown to be feasible for a detector by Norton (1976a, b) and Norton et al. (1977). Arsenic, phosphorus, and antimony dopants give rather similar spectral responses, with significant response stretching to about 0.5 mm at a working temperature of ~ 2 K. Although the estimated video NEP is not outstanding (Table VII), such devices may be useful for applications requiring high speed or as heterodyne mixers.

High purity epitaxial n-type gallium arsenide exhibits extrinsic photoconductivity in the range between about 50 μm and 0.35 mm. The details of the mechanisms involved and the performance have been discussed by Stillman et al. (1977). Good video NEPs are possible in low-background conditions (see Table VII). While the intrinsic response time is probably about

10 ns, the usual impedance problems make it difficult to approach this limit without significant deterioration in responsivity and NEP. However, using a cryogenic amplifier, Kunz and Madey (1974) achieved a best video NEP of 6×10^{-13} W Hz$^{-1/2}$ with 40 kHz bandwidth, increasing by about a factor of 6 with 40 MHz bandwidth. Gallium arsenide can also exhibit a photo-conductive response at millimeter and longer wavelengths, via mechanisms other than the usual extrinsic photoconductivity (Stillman et al., 1977). However, these mechanisms appear to require rather large radiation powers or do not provide sufficiently good NEPs to be competitive, and have not been developed into practical detectors.

Another III-V semiconductor, indium phosphide, has recently been investigated as a detector (Wilson and Epton, 1978) and gives good response to wavelengths as long as 220–250 μm. At 4.2 K, NEP is not presently out-standing (see Table VII), but may improve if the detector is operated at a slightly higher temperature.

3. Hot-Electron Photoconductors

Photoconductive detectors based on n-type indium antimonide are now well established as hot-electron "bolometers" at the longest wavelengths of the SMSMR. They may be used in various ways (Putley, 1977c; and Table VII). In most of the methods of use, performance is best at wavelengths of about 1 mm and longer, with a decrease in responsivity below 1 mm; but use-ful operation is possible down to a few hundred microns. By use of applied magnetic fields, tunable response at shorter wavelengths is possible (Table VII).

At 4.2 K, the intrinsic response time of the photoconductive effect is a few hundred nanoseconds. Unlike the photoconductors discussed above, the resistance of the indium antimonide element is low, down to a few tens of ohms, and the provision of low-noise postdetector amplification can be a problem when working from this low resistance. For video detectors the problem has been approached in three main ways. A cooled transformer can be used to match the detector crystal to a higher resistance amplifier, al-though to be most effective, the transformer circuit is usually made resonant, with consequently rather small instantaneous modulation bandwidth (Kinch and Rollin, 1963; Clegg and Huizinga, 1971, 1972). The second method is to increase the crystal's resistance by application of a magnetic field. This changes the mechanism of response to some extent, and requires a working temperature of $\gtrsim 2$ K for the best performance (Putley, 1977c). Third, by the use of high-performance transistor amplifiers, and/or specially shaped high-purity crystals of rather higher resistance, Ade (1978) and Naka-jima et al. (1978) have achieved good NEPs and bandwidths up to about

1 MHz without the need for transformers or magnetic fields. This third form of indium antimonide detector is now probably the preferred type for most purposes.

By operation at higher temperatures, the relaxation time associated with the hot-electron photoconductive effect can be reduced. Putley (1978b) estimates that at 20 K, a video NEP of 4×10^{-11} W Hz$^{-1/2}$ should be possible with a response time of about 3 ns, and that the device could be particularly attractive for heterodyne mixing.

Hot-electron effects do occur in other semiconducting materials, but as regards photoconductivity, the effects are often rather weak. At present, only indium antimonide has been developed as a practical hot-electron detector in the SMSMR.

E. MISCELLANEOUS DETECTORS

Of the various types of detection devices that do not readily fit into the categories already discussed, three which may be useful or potentially useful in the SMSMR are worth mentioning.

The "photon-drag" detector, which depends on the momentum transfer from photons to the free carriers in a semiconductor, has been used for some time at 10 μm wavelength, particularly for detection of fast laser pulses. This fast ($\gtrsim 1$ ns), room-temperature detector can work, with slightly improved response, throughout the SMSMR, although responsivity is still low. Recent work (Kimmitt, 1977) indicated a responsivity of about 10^{-5} VW^{-1} and an NEP of about 10^{-4} W Hz$^{-1/2}$ at 500 μm wavelength in a p-type germanium device. However, these rugged and stable devices may well be useful for monitoring high-power cw and pulsed sources in the SMSMR.

The influence of microwave radiation on the properties of dc glow-discharge tubes has been studied as the basis of a detector at millimeter wavelengths (Kopeika and Farhat, 1975; Farhat and Politch, 1979). At 70 GHz, video NEPs down to 10^{-9} W Hz$^{-1/2}$ have been measured, with response times of a few microseconds. The detection effect is much enhanced by the application of a magnetic field to produce cyclotron resonance between the discharge electrons and the radiation frequency (Opher et al., 1978). This form of detection should extend in the SMSMR, and the devices, being inexpensive and quite rugged, may be of interest for the detection of modestly powerful sources or possibly for heterodyne work.

At high quantum numbers (say, $\gtrsim 20$), the so-called Rydberg states of hydrogenic atoms are spaced by energies corresponding to quanta in the SMSMR. Changes in such states, brought about, say, by incident radiation, can be sensed by observing the change of ionization potential of the atom. A quantum detector based on this principle has been proposed by Kleppner and Ducas (1976), who suggest that a turnable device of high quantum efficiency and low noise should be possible. Goy *et al.* (1979) have reported the detection of millimeter-wave powers of 10^{-16} W, using the effect in beams of alkali atoms.

VI. Heterodyne Receivers

A. GENERAL COMMENTS

Except at 100 GHz and slightly higher frequencies, SMSMR heterodyne receivers are still at an early stage of development and some considerable improvements are desirable and possible. There is presently most interest in such receivers in the range 100–1000 GHz for applications in astronomy and plasma diagnostics, in both of which high receiver sensitivity is a leading requirement.

As discussed in Section IV.D, the performance of a heterodyne system depends not only on the properties of the detector/mixer and the signal-input coupling efficiency, as does video performance, but also rather critically on the performance of the local oscillator (power, tunability, purity of spectrum) and IF amplifier (noise level and bandwidth). We shall generally restrict the detailed discussion to the mixer performance, but it is worth making some brief remarks on local oscillators.

The provision of suitable local oscillator (LO) sources becomes increasingly difficult as the frequency increases, particularly for those mixers (e.g., Schottky-barrier diodes) for which a relatively high power (say, $\gtrsim 10$ mW) is required. Some of the LOs which have been used are indicated in Table VIII, and include klystrons, harmonics of klystrons, backward-wave oscillators, and fixed-frequency lasers. To make best use of the LO power available, it must be efficiently coupled to the mixer, simultaneously with the signal power. In the SMSMR it is becoming increasingly common to combine the LO and signal beams in free-space "quasi-optical" diplexers, although this can usually only be done effectively if the IF is above about 1 GHz. Various interferometric diplexers have been described, (Goldsmith, 1977; Wrixon and Kelly, 1978; Payne and Wordeman, 1978) and usually have the added advantage of rejecting LO noise at the signal frequency.

At the longer wavelengths of the SMSMR, rectifying types of mixers are currently of most interest, mainly because of their high IF bandwidth and impedances suitable for matching to IF amplifiers. The properties of receivers based on them are described in Section B. The use of thermal and photoconductive mixers is discussed in Section C.

B. Heterodyne Receivers Using Rectifier-Type Mixers

1. Schottky and Other Diode Mixers

At SMSMR wavelengths the metal–semiconductor and metal–insulator–metal diodes can be regarded as nonlinear resistive mixers. For fundamental reasons, such devices are usually limited in their mixer conversion efficiency (see Section IV) to 50% or less (Torrey and Whitmer, 1948; Page, 1956).

The metal–semiconductor types of mixer have attracted much more interest than the metal–insulator–metal types, although it may be that at frequencies above around 1000 GHz, the latter type could offer a competitive performance and be capable of development in the future (Abrams and Gandrud, 1970; Blaney et al., 1974; Kobzev et al., 1975).

Of the metal–semiconductor types, the point contact form of device has been investigated and used in the past at short millimeter and longer submillimeter wavelengths (e.g., see Blaney, 1978a, for references). However, for practical receivers, the small-area lithographically produced diodes (often with a whisker contact) are now usually preferred, and appear to offer an electrical performance at least as good as the point contacts. Table VIII lists some of the results obtained with lithographically prepared Schottky-type diodes, which will be discussed in the following paragraphs.

A considerable amount of receiver development work has gone on around 100 GHz, but the available data decrease rather rapidly as the frequency increases. Several groups have been studying, both theoretically and experimentally, the noise and frequency conversion properties of diodes at high frequencies (Wrixon 1976; Oxley 1976; McColl, 1977; Clifton, 1977; Mattauch and Fei, 1977; Schneider et al., 1977; Keen et al., 1978; Held and Kerr, 1978). Mixer temperatures down to less than 500 K have been reported near 100 GHz in room-temperature Schottky diodes (Cong et al., 1979; Haas 1979), although to achieve a good overall receiver noise temperature, low-noise IF amplifiers are required, as mixer conversion efficiency is usually 25% or less. By cooling the diodes down to as low as 20 K, mixer noise temperatures improve by up to a factor of about two, although conversion efficiency is often marginally worse (Kerr, 1975; Schneider et al., 1977; Haas, 1979). Strictly speaking, the metal–semiconductor barrier in some of these diodes is of the Mott rather than the Schottky type (Irvin and Vanderwal, 1969). It is interesting to note that a Mott diode operating at 20 K, at 115 GHz, required only 150 μW of LO power, compared with the several milliwatts at least usually required by Schottky devices.

"Subharmonically pumped" mixers, operating with signals near 100 GHz, at room temperature, and with sensitivity very similar to that of good conventional room-temperature mixers have been described by Carlson

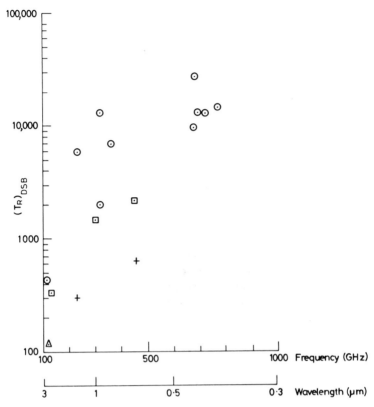

Fɪɢ. 9   A comparison of the available experimental data for heterodyne-system noise temperatures (double sideband) in the short millimeter- and long submillimeter-wavelength range. The data are drawn from references given in the text and Table VIII. Key: ⊙ Schottky, ⊡ Josephson, + InSb, △ SIS.

et al. (1978). This latter system used two diodes coupled to the signal etc. by strip lines. However, hollow waveguide mixer mounts are usual near 100 GHz, although antenna-type coupling has been tried with some success (Kerr *et al.*, 1977b).

Fundamental-mode waveguide mounts have been used with Schottky mixers to beyond 200 GHz (Goldsmith and Plambeck, 1976; Kerr *et al.*, 1977a). However, at these and higher frequencies, overmoded waveguide mounts (Zuidberg and Dymanus, 1978), whisker antennas with reflectors (Fetterman *et al.*, 1978), and biconical cavities (Gustincic, 1977) are more favored. As indicated in Table VIII and Fig. 9, the present performance achieved at these higher frequencies is much poorer than at 100 GHz and is probably of little practical use at wavelengths less than 0.3 mm. However, the results achieved by Fetterman *et al.* (1978) between 1 and 0.4 mm are very promising, while even their result at 0.17 mm (a double-sideband receiver noise temperature of 370,000 K) can almost certainly be improved substantially.

2. *Josephson and Related Mixers*

Josephson mixers can be used with the internal voltage-tunable oscillations as the LO (Avakjan *et al.*, 1975). Although this might be of some interest for wide-bandwidth radiometry in the future, the use of the devices with an external LO is currently of most interest. The Josephson device is not a simple resistive mixer when used with an external LO. Conversion gain is possible, although in the SMSMR, conversion loss is usual in devices of low noise. The broad mechanisms of Josephson mixers are fairly well understood, and there is now a reasonable quantitative agreement between theory and experiment as regards mixer noise temperature and other parameters (Claassen and Richards, 1978; Richards, 1978a). The theory would indicate a mixer noise temperature increasing linearly with frequency above a few hundred gigahertz. Reasonably good performance would be obtainable to about 1000 GHz, where the theoretical mixer noise temperature would be about 1000 K. However, detailed quantitative results are presently available to only 450 GHz (Blaney, 1978b).

The particular assets of Josephson mixers are very good mixer noise temperatures (see Table VIII and Fig. 9), low conversion loss relative to devices such as Schottky diodes (at least up to 450 GHz, and very low LO power requirements (typically microwatts or less). The present major drawback is the dependence on point-contact devices, although this problem is being reduced and possibly solved as new junction devices are developed (Taur and Kerr, 1978; and Section V.C.5). The high broadband sensitivity may also be a problem in some circumstances, possibly making filtering out of unwanted radiation a necessity.

So far, Josephson heterodyne receivers have been used in virtually no routine measurements of low-level signals. However, given the effort to develop a sufficiently rugged form of Josephson junction with good high-frequency properties, they can almost certainly offer a very competitive performance up to 1000 GHz when good sensitivity is required.

Superconductor–insulator–superconductor (SIS) structures, which in suitable circumstances can be Josephson junctions, show resistive non-linearities due to quasi-particle tunneling. This has been investigated as the basis for a "classical" resistive mixer at 30 GHz (Richards *et al.*, 1979) and 115 GHz (Dolan *et al.*, 1979). The latter result gives a promising mixer noise temperature (< 100 K) with only 10 nW of LO power required, but the

TABLE VIII

Heterodyne Mixer and Receiver Performance

| Type of mixer | Signal wavelength | LO | Example of performance[a] | | | | Reference |
			Signal coupling	$(T_M)_{SSB}$	Conversion loss	$(T_R)_{DSB}$	
GaAs Schottky diode (300 K)	2.6 mm (114 GHz)	Klystron	Fundamental-mode hollow waveguide	440 K	5.3 dB	430 K	Cong et al. (1979)
	1.76 mm (170 GHz)	Klystron	Fundamental-mode hollow waveguide	1100 K –1300 K	6.2 dB –7.2 dB	—	Kerr et al. (1977a)
	0.94 mm (318 GHz)	Carcinotron (~10 mW)	Fundamental-mode hollow waveguide	3100 K	9.3 dB	2000 K	Erickson (1978)
	0.83 mm (360 GHz)	Carcinotron (~12 mW)	Biconical cavity	~10,000 K[a]	~14 dB[a]	~7000 K	Gustincic et al. (1977)
	0.45 mm (671 GHz)	Laser (~30 mW)	Biconical cavity	~32,000 K[a]	~19 dB[a]	27,000 K	Gustincic et al. (1977)
	0.39 mm (760 GHz)	Laser (~30 mW)	Whisker antenna with reflector	15,200 K[a]	15.3 dB[a]	14,500 K	Fetterman et al. (1978)
GaAs "Mott" diode (20 K)	2.7 mm (111 GHz)	Klystron (150 µW)	Fundamental-mode hollow waveguide	200 K	~6 dB	—	Keen et al. (1978)
Josephson junction Nb-Nb point contact (~4 K)	2.6 mm (115 GHz)	Klystron (3 nW required)	Fundamental-mode hollow waveguide	120 K	0 dB	—	Taur and Kerr (1978)
	~1 mm (300 GHz)	Backward-wave oscillator (<10 µW required)	Fundamental-mode hollow waveguide	~220 K	9.5 dB	~1500 K	Edrich et al. (1977)
	~0.66 mm (450 GHz)	Laser (25–100 nW typically required)	Fundamental-mode hollow waveguide	350–1000 K	Down to 5 dB	2100 K	Blaney, Cross, and Jones (unpublished); Blaney (1978b)

				Down to ~1200 K estimated	Down to ~10 dB estimated	—	Blaney, Cross, and Jones (unpublished)
SIS mixer (see text) (~1.5 K)	0.337 mm (890 GHz)	Laser	Wire antenna, not optimized				
	2.6 mm (115 GHz)	Klystron (~10 nW required)	Fundamental-mode hollow waveguide	<100 K	~10 dB	—	Dolan et al. (1979)
Indium antimonide (hot-electron) (~4 K)	1.3 mm	115 GHz Klystron, externally doubled	Fundamental-mode hollow waveguide	~500 K	~10 dB	300 K	Phillips and Jefferts (1974)
	0.65 mm	Carcinotron	Oversized waveguide	—	—	650 K	Van Vliet et al. (1978)
Pyroelectric, TGS in various types of mount	0.337 mm	Laser (5×10^{-5} W)	Flat element, 0.25 mm^2	Het. NEP $\sim 10^{-12}$ W Hz^{-1} (800 Hz IF)			
	10.6 μm	Laser (7×10^{-3} W)	Flat element, 0.8 mm^2	Het. NEP $\sim 2 \times 10^{-15}$ W Hz^{-1} (10^3 Hz IF)			Baynham et al. (1974)
	10.6 μm	Laser (4×10^{-2} W)	Flat element, 0.8 mm^2	Het. NEP 5×10^{-14} W Hz^{-1} (10^6 Hz IF)			

[a] The performance figures quoted are usually the best available in the literature. The mixer noise temperatures, T_M, and conversion loss figures (which are given for single sideband operation) in most cases do not include the signal input coupling loss, or the latter is relatively small compared to the intrinsic conversion loss. However, the figures for Schottky diodes at 0.83 mm and shorter wavelengths include the input coupling loss. The receiver noise temperatures, T_R, are for double-sideband operation, and include all losses and IF amplifier noise contributions, which vary considerably in the various examples.

conversion loss was 10 dB. However, the evaporated-film devices used are durable, and with improvement and extension to higher frequencies, this may be a very useful mixer.

The super-Schottky diode (see Section V.C.3) has shown a good mixing performance at 30 GHz (McColl *et al.*, 1979), but no data are yet available at higher frequencies.

C. HETERODYNE RECEIVERS USING THERMAL AND PHOTOCONDUCTIVE MIXERS

In general, when a thermal or photoconductive detector is used for heterodyne mixing, the mounting geometry and optimization of device parameters might be considerably different than in the video case. For example, the use of an integrating cavity is not generally desirable for mixing because of the radiation phase scrambling it would produce. Also, the use of the detector at relatively high LO powers may considerably change its characteristics when compared with video operation. Thus, the video characteristics cannot readily be translated into heterodyne performance.

Thermal detectors are usually too slow for heterodyne work. However, pyroelectrics have good responsivity and might conceivably be used at IFs up to ~ 1 MHz throughout the SMSMR. The few results so far (Table VIII) are disappointing, and even if the projected NEPs of $\sim 10^{-16}$ W Hz^{-1} are obtained (Baynham *et al.*, 1974), their range of application is likely to be rather limited. Antenna-coupled microbolometers (Hwang *et al.*, 1979b) are probably sufficiently fast for effective mixing performance, but no heterodyne results are yet available.

Of the photoconductive mixers, "hot-electron" indium antimonide has provided a very useful performance, particularly for astronomy, over the past few years (Phillips and Jefferts, 1973: van Vliet *et al.*, 1978). They give outstanding receiver noise temperatures (see Table VIII) with little LO power required ($\gtrsim 1 \mu$W), but the IF limit in the present systems (~ 2 MHz) is a great disadvantage in practice. However, Putley (1978b) suggests that wider IF bandwidth will be possible if a higher operating temperature is used, the major penalty being greater LO power requirements. Indium antimonide mixers could probably be used at wavelengths down to ~ 0.3 mm.

Germanium photoconductors have certainly provided good heterodyne performance at 10 μm wavelengths (De Graauw, 1975) where powerful laser local oscillators are available. However, a published result at 0.12 mm using Ge–Ga (Seib, 1974) is relatively poor, almost certainly because of insufficient LO power.

Mixing in extrinsic silicon at 0.12 mm using the D$^-$ states detection effect (see Section V.D.2) has yielded heterodyne NEPs of $\gtrsim 10^{-16}$ W Hz^{-1} (Norton *et al.*, 1977) and improvements of two to three orders of magnitude could be possible. An experiment at 0.337 mm using gallium arsenide gave a heterodyne NEP of $\sim 10^{-14}$ W Hz^{-1} (De Graauw et al., 1976) with only 3% quantum efficiency. Silicon and gallium arsenide presently cover approximately the same region of the SMSMR (see Table VII), but it is not clear from the results so far if either will provide a satisfactory heterodyne performance.

VII. Imaging Receivers

The purpose of an imaging receiver, as defined here, is to be able to display the spatial intensity variation of radiation falling on a target, usually in two dimensions, but possibly in just one dimension. The tasks for which such a device might be used in the SMSMR would include the measurement of the shape or position of a laser beam, or the structure of an interference pattern. We do not specifically consider the problem of building up an image of a physical object that is remote from the receiver, although obviously the imaging receivers we have defined may well be used to do so.

In addition to the detector parameters discussed in previous sections, an imaging system is also characterized by the spatial resolution which it can provide, as well as the total number of resolved units available in the "picture." By the nature of radiation, a single mode of the radiation field must occupy an area of at least $\sim \lambda^2$, so that this is the resolution limit.

The range of different schemes by which the spatial variation of intensity may be recorded is obviously large. In the SMSMR the number tried in practice is relatively small, and most of these have had only preliminary investigation. Blaney (1978a) has briefly reviewed the published experimental work in the SMSMR and little progress has been reported since then.

One way of building up a picture of the spatial variation of intensity is to use an array of the sort of detectors already discussed, or to spatially scan a single detector and record the variation of intensity. For example, a scanning system has been described by Zav'yalov and Voronin (1976) for displaying the power distribution in submillimeter-wavelength laser beams. Arrays of pyrolectric detectors (Watton, 1976; Roundy, 1979) could be particularly useful, but no work on their use in the SMSMR has been published.

So far, the technique that has been most used in the SMSMR is that in which the radiation is absorbed to produce a thermal image, which is then sensed by a thermal detection effect. At rather high power densities, the discoloration of heat-sensitive paper, temperature-sensitive liquid crystals on suitable absorbing substrates, and thermographic phosphor screens have all been used. For better performance, liquid-crystal cameras have been described and used at 0.337 mm wavelength (Keilmann and Renk, 1971; Lesieur et al., 1972), "evapographic" devices used at 0.5 mm and shorter wavelengths (Dodel *et al.*, 1976; Dodel and Kunz, 1977), and pyroelectric vidicons at 0.337 mm and shorter wavelengths (Fuller *et al.*, 1972; Hadni, 1974; Felix *et al.*, 1976; Dodel and Kunz, 1977). The published quantitative data for these devices in the SMSMR have been compared by Blaney (1978a).

The pyroelectric vidicon (Holeman and Wreathall, 1971; Watton, 1978) shows particularly good potential, but the present devices are usually designed for 10 μm or shorter wavelengths, so that target areas tend to be rather small for SMSMR work, particularly at the longer wavelengths. As with any device that uses a thermal image in a continuous sheet of material, the early pyroelectric vidicons could suffer from thermal spread of the image and hence some degradation of spatial resolution. However, reticulated layers of pyroelectric material that considerably reduce the problem, can now be produced (Watton, 1976, 1978), although no devices specifically for SMSMR use appear to have been produced.

VIII.   Calibration and Characterization of Receiver Performance

A.   General Remarks

It was remarked earlier that proper standards and calibration techniques are lacking for some aspects of SMSMR measurements. Receiver performance is one of these. In practice, various ad hoc techniques have been used for calibration, and we do not intend to give a comprehensive discussion of these here. Much of the discussion will just restate the difficulties involved.

Tables II and III list the properties that should be known if a receiver system is to be properly characterized, although we shall restrict attention here to the properties that are currently most difficult to measure. These are spectral response, responsivity, NEP, angular variation of response, and response speed, as well as other more detailed factors, such as conversion loss for a heterodyne mixer. Several of these require knowledge or a measurement of radiation power, and in Section C there is a short discussion of power-measuring detectors. It is worth reemphasizing that many of the most difficult calibration problems exist at short millimeter and long submillimeter wavelengths in free-space systems because the wavelengths are often not negligible compared with aperture size, nor can the simplicities of geometrical optics be assumed. This is compounded by the weakness of blackbody sources at these long wavelengths. Calibration problems are considerably eased as the wavelength shortens below 100 μm.

B.   Calibration and Measurement Methods

As an obvious general rule, it is desirable to calibrate a receiver system under conditions as close as possible to those in which it will be used, e.g., similar power levels; wavelength ranges; background levels; size, type, and distance of sources; and output bandwidths. Spectral response in the SMSMR is often measured with a broadband source, with a spectrometer, or possibly a set of filters, to provide spectral resolution. If the spectral characteristics of the spectrometer are known, or if the combined performance of the spectrometer plus receiver are required, then spectral response is most readily measured by use of a "standard" source. This would usually be a blackbody. At 300 K and lower temperatures, absorbers based on carbon or metallic-black paints, or on iron-loaded epoxy foams are often used as the basis of SMSMR black bodies. These are adequate for good, cooled detectors, but for detectors of more modest sensitivity, the signal-to-noise ratio they provide is usually insufficient for useful measurements. Source power increases with higher temperature (although only approximately linearly with temperature at the longer SMSMR wavelengths), but relatively little work has been done on high-temperature blackbody sources for long wavelengths (but see, for example, Lichtenberg and Sesnic, 1966). A possibility is to calibrate the spectrum of a brighter source (e.g., a mercury lamp) against a blackbody, although the introduction of this extra step is obviously reflected in poorer overall accuracy.

A second method for spectral response is to use monochromatic sources, of measured relative power, to give a series of "spot" measurements across the spectral bandwidth of interest. This method normally requires a detector, or detectors, that can give the relative power levels at the different wavelengths. This is also true of the third method, in which the detector under test is ratioed directly against a detector of known, or assumed, spectral response, such as a Golay cell or bolometer. Many authors are content to give the spectrum a detector under test relative to a more commonly available device, such as a Golay cell, without making any assumption about the latter's spectral response.

For the measurement of responsivity, the incident power from the source must be known absolutely. If sufficiently powerful monochromatic sources are available, they may be calibrated by the sorts of methods discussed in Section C, and then possibly attenuated to give a power level suitable for detector tests. Daehler (1976) has discussed a calorimeter system, operating at powers down to about 1 μW, specifically for calibrating SMSMR detectors. If a blackbody source is used, then the detector's area, angular variation of sensitivity, and spectral response should be known, as well as the area of the source and the geometry of the test apparatus. For some wavelength ranges and for some types of detectors, these measurements may be difficult to make if an absolute value for responsivity is required.

The measurement of NEP requires the same sorts of measurements as for responsivity, together with measurements of the detection system's noise level. It may be important to ensure that the background and signal levels under calibration are similar to those in which the detector is usually employed. Thus, for example, for a good, cooled bolometer, a low-temperature blackbody calibrating source may be most appropriate (Nolt et al., 1977).

For heterodyne receivers, the single-mode response means that assumptions can be made regarding the étendue, and as long as the field of view is filled by a blackbody source, the effective incident power can be calculated. Where appropriate, the noise temperature of the system can be measured by the Y-factor method (Mumford and Scheibe, 1968). If the system is sufficiently sensitive, room temperature and cryogenic black bodies can be used (Goldsmith, 1977; Fetterman et al., 1978). To measure overall system conversion efficiency, a small, known, monochromatic signal may be used and compared with the IF output, or alternatively, modulation of the local oscillator by a known amount can provide a calibration (Kurbatov and Penin, 1976; Sauter and Schultz, 1977).

A knowledge of the variation of receiver sensitivity with the angle of incidence of the signal radiation is obviously important for some applications. The radiation patterns of small SMSMR antennas and single-mode receivers can be measured fairly readily using monochromatic sources (Krautle et al., 1978), although a accurate knowledge of the characteristics of the probe beam is required for precise measurements. Measurements on incoherent detection systems tend to be more difficult, particularly if measurements of small side lobes on the receiver's radiation pattern are necessary. The use of monochromatic coherent sources for such measurements can be misleading because of interference effects unless the spatial coherence is destroyed by a diffuser. Even with incoherent test sources, a knowledge of the mode structure of the propagating radiation and the application of diffraction corrections may be necessary (Richards, 1978b).

The measurement of receiver linearity has not received much attention although it may be important, for example, in spectrometric measurements requiring a large dynamic range. Linearity measurements require accurate attenuators, although particularly for free-space SMSMR systems, little work has been done on standard attenuators. For detectors of relatively slow response, chopping attenuators, which effectively change the duty cycle of the single source are probably the most simple (Hennerich et al., 1966) and can be used to calibrate other types of attenuator.

For measurement of receiver response time, mechanical or other types of amplitude modulator (Melngailis and Tannenwald, 1969; Birch and Jones, 1970; Fan et al., 1977) may be used up to 1 MHz and possibly beyond. For higher frequencies the detection system can be tested by using it as a mixer of two monochromatic sources of suitable frequency difference, or short rise-time pulsed sources may be used if available (Kohl et al., 1978).

C. POWER-MEASURING DETECTORS

A common route to providing a radiation-power standard is that of the electrically calibrated thermal detector, i.e., a device in which radiative and electrical heating can be directly compared. The absolute accuracy achieved with such systems depends on such factors as how well the effective radiation absorption factor is known and how well the quantitative effects of the different thermal distributions produced by radiative and electrical heating can be estimated.

At microwave frequencies up to around 100 GHz, good power standards are available for radiation propagating in single-mode hollow waveguides, the absolute instruments usually being dry-load calorimeters. Commercial calorimeters giving absolute accuracies of a few percent or better at a few milliwatts and higher power are available. Above 100 GHz, formal power standards are not yet available, although electrically calibrated calorimeters operating in fundamental waveguides (Stevens, 1979) and overmoded waveguides (Keen, 1978; Clarici and Keen, 1979) have been described for use up to around 300 GHz. At 300 GHz, such devices typically have an accuracy of $\gtrsim 10\%$ for cw power levels of $\gtrsim 10$ mW with thermal time constants of at least a few tens of seconds. Water-flow calorimeters, usually in overmoded waveguides, have been used for measurements (of $\gtrsim 10$ mW) up to around 1000 GHz (Schaer, 1970).

For power measurement of radiation propagating in free space, there has been relatively little effort to produce "standard" detectors in the SMSMR. At the infrared end of the range, electrically calibrated thermopiles (Gillham, 1962) or pyroelectric detectors (Doyle et al., 1976) may sometimes be usable, depending on the effectiveness of the black paint or other coating as an absorber of radiation. Commercially available power meters for both continuous-wave and pulse applications are often claimed to be accurate to wavelengths as long as several tens of microns. By addition of extra layers of absorbing paint, combined with measurements of the reflectivity of the absorber so formed, thermopiles have been used to wavelengths as long as about 500 μm, giving accuracies of 20% or better at cw power levels of around 1 mW (D. W. E. Fuller, private communication).

A few workers have devised electrically calibrated detectors specifically for the long-wavelength submillimeter region. A thermopile device, with a metal film on a glass substrate serving as both the radiation absorber and electrical heater, has been described by Stone et al. (1975). This prototype standard had a provisional uncertainty of about 15% at 1 mW and 0.337 mm, but improvement in accuracy and extension to both shorter and longer wavelengths should be possible. Evenson et al. (1977a) have demonstrated a thermocouple detector, using cone-absorber geometry and with an estimated accuracy of 10% at 1 mW for wavelengths of about 500 μm and shorter.

Daehler (1976) has described a cryogenic calorimeter, operating at wavelengths from 100 μm to 5 mm and at powers down to about 1 μW, and specifically designed for detector calibrations.

A means of measuring source power, which might be particularly useful for the 3 mm to 300 μm wavelength range has been discussed by Llewellyn–Jones and Gebbie (1979). This uses a heavily overmoded cavity that allows the emissivity of a thermal source to be determined and its radiation compared with that of the source to be calibrated.

Overall, the availability of satisfactory power meters in the SMSMR for both continuous wave and pulsed measurements is presently not good. However, the means exist for solving many of the problems and considerable progress can be expected in the next few years.

IX.   Conclusion

There are many improvements to SMSMR receivers that can be foreseen as being both desirable and probable in the future.

As regards video systems, the present sensitivities of the best photoconductive devices, particularly at wavelengths of less than about 100 μm, and of the best bolometers are unlikely to be much improved. However, specific types within the latter group of devices are capable of improvement in sensitivity and speed of response. The room-temperature devices, such as Golay cells and pyroelectric detectors, will continue to be widely used, but as cryogenic systems become more convenient and less expensive, cooled devices may well be used much more routinely than at present. Cooled electronics will become increasingly common in conjunction with cooled detectors. As mentioned throughout the text, there are several video devices that could be further developed for specific requirements.

As regards heterodyne systems, the main requirements in the immediate future will be at short-millimeter wavelengths, with considerable activity at wavelengths down to around 300 μm. Further significant improvements in heterodyne receiver sensitivity are possible, particularly at wavelengths of 2 mm and less. As local oscillators improve, convenience and compactness in design will become even more important, with considerable emphasis on evaporated-film components, integrated-circuit techniques, and device arrays. There is a requirement for considerable further work in the design, fabrication, and characterization of antennas and other coupling systems for rectifier-type mixers, while improved types of waveguide and propagation techniques will make new demands on receiver design. Schottky-barrier diodes will almost certainly be the most-used mixers in the foreseeable future, although other types may be important where the very best sensitivity is required.

For many of the SMSMR receivers, including some of those in quite widespread use, there is a need for much better commercial availability of devices approaching state-of-the-art performance, and for greater ruggedness, reliability, and convenience in use. While cryogenic techniques have improved in convenience in recent years, compact refrigerators, preferably closed cycle and providing temperatures of $\gtrsim 10$ K with the rather modest refrigeration powers required by many detectors, would be a great advantage for the cooled detectors. Particularly at the longer wavelengths of the SMSMR, where commercial activity is likely to be greatest, reliability and cost effectiveness of devices will obviously be important. Commercial pressure will add to the already great need for much improved measurement practice and calibration techniques in this region of the spectrum.

REFERENCES

Abbas, M. M., Kostiuk, T., and Ogilvie, K. W. (1976). *Appl. Opt.* **15**, 961–70.
Abrams, R. L., and Gandrud, W. B. (1970). *Appl. Phys. Lett.* **17**, 150–152.
Ade, P. A. (1978). Private communication.
Allen, C., Abrams, F. R., Wang, M., and Bradley, C. C. (1969). *Appl. Opt.* **8**, 813–817.
Adronov, A. A. *et al.* (1978). *Infrared Phys.* **18**, 385–393.
Annis, A. D., and Simpson, G. (1974). *Infrared Phys.* **14**, 199–205.
Arams, F. R. (1973). "Infrared to Millimeter Wavelength Detectors." Artech, Dedham, Massachusetts.
Astheimer, R., and Weiner, S. (1964). *Appl. Opt.* **3**, 493–500.
Avakjan, R. S., Vystavkin, A. N., Gubankov, V. N., Migulin, V. V., and Shtykov, V. D. (1975). *IEEE Trans. Magn.* **MAG-11**, 838–840.
Bachmann, R., Kirsch, H. C., and Geballe, T. H. (1970). *Rev. Sci. Instrum.* **41**, 547–549.
Baker, H. J. (1975). *J. Phys. E: Sci. Instrum.* **8**, 261–262.
Baker, G., Charlton, D. E., and Lock, P. J. (1972). *Radio Electron. Eng.* **42**, 260–264.
Baynham, A. C., Elliot, C. T., Shaw, N., and Wilson, D. J. (1974). *Digest Conf. Submillimeter Waves Their Appl., Atlanta* IEEE Cat. No. 74 CHO 856-MTT, pp. 153–155.
Becklake, E. J., Payne, C. D., and Prewer, B. E. (1970). *J. Phys. D: Appl. Phys.* **3**, 473–481.
Beerman, H. P. (1975). *Infrared Phys.* **15**, 225–231.
Bertin, C. L., and Rose, K. (1968). *J. Appl. Phys.* **39**, 2561–2568.
Bertin, C. L., and Rose, K. (1971). *J. Appl. Phys.* **42**, 163–166.
Birch, J. R., and Jones, R. G. (1970). *Infrared Phys.* **10**, 217–224.
Blaney, T. G. (1971). *Phys. Lett.* **37A**, 19–20.
Blaney, T. G. (1975). *Space Sci. Rev.* **17**, 691–702.
Blaney, T. G. (1978a). *J. Phys. E: Sci. Instrum.* **11**, 856–881.
Blaney, T. G. (1978b). *In* "Future Trends in Superconductive Electronics" (B. S. Deaver, C. M. Falco, J. H. Harris, and S. A. Wolf, eds.), pp. 230–238. American Institute of Physics, New York.
Blaney, T. G., and Cross, N. R. (1980). To be published.
Blaney, T. G., Bradley, C. C., Edwards, G. J., and Knight, D. J. E. (1974). *Digest Conf. Precision Electromagn. Measurements, London* IEE Conf. Publ. No. 113, IEEE. Cat. No. 74 CHO 770-8 IM, pp. 99–100.

Block, W. H., and Gaddy, O. L. (1973). *IEEE J. Quantum Electron.* **QE-9**, 1044–53.

Bloembergen, N. (1959). *Phys. Rev. Lett.* **2**, 84–85.

Boissel, P. (1975). *J. Phys.* **36**, 1023–1028.

Boyd, R. W. (1977). *Opt. Eng.* **16**, 563–568.

Boyle, W. S., and Rodgers, K. F. (1959). *J. Opt. Soc. Am.* **49**, 66–69.

Bratt, P. R. (1977). *In* "Semiconductors and Semimetals" (R. K. Willardson and A. C. Beer, eds.), Vol. 12, Infrared Detectors II, pp. 39–142. Academic Press, New York.

Brown, M. A. C. S., and Kimmitt, M. F. (1965). *Infrared Phys.* **5**, 93–97.

Burrus, C. A. (1966). *Proc. IEEE* **54**, 575–587.

Busse, G. (1978). *Proc. Int. Conf. Infrared Phys., 2nd, Zurich* (E. Affolter and F. Kneubühl, eds.), pp. 340–341.

Byer, N. E., Stokowski, S. E., and Venables, J. D. (1975). *Appl. Phys. Lett.* **27**, 639–641.

Carlson, E. R., Schneider, M. V., and McMaster, T. F. (1978). *IEEE Trans. Microwave Theory Tech.* **MTT-26**, 706–715.

Champlin, K. S., and Eisenstein, G. (1977). *Appl. Phys. Lett.* **31**, 221–223.

Champlin, K. S., and Eisenstein, G. (1978). *IEEE Trans. Microwave Theory Tech.* **MTT-26**, 31–34.

Chanin, G. (1972). *In* "Infrared Detection Techniques for Space Research" (V. Manno and J. Ring, eds.), pp. 114–120. Riedel, Dordrecht.

Chanin, G., Torre, J. P., and Peccoud, L. (1978). *Infrared Phys.* **18**, 657–662.

Chantry, G. W. (1971). "Submillimetre Spectroscopy." Academic Press, New York.

Chatanier, M., and Gauffre, G. (1972). *In* "Infrared Detection Techniques for Space Research" (V. Manno and J. Ring, eds.), pp. 141–146. Riedel, Dordrecht.

Chiao, R. Y. (1979). *IEEE Trans. Magn.* **MAG-15**, 446–449.

Claassen, J. H., and Richards, P. L. (1978). *J. Appl. Phys.* **49**, 4117–4129, 4130–4140.

Clarici, J. B., and Keen, N. J. (1979). *Colloq. Measurement Power Higher Microwave Frequencies, IEE, London* (unpublished).

Clarke, J., Hoffer, G. I., and Richards, P. L. (1974). *Rev. Phys. Appl.* **9**. 69–71.

Clarke, J., Hoffer, G. I., Richards, P. L., and Yeh, N. H. (1977). *J. Appl. Phys.* **48**, 4865–4879.

Clegg, P. E., and Huizinga, J. S. (1971). *Proc. Conf. Infrared Tech., Reading* IERE Conf. Proc. No. 22, pp. 21–30.

Clegg, P. E., and Huizinga, J. S. (1972). *In* "Infrared Detection Techniques for Space Research" (V. Manno and J. Ring, eds.), pp. 132–140. Riedel, Dordrecht.

Clifton, B. J. (1977). *IEEE Trans. Microwave Theory Tech.* **MTT-25**, 457–463.

Cong, H., Kerr. A. R., and Mattauch, R. J. (1979). *IEEE Trans. Microwave Theory Tech.* **MTT-27**, 246–248.

Coron, N., Dambier, G., and Leblanc, J. (1972). *In* "Infrared Detection Techniques for Space Research" (V. Manno and J. Ring, eds.), pp. 121–131. Riedel, Dordrecht.

Corsi, S., Dall'Oglio, G., Fanton, G., and Melchiorri, F. (1973). *Infrared Phys.* **13**, 253–272.

Costley, A. E. (1979). *In* "Trends in Physics, 1978" (M. M. Woolfson, ed.), pp. 351–360. Adam Hilger, Bristol.

Costley, A. E., How, J. A., and Vizard, D. R. (1978). National Physical Laboratory Rep. No. DES 54.

Crenn, J. P. (1979). *IEEE Trans. Microwave Theory Tech.* **MTT-27**, 573–577.

Cross, N. R., and Blaney, T. G. (1977). *J. Phys. E: Sci. Instrum.* **10**, 146–149.

Crouch, J. N. (1978). *Infrared Phys.* **18**, 89–98.

Crowley, J. D., Wilson, W. L., Tittel, F. K., and Rabson, T. A. (1976). *Infrared Phys.* **16**, 225–232.

Daalmans, G. M., and Zwier, J. (1978). *In* "Future Trends in Superconductive Electronics"

(B. S. Deaver, C. M. Falco, J. H. Harris, and S. A. Wolf, eds.), pp. 312–316. American Institute of Physics, New York.

Daehler, M. (1976). Rep. No. 7976, Naval Research Laboratory, Washington, D.C.

Daiku, Y., Mizuno, K., and Ono, S. (1978). *Infrared Phys.* **18**, 679–682.

Dall'Oglio, G., Melchiorri, B., Melchiorri, F., and Natale, V. (1974). *Infrared Phys.* **14**, 347–350.

De Graauw, Th. (1975). *Space Sci. Rev.* **17**, 709–719.

De Graauw, Th., and Norton, P. (1976). *Infrared Phys.* **16**, 51–54.

De Graauw, Th., Van de Stadt, H., Bicanic, D., Zuidberg, D., and Hugenholtz, A. (1976). *Infrared Phys.* **16**, 233–235.

Dicke, R. H. (1946). *Rev. Sci. Instrum.* **17**, 268–275.

Divin, Yu Ya, and Nad, F. Ya (1973). *Radio Eng. Electron. Phys.* **18**, No. 4, 642–645.

Divin, Yu Ya, and Nad, F. Ya (1979). *IEEE Trans. Magn.* **MAG-15**, 450–453.

Dodel, G., and Kunz, W. (1977). *J. Opt. Soc. Am.* **67**, 975–978.

Dodel, G., Krautter, J., and Haglsperger, H. (1976). *Infrared Phys.* **16**, 237–242.

Dolan, G. J., Phillips, T. G., and Woody, D. P. (1979). *Appl. Phys. Lett.* **34**, 347–349.

Doyle, W. M., McIntosh, B. C., and Geist, J. (1976). *Opt. Eng.* **15**, 541–548.

Draine, B. T., and Sievers, A. J. (1976). *Opt. Commun.* **16**, 425–428.

Drew, H. D., and Sievers, A. J. (1969). *Appl. Opt.* **8**, 2067–2071.

Dryagin, Yu A., and Fedoseev, L. I. (1969). *Izv. VVZ. Radiofiz.* **12**, 813–818.

Dyubko, S. F., and Efimenko, M. N. (1971). *Sov. Phys.—JETP Lett.* **13**, 379–380.

Edrich, J., Sullivan, D. B., and McDonald, D. G. (1977). *IEEE Trans. Microwave Theory Tech.* **MTT-25**, 476–479.

El-Atawy, S. A. (1979). *Infrared Phys.* **19**, 43–47.

Erickson, N. R. (1978). *IEEE MTT-S Microwave Symp. Digest* (J. Y. Wong, ed.), IEEE Cat. 78 CH 1355-7 MTT, pp. 438–439.

Evans, D. E. (1976). *Physica* **82C**, 27–42.

Evenson, K. M. *et al.* (1977a). *IEEE J. Quantum Electron.* **QE-13**, 442–444.

Evenson, K. M., Jennings, D. A., Petersen, F. R., and Wells, J. S. (1977b). *In* "Laser Spectroscopy" (J. L. Hall and J. L. Carsten, eds.), Vol. III, pp. 56–68. Springer, Berlin and New York.

Fan, J. C. C., Fetterman, H. R., Bachner, F. J., Zavracky, P. M., and Parker, C. R. (1977). *Appl. Phys. Lett.* **31**, 11–13.

Farhat, N. H., and Politch, J. (1979). *J. Phys. E: Sci. Instrum.* **12**, 89–90.

Fedoseyev, L. I., and Shabanov, V. N. (1973). *Radio Engng. Electron. Phys.* **18**, No. 3, 469–70.

Feldman, P. D., and McNutt, D. P. (1969). *Appl. Opt.* **8**, 2205–10.

Felix, P. *et al.* (1976). *Opt. Laser Technol.* **8**, 75–80.

Fetterman, H. R., Clifton, B. J., Tannenwald, P. E., and Parker, C. D. (1974a). *Appl. Phys. Lett.* **24**, 70–72.

Fetterman, H. R., Clifton, B. J., Tannenwald, P. E., Parker, C. D., and Penfield, H. (1974b). *IEEE Trans. Microwave Theory Tech.* **MTT-22**, 1013–1015.

Fetterman, H. R., Tannenwald, P. E., Clifton, B. J., Parker, C. D., Fitzgerald, W. D., and Erickson, N. R. (1978). *Appl. Phys. Lett.* **33**, 151–154.

Firth, J. R., and Davies, L. B. (1971). *Proc. Conf. Infrared Tech.*, *Reading* IERE Conf. Proc. No. 22, pp. 227–230.

Frayne, P. G., Chandler, N., and Booton, M. W. (1978). *J. Phys. D: Appl. Phys.* **11**, 2391–2400.

Fuller, D. W. E., Cross, N. R., Wreathall, W. M., and Chaplin, L. I. (1972). *Electron Lett.* **8**, 44–45.

Gabriel, F. C. (1974). *Appl. Opt.* **13**, 1294–1295.

Galantowicz, T. A. (1977). *IEEE J. Quantum Electron.* **QE-13**, 459–461.

Gallinaro, G., and Varone, R. (1975). *Crogenics* **15**, 292–293.

Gehre, O. (1974). *IEEE Trans. Microwave Theory Tech.* **MTT-22**, 1061–1064.

Geist, J., Dewey, H. J., and Lind, H. A. (1976). *Appl. Phys. Lett.* **28**, 171–173.

Gelbwachs, J. A., Klein, C. F., and Wessel, J. E. (1978). *IEEE J. Quantum Electron.* **QE-14**, 77–79.

Gelinas, R. W., and Genoud, R. H. (1959). Rand Corporation Rep. No. P-1697.

Gillham, E. J. (1962). *Proc. R. Soc. London Ser. A* **269**, 249–256.

Ginsberg, D. M., and Tinkham, M. (1960). *Phys. Rev.* **118**, 900–1000.

Golay, M. J. E. (1947). *Rev. Sci. Instrum.* **18**, 347–362.

Golay, M. J. E. (1949). *Rev. Sci. Instrum.* **20**, 816–820.

Goldsmith, P. F. (1977). *Bell. Syst. Tech. J.* **56**, 1483–1501.

Goldsmith, P. F., and Plambeck, R. L. (1976). *IEEE Trans. Microwave Theory Tech.* **MTT-24**, 859–861.

Goy, P., Fabre, C., Haroche, S., Gross, M., and Raimond, J. M. (1979). Paper given at British Radio Frequency Spectroscopy Group Meeting, Keele, March (unpublished).

Green, M. R. (1978). *Phys. Bull.* **29**, 254–255.

Green, M. R., Marsden, P. L., Donaldson, G. B., Cash, A., and Cawthraw, M. J. (1979). *Proc. Int. Conf. Infrared Phys., 2nd, Zurich* (E. Affolter and F. Kneubühl, eds.), pp. 314–316.

Green, S. I., Coleman, P. D., and Baird, J. R. (1971). *Proc. Symp. Submillimeter Waves, New York, 1970* (J. Fox, ed.), pp. 369–389. Polytechnic Press, Brooklyn, New York.

Gundersen, M. (1974). *Appl. Phys. Lett.* **24**, 591–592.

Gustincic, J. J. (1977). *Proc. Soc. Photo-Opt. Instrum. Eng.* **105**, 40–43.

Gustincic, J. J., De Graauw, Th., Hodges, D. T., and Luhmann, N. C. (1977). Unpublished report.

Haas, R. W. (1979). *Electron. Lett.* **15**, 115–116.

Hadni, A. (1969). *Opt. Commun.* **1**, 251–253.

Hadni, A. (1974). *IEEE Trans. Microwave Theory Tech.* **MTT-22**, 1016–1018.

Hadni, A., Thomas, R., Mangin, J., and Bagard, M. (1978). *Infrared Phys.* **18**, 663–668.

Haller, E. E., Hueschen, M. R., and Richards, P. L. (1979). *Appl. Phys. Lett.* **34**, 495–497.

Harper, D. A., Hildebrand, R. H., Stiening, R., and Winston, R. (1976). *Appl. Opt.* **15**, 53–60.

Harris, D. J., Lee K. W., and Batt, R. J. (1978). *Infrared Phys.* **18**, 741–747.

Hartfuss, H. J., Kadlec, J., and Gundlach, K. H. (1979). *Proc. Int. Conf. Infrared Phys., Zurich* (E. Affolter and F. Kneubühl, eds.), pp. 319–321.

Hauser, M. G., and Notarys, H. A. (1975). *Bull. Am. Astron. Soc.* **7**, 409.

Heiblum, M., Wang, S., Whinnery, J. R., and Gustafson, T. K. (1978). *IEEE J. Quantum Electron.* **QE-14**, 159–169.

Held, D. N., and Kerr, A. R. (1978). *IEEE Trans. Microwave Theory Tech.* **MTT-26**, 49–55, 55–61.

Hennerich, K., Lahmann, W., and Witte, W. (1966). *Infrared Phys.* **6**, 123–128.

Hickey, J. R., and Daniels, D. B. (1969). *Rev. Sci. Instrum.* **40**, 732–733.

Hodges, D. T. (1978). *Infrared Phys.* **18**, 375–384.

Holeman, B. R., and Wreathall, W. M. (1971). *J. Phys. D: Appl. Phys.* **4**, 1898–1909.

Hu, E. L., Jackel, L. D., Epworth, R. W., and Fetter, L. A. (1979). *IEEE Trans. Magn.* **MAG-15**, 585–588.

Hwang, T., Rutledge, D. B., and Schwarz, S. E. (1979a). *Appl. Phys. Lett.* **34**, 9–11.

Hwang, T., Schwarz, S. E., and Rutledge, D. B. (1979b). *Appl. Phys. Lett.* **34**, 733–736.

Irvin, J. C., and Vanderwal, N. C. (1969). *In* "Microwave Semiconductor Devices and their Circuit Applications" (H. A. Watson, ed.), pp. 340–369. McGraw-Hill, New York.

Jakes, W. C. (1961). *In* "Antenna Engineering Handbook" (H. Jasik, ed.), pp. 10-1–10-18. McGraw-Hill, New York.

Jasik, H. (1961). *In* "Antenna Engineering Handbook" (H. Jasik, ed.), pp. 2-1–2-53. McGraw-Hill, New York.

Jeffers, W. Q., and Johnson, C. J. (1968). *Appl. Opt.* **7**, 1859–1860.

Jones, R. C. (1953a), *Adv. Electron.* **5**, 1–96.

Jones, R. C. (1953b). *J. Opt. Soc. Am.* **43**, 1–14.

Kadlec, J., Gundlach, K. H., and Hartfuss, H. J. (1979). *Proc. Int. Conf. Infrared Phys.*, *2nd*, *Zurich* (E. Affolter and F. Kneubühl, eds.), pp. 322–324.

Kantorowicz, G., Palluel, P., and Portvianne, J. (1979). *Microwave J.* **22**, No. 2, 57–59.

Kastler, A. (1964). *In* "Quantum Electronics" (*Proc. Int. Congr., 3rd*) (P. Grivet and N. Bloembergen, eds.), Vol. 1, pp. 3–11. Dunod, Paris.

Keen, N. (1978). Unpublished report.

Keen, N., Haas, R., and Perchtold, E. (1978). *Electron. Lett.* **14**, 825–826.

Keene, J., Hildebrand, R. H., Whitcomb, S. E., and Winston, R. (1978). *Appl. Opt.* **17**, 1107–1109.

Keilmann, F., and Renk, K. F. (1971). *Appl. Phys. Lett.* **18**, 452–454.

Kerecman, A. J. (1973). *Int. Microwave Symp. Dig.* IEEE Cat. No. 73 CHO 736-9 MMT, p. 30–34.

Kerr, A. R. (1975). *IEEE Trans. Microwave Theory Tech.* **MTT-23**, 781–787.

Kerr. A. R., Mattauch, R. J., and Grange, J. A. (1977a). *IEEE Trans. Microwave Theory Tech.* **MTT-25**, 399–401.

Kerr, A. R., Siegel, P. H., and Mattauch, R. J. (1977b). *Int. Microwave Symp. Dig.* IEEE Cat. No. 77 CH 1219-5 MTT, pp. 96–98.

Keve, E. T. (1975). *Philips Tech. Rev.* **35**, 247–257.

Kimmitt, M. F. (1970). "Far-Infrared Techniques." Pion, London.

Kimmitt, M. F. (1977). *Infrared Phys.* **17**, 459–466.

Kinch, M. A. (1971). *J. Appl. Phys.* **42**, 5861–5863.

Kinch, M. A., and Rollin, B. V. (1963). *Br. J. Appl. Phys.* **14**, 672–676.

Kleinmann, D. E. (1976). *In* "Far-Infrared Astronomy" (M. Rowan-Robinson, ed.), supplement to "Vistas in Astronomy," pp. 33–45. Pergamon, Oxford.

Kleppner, D., and Ducas, T. W. (1976). *Bull. Am. Phys. Soc.* **21**, 600.

Knight, D. J. E. (1979). National Physical Lab. Rep. No. Qu 45.

Knight, D. J. E., and Woods, P. T. (1976). *J. Phys. E: Sci. Instrum.* **9**, 898–916.

Knotek, S., and Gush, H. P. (1977). *Rev. Sci. Instrum.* **48**, 1223–1224.

Kobzev, V. V., Rivlin, A. A., and Solov'ev, V. S. (1975). *Sov. J. Quantum Electron.* **4**, 1146–1147.

Kohl, F., Müller, W., and Gornik, E. (1978). *Infrared Phys.* **18**, 697–704.

Kopeika, N. S., and Farhat, N. H. (1975). *IEEE Trans. Electron Devices* **ED-22**, 534–548.

Krautle, H., Sauter, E., and Schultz, G. V. (1977). *Infrared Phys.* **17**, 477–483.

Krautle, H., Sauter, E., and Schultz, G. V. (1978). *Infrared Phys.* **18**, 705–712.

Kremenchugskii, L. S., and Roitsina, O. V. (1976). *Instrum. Exp. Tech.* **19**, 603–621.

Krishnan, K. S., Ostrem, J. S., and Stappaerts, E. A. (1978). *Opt. Eng.* **17**, 108–113.

Kunz, L. W., and Madey, J. M. (1974). *Dig. Conf. Submillimeter Waves Their Appl.*, *Atlanta* IEEE Cat. No. 74 CHO 856-MTT, pp. 69–71.

Kurbatov, V. A., and Penin, N. A. (1976). *Sov. J. Quantum Electron.* **6**, 1041–1044.

Lengfellner, H., and Renk, K. F. (1977). *IEEE J. Quantum Electron.* **QE-13**, 421–424.

Lesieur, J. P., Sexton, M. C., and Veron, D. (1972). *J. Phys. D: Appl. Phys.* **5**, 1212–1217.

Levinstein, H. (1965). *Appl. Opt.* **4**, 639–647.

Lewis, W. B. (1947). *Proc. Phys. Soc.* **59**, 34–40.

Lichtenberg, A. J., and Sesnic, S. (1966). *J. Opt. Soc. Am.* **56**, 75-79.
Liu, S. T., and Long, D. (1978). *Proc. IEEE* **66**, 14-26.
Llewellyn-Jones, D. T., and Gebbie, H. A. (1979). *Colloq. Measurement Power at Higher Microwave Frequencies, IEE, London* (unpublished).
Low, F. J. (1961). *J. Opt. Soc. Am.* **51**, 1300-1304.
Low, F. J. (1965). *Proc. IEEE* **53**, 516.
Low, F. J. (1966). *Proc. IEEE* **54**, 477-484.
Mandel, L. (1964). *In* "Quantum Electronics" (*Proc. Int. Congr., 3rd*), (P. Grivet and N. Bloembergen eds.) Vol. 1, pp. 101-109. Dunod, Paris.
Martin, D. H., and Bloor, D. (1961). *Cryogenics* **1**, 159-163.
Martin, D. H., and Lesurf, J. (1978). *Infrared Phys.* **18**, 405-412.
Martin, D. H., and Mizuno, K. (1976). *Adv. Phys.* **25**, 211-246.
Matarrese, L. M., and Evenson, K. M. (1970). *Appl. Phys. Lett.* **17**, 8-11.
Mattauch, R. J., and Fei, F. S. (1977). *Electron. Lett.* **13**, 22-23.
Maul, M. K., and Strandberg, M. W. P. (1969). *J. Appl. Phys.* **40**, 2822-2827.
McColl, M. (1977). *IEEE Trans. Microwave Theory Tech.* **MTT-25**, 54-59.
McColl, M., Garber, W. A., and Millea, M. F. (1972). *Proc. IEEE* **60**, 1446-1447.
McColl, M., Hodges, D. T., and Garber, W. A. (1977). *IEEE Trans. Microwave Theory Tech.* **MTT-25**, 463-467.
McColl, M., Bottjer, M. F., and Chase, A. B. (1979). *IEEE Trans. Magn.* **MAG-15**, 468-470.
McLean, T. P., and Putley, E. H. (1965). *RRE J.* **52**, 5-34.
Melngailis, I., and Tannenwald, P. E. (1969). *Proc. IEEE* **57**, 806-807.
Merat, F. L., Claspy, P. C., and Pao, Y-H. (1976). *Dig. Conf. Submillimeter Waves Their Appl., Puerto Rico* IEEE Cat. No. 76 CHM 1152-8 MTT, pp. 81-82.
Meredith, R., Warner, F. L., Davis, Q. V., and Clarke, J. L. (1964). *Proc. IEE* **111**, 241-256.
Messenger, G. C., and McCoy, C. T. (1957). *Proc. IRE* **45**, 1269-1283.
Mizuno, K., Kuwakara, R., and Ono, S. (1975). *Appl. Phys. Lett.* **26**, 605-607.
Moore, W. J. (1976). *Dig. Conf. Submillimeter Waves Their Appl., Puerto Rico* IEEE Cat. No. 76 CHH 1152-8 MTT, pp. 68-69.
Moore, W. J., and Shenker, H. (1965). *Infrared Phys.* **5**, 99-106.
Mumford, W. W., and Scheibe, E. H. (1968). "Noise Performance Factors in Communications Systems." Horizon House-Microwave, Dedham.
Murphy, R. A. *et al.* (1977). *IEEE Trans. Microwave Theory Tech.* **MTT-25**, 494-495.
Nad, F. Ya, (1975). *Instrum. Exp. Tech.* **18**, 1-15.
Nakajima, F., Kobayashi, M., and Narita, S. (1978). *Jpn. J. Appl. Phys.* **17**, 149-153.
Nayar, P. S., and Hamilton, W. O. (1977). *Appl. Opt.* **16**, 2942-2944.
Nishioka, N. S., Richards, P. L., and Woody, D. P. (1978). *Appl. Opt.* **17**, 1562-1567.
Nokagawa, Y., and Yoshinaga, H. (1970). *Jpn. J. Appl. Phys.* **9**, 125-131.
Nolt, I. G., Radostitz, J. V., Kittel, P., and Donelly, R. J. (1977). *Rev. Sci. Instrum.* **48**, 700-702.
Norton, P. (1976a). *J. Appl. Phys.* **47**, 308-320.
Norton, P. (1976b). *Phys. Rev. Lett.* **37**, 164-168.
Norton, P., Slusher, R. E., and Sturge, M. D. (1977). *Appl. Phys. Lett.* **30**, 446-448.
Oka, Y., Nagasaka, K., and Narita, S. (1968). *Jpn. J. Appl. Phys.* **7**, 611-618.
Okamura, S., and Ijichi, K. (1976). *IEEE Trans. Instrum. Measurement* **IM-25**, 437-440.
Opher, R., Politch, J., and Felsteiner, J. (1978). *Appl. Phys. Lett.* **32**, 701-702.
Oxley, T. H. (1976). *Microwave J.* **19**, No. 12, 46-47.
Page, C. H. (1956). *J. Res. Nat. Bur. Std.* **56**, 178-182.
Pankratov, N. A., Kulikov, Yu. V., and Shchetinina, N. V. (1978). *Sov. J. Opt. Technol.* **45**, 435-437.
Payne, J. M., and Wordeman, M. R. (1978). *Rev. Sci. Instrum.* **49**,,1741-1743.

Phillips, T. G., and Jefferts, K. B. (1973). *Rev. Sci. Instrum.* **44**, 1009–1014.

Phillips, T. G., and Jefferts, K. B. (1974). *IEEE Trans. Microwave Theory Tech.* **MTT-22**, 1290–1292.

Poorter, T., and Tolner, H. (1979). *Proc. Int. Conf. Infrared Phys.*, *2nd, Zurich* (E. Affolter and F. Kneubühl, eds.), pp. 317–318.

Putley, E. H. (1964). *Infrared Phys.* **4**, 1–8.

Putley, E. H. (1965). *Appl. Opt.* **4**, 649–657.

Putley, E. H. (1970). *In* "Semiconductors and Semimetals" (R. K. Willardson and A. C. Beer, eds.) Vol. 5, pp. 259–285. Academic Press, New York.

Putley, E. H. (1971). *Opt. Laser Technol.* **3**, 150–156.

Putley, E. H. (1973). *Phys. Technol.* **4**, 202–222.

Putley, E. H. (1977a). *In* "Semiconductors and Semimetals" (R. K. Willardson and A. C. Beer, eds.), Vol. 12, pp. 441–449. Academic Press, New York.

Putley, E. H. (1977b). *In* "Topics in Applied Physics" (R. J. Keyes, ed.), Vol. 19, Optical and Infrared Detectors, pp. 71–100. Springer, Berlin and New York.

Putley, E. H. (1977c). See Putley (1977a, pp. 143–168).

Putley, E. H. (1978a). *Infrared Phys.* **18**, 373–374.

Putley, E. H. (1978b). *Infrared Phys.* **18**, 371.

Putley, E. H., and Martin, D. H. (1967). In "Spectroscopic Techniques for Far Infrared, Submillimetre and Millimetre Waves" (D. H. Martin, ed.), pp. 113–151. North-Holland Publ., Amsterdam.

Radostitz, J. V., Nolt, I. G., Kittel, P., and Donnelly, R. J. (1978). *Rev. Sci. Instrum.* **49**, 86–88.

Richards, P. L. (1970). *In* "Far-Infrared Properties of Solids" (S. S. Mitra and S. Nudelman, eds.), pp. 103–120. Plenum Press, New York.

Richards, P. L. (1977a). *In* "Semiconductors and Semimetals" (R. K. Willardson and A. C. Beer, eds.), Vol. 12, pp. 395–440. Academic Press, New York.

Richards, P. L. (1977b). *Infrared Phys.* **17**, 241–244.

Richards, P. L. (1978a). *In* "Future Trends in Superconductive Electronics" (B. S. Deaver, C. M. Falco, J. H. Harris, and S. A. Wolf, eds.), pp. 223–229. American Institute of Physics, New York.

Richards, P. L. (1978b). Paper given at U.S. National Radio Science Meeting, Boulder, Colorado, November (unpublished).

Richards, P. L., and Tinkham, M. (1960). *Phys. Rev.* **119**, 575–590.

Richards, P. L., Shen, T. M., Harris, R. E., and Lloyd, F. L. (1979). *Appl. Phys. Lett.* **34**, 345–347.

Robinson, L. C. (1973). "Physical Principles of Far-Infrared Radiation." Academic Press, New York.

Rogers, W. M., and Powell, R. L. (1958). National Bureau of Standards Circular No. 595.

Ross, M. (1966). "Laser Receivers." Wiley, New York.

Roundy, C. B. (1979). *Appl. Opt.* **18**, 943–945.

Roundy, C. B., and Byer, R. L. (1972). *Appl. Phys. Lett.* **21**, 512–515.

Rutledge, D. B., Schwarz, S. E., and Adams, A. T. (1978). *Infrared Phys.* **18**, 713–729.

Sakuma, E., and Evenson, K. M. (1974). *IEEE J. Quantum Electron.* **QE-10**, 599–603.

Saur, W. (1968). *Infrared Phys.* **8**, 255–258.

Sauter, E., and Schultz, G. V. (1977). *IEEE Trans. Microwave Theory Tech.* **MTT-25**, 468–470.

Schaer, A. (1970). *Microwave J.* **13**, No. 4, 69–72.

Schneider, M. V., Linke, R. A., and Cho, A. Y. (1977). *Appl. Phys. Lett.* **31**, 219–221.

Schwarz, S. E., and Ulrich, B. T. (1977). *J. Appl. Phys.* **48**, 1870–1873.

Seib, D. H. (1974). *IEEE J. Quantum Electron.* **QE-10**, 130–131.

Sher, A., Foles, C. L., and Stubblefield, J. F. (1976). *Appl. Phys. Lett.* **28**, 676–678.

Shivanandan, K., McNutt, D. P., Daehler, M., and Feldman, P. D. (1975). *Proc. Soc. Photo. Opt. Inst. Eng.* **67**, 48–52.

Siegman, A. E. (1966). *Proc. IEEE* **54**, 1350–1356.

Silberg, P. A. (1957). *J. Opt. Soc. Am.* **47**, 575–578.

Silver, A. H., Chase, A. B., McColl, M., and Millea, M. F. (1978). *In* "Future Trends in Superconductive Electronics" (B. S. Deaver, C. M. Falco, J. H. Harris, and S. A. Wolf, eds.), pp. 364–379. American Institute of Physics, New York.

Smith, R. A. (1951). *Proc. IEE* **98**, 43–54.

Stevens, E. (1979). *Colloq. Measurement Power at Higher Microwave Frequencies, IEE, London* (unpublished).

Stevens, N. B. (1970). *In* "Semiconductors and Semimetals" (R. K. Willardson and A. C. Beer, eds.), Vol. 5, pp. 287–318. Academic Press, New York.

Stillman, G. E., Wolfe, C. M., and Dimmock, J. O. (1977). *In* "Semiconductors and Semimetals" (R. K. Willardson and A. C. Beer, eds.), Vol. 12, pp. 169–290. Academic Press, New York.

Stokowski, S. E. (1976). *Appl. Phys. Lett.* **29**, 393–395.

Stone, N. W. B. *et al.* (1975). *Proc. IEE: IEE Rev.* **122**, 1054–1070.

Summers, C. J., and Zwerdling, S. (1974). *IEEE Trans. Microwave Theory Tech.* **MTT-22**, 1009–1013.

Taur, Y., and Kerr, A. R. (1978). *Appl. Phys. Lett.* **32**, 775–777.

Tiuri, M. E. (1964). *IEEE Trans. Antennas Propag.* **AP-12**, 930–938.

Tiuri, M. E. (1966). *In* "Radio Astronomy" (J. D. Kraus, ed.), Chapter 7. McGraw-Hill, New York.

Tolner, H. (1977). *J. Appl. Phys.* **48**, 691–701.

Tolner, H., Andriesse, C. D., and Schaeffer, H. H. A. (1976). *Infrared Phys.* **16**, 213–223.

Torrey, H. C., and Whitmer, C. A. (1948). "Crystal Rectifiers," Radiation Laboratory Ser., Vol. 15. McGraw-Hill, New York.

Troup, G. J., and Turner, R. G. (1974). *Rep. Prog. Phys.* **37**, 771–816.

Tsang, D., and Schwarz, S. E. (1977). *Appl. Phys. Lett.* **30**, 263–265.

Twu, B., and Schwarz, S. E. (1975). *Appl. Phys. Lett.* **26**, 672–675.

Van der Ziel, A. (1976). *J. Appl. Phys.* **47**, 2059–2068.

Van Vliet, A. H. F., De Graauw, Th., and Schotzau, H. J. (1978). *Digest Int. Conf. Submillimeter Waves Their Appl., 3rd, Guildford* pp. 248–250.

Von Ortenberg, M., Link, J., and Helbig, R. (1977). *J. Opt. Soc. Am.* **67**, 968–970.

Voss, R. F., Laibowitz, R. B., Raider, S. I., Grobman, W. D., and Clarke, J. (1979). *Bull. Am. Phys. Soc.* **24**, 264–265.

Wahlsten, S., Rudner, S., and Claeson, T. (1978). *J. Appl. Phys.* **49**, 4248–4263.

Waldram, J. R. (1976). *Rep. Prog. Phys.* **39**, 751–821.

Walser, R. M., Bené, R. W., and Carruthers, R. E. (1971). *IEEE Trans. Electron Devices* **ED-18**, 309–315.

Warner, F. L. (1969). *In* "Millimetre and Submillimetre Waves" (F. A. Benson, ed.), pp. 453–511. Iliffe, London.

Warner, J. (1971). *Opto-Electron.* **3**, 37–48.

Watton, R. (1976). *Ferroelectrics* **10**, 91–98.

Watton, R. (1978). *Infrared Phys.* **18**, 73–87.

Weinreb, S., and Kerr, A. R. (1973). *IEEE J. Solid State Circuits* **SC-8**, 58–63.

Weitz, D. A., Skocpol, W. J., and Tinkham, M. (1978a). *Infrared Phys.* **18**, 647–656.

Weitz, D. A., Skocpol, W. J., and Tinkham, M. (1978b). *Phys. Rev. B* **18**, 3282–3292.

Wiesendanger, E., and Kneubühl, F. (1977). *Appl. Phys.* **13**, 343–349.

Wilson, W. L., and Epton, P. J. (1978). *Infrared Phys.* **18**, 669–673.

Winston, R. (1970). *J. Opt. Soc. Am.* **60**, 245–247.

Woody, D. P., Nishioka, N. S., and Richards, P. L. (1978). *J. Phys.* **39**, Coll. C-6, Suppl. to No. 8. pp. C6-1629–C6-1630.

Wrixon, G. T. (1974). *IEEE Trans. Microwave Theory Tech.* **MTT-22**, 1159–1165.

Wrixon, G. T. (1976). *IEEE Trans. Microwave Theory Tech.* **MTT-24**, 702–706.

Wrixon, G. T., and Kelly, W. M. (1978). *Infrared Phys.* **18**, 413–428.

Wyatt, C. L., Baker, D. J., and Frodsham, D. G. (1974). *Infrared Phys.* **14**, 165–176.

Zav'yalov, V. V., and Voronin, V. I. (1976). *Instrum. Exp. Tech.* **19**, 1702–1704 (English translation 1977).

Zuidberg, B. F. J., and Dymanus, A. (1978). *Appl. Phys.* **16**, 375–379.

Zwerdling, S., Smith, R. A., and Theriault, J. P. (1968). *Infrared Phys.* **8**, 271–336.

CHAPTER 2

Optimization of Schottky-Barrier Diodes for Low-Noise, Low-Conversion Loss Operation at Near-Millimeter Wavelengths

W. M. Kelly

Microelectronics Research Center
University College
Cork, Ireland

and

G. T. Wrixon

Microelectronics Research Center
University College
Cork, Ireland

I. Summary

If Schottky-barrier diodes are to retain the low-loss, low-noise properties they possess at microwave and millimeter wavelengths as the wavelength decreases to 1 mm and less, special constraints are necessary in their design. This chapter attempts to list these constraints and show how they may be optimized to obtain the best possible performance.

In Section II a brief introduction to Schottky-barrier diodes and their role in near-millimeter superheterodyne detection is given. In Section III, following a review of conduction mechanisms in diodes, the noise behavior in diodes is discussed. The overriding importance of the conversion loss as opposed to intrinsic noise-generation mechanisms in determining diode performance in the near-millimeter is stressed. Section IV looks at conversion loss in detail, especially at the contribution of the parasitic loss elements, which become especially important at short wavelengths. Finally, in light of criteria discussed in Section IV, Section V sets out the optimization techniques that are possible when constructing Schottky-barrier diodes to operate at near-millimeter wavelengths.

II. Introduction

In recent years low-noise superheterodyne mixing techniques have been extended from the millimeter to the submillimeter and far-infrared wavelength regions. This has placed the benefits of coherent detection and, in particular, the availability of broadband multichannel spectrometers of very high sensitivity at the disposal of workers in this wavelength region.

The most widely used mixer element at microwave and millimeter wavelengths is the Schottky-barrier diode. It is a wide bandwidth device having high sensitivity and good mechanical stability. It can be operated satisfactorily at room temperature, but its sensitivity can increase by a factor of 3–7 when it is cooled cryogenically to temperatures of about 15 K. It consists of a small metal contact deposited on an epitaxial semiconductor (usually GaAs), with the contact area playing a vital role in determining the device capacitance and consequently its performance at submillimeter wavelengths. The diode is mounted in a mixer circuit whose basic function is to match the incoming radiation into the diode and match the diode into the IF amplifier.

A basic property of a Schottky diode is that compared to superheterodyne devices operating at liquid helium temperatures (e.g., MOM devices, Josephson junctions, InSb bolometer), the LO power requirements are comparatively high, being of the order of 1–2 mW. Since one of the difficulties of operation in the near-millimeter range is the dearth of suitable sources of LO power, various schemes such as harmonic mixing with a single diode (Zimmerman and Mattauch, 1979) or subharmonic mixing using two diodes (Carlson et al., 1978) have been successfully used to allow the mixer to work at the second harmonic of the LO frequency.

Recently Schottky diodes have been developed with LO requirements of an order of magnitude less than that quoted above (Vizard et al., 1979). The reason for the low LO power requirement is not yet fully understood (see Section V), but a development such as this will clearly be of enormous importance in helping to extend the range of mixers using Schottky diodes further into the submillimeter range.

In this chapter we will confine our attention to the Schottky diode itself, and we shall not treat the mixer circuit. Obviously, optimization of the performance of the Schottky diode will not be of much consequence unless it is accompanied by similar optimization and attention to the mixer circuit. Likewise, optimization of the mixer structure without a similar treatment of the diode will prove fruitless. Thus, it is important to view this chapter as a necessary but not sufficient step toward the achievement of the overall goal of high performance, superheterodyne receivers working in the near-millimeter range.

III. Diode Noise

A. CONDUCTION MECHANISMS IN DIODES

In general, conduction in Schottky-barrier diodes is due to a combination of two different types of electronic processes (i) thermionic emission and (ii) field emission. The former consists of electrons with sufficient thermal energy to cross the barrier, and the latter consists of electrons that tunnel through the barrier. The relative proportion of electrons engaged in these two processes depends on the physical temperature of the diode (T), the doping density of the semiconductor (N), and the effective mass of the majority carriers (m^*). In general, a high T will increase the proportion of electrons with energy sufficient to overcome the barrier. On the other hand, image force lowering and narrowing of the potential barrier increases as the ratio of N/m^*. Initially, the barrier becomes thin enough so that thermally excited carriers can tunnel through near the top of the barrier. This temperature-dependent mode of current transport is called thermionic-field emission or thermally assisted tunneling. As the impurity concentration is further increased, the barrier becomes so thin that significant numbers of carriers can tunnel through even at the base of the barrier. This is field emission tunneling and is temperature independent (Rideout, 1975).

For $I \gg I_s$ the $I-V$ characteristic of a Schottky diode may be written

$$I = I_s \exp(V/V_0). \tag{1}$$

Padovani and Stratton (1966) have shown that

$$V_0 = (E_{00}/q) \coth(E_{00}/kT), \tag{2}$$

where

$$E_{00} = qh(N/4\varepsilon m^*)^{1/2}. \tag{3}$$

For the field emission case, i.e., high N and/or low T, $\coth(E_{00}/kT) \to 1$ and

$$V_0 = E_{00}/q, \tag{4}$$

which is temperature independent.

For the thermionic emission case, i.e., high T and/or low N,

$$\coth(E_{00}/kT) \to kT/E_{00}$$

and

$$V_0 = kT/q. \tag{5}$$

Often, the diode $I-V$ characteristic is written ($I \gg I_s$)

$$I = I_s \exp(qV/\eta kT). \tag{6}$$

For pure thermionic emission $\eta = 1$, and the deviation of the η value from unity may be used as a measure of the relative contribution of tunneling to conduction.

At room temperature, deviations from purely thermionic emission occur in GaAs at doping densities as low as $1 - 2 \times 10^{17}/\text{cm}^3$. Thus for doping densities higher than this, values of η greater than unity can be expected. This measure must be cautiously applied, however, as imperfections in diode processing, which are especially difficult to eliminate given the small size of near-millimeter diodes, may lead to the presence of an interfacial layer or impurity diffusion, both of which also tend to increase η.

At near-millimeter wavelengths Schottky diodes are usually made on GaAs of doping $2 \times 10^{16} - 2 \times 10^{17}/\text{cm}^3$ (see Section V). When a diode such as this is used in a cooled mixer, the onset of full-field emission occurs at temperatures less than ~ 70 K. At full-field emission it is seen from Eqs. (1), (4), and (6) that the effective value of η is E_{00}/kT, which from Eq. (3) is seen to vary as $N^{1/2}$. The importance of η in determining the noise properties of the diode is now examined.

B.   DIODE NOISE BEHAVIOR

At near-millimeter wavelengths, because of the physical difficulties involved in tuning out one sideband, most mixers using Schottky-barrier diodes will be of the "dual-response" type, where the conversion losses for the signal and image, L_s and L_i, need not necessarily be equal (Carlson et al.,

1978). For a receiver containing such a mixer, assuming the signal of interest appears in the signal sideband only, the appropriate measure of receiver noise is the single-sideband receiver noise temperature, given by

$$T_r(\text{SSB}) = T_m(\text{SSB}) + L_s T_{\text{IF}}, \tag{7}$$

where $T_m(\text{SSB})$ is the single-sideband mixer noise temperature referred to the input and T_{IF} is the noise temperature of the IF amplifier.

Under most conditions the principal source of noise in a Schottky-barrier diode is shot noise. For an ideal exponential diode having full-shot noise, negligible series resistance, and nonlinear junction capacitance, it has been shown (Dragone, 1968; Saleh, 1971b; Kerr, 1979) that when operating as a mixer, it is equivalent to a lossy network at a temperature T_A, where

$$T_A = \eta T/2, \tag{8}$$

and T is the physical temperature of the diode.

$T_m(\text{SSB})$ is related to T_A by (Kerr, 1979)

$$T_m(\text{SSB}) = T_A(L_s - 1 - L_s/L_i); \tag{9}$$

thus

$$T_m(\text{SSB}) = \tfrac{1}{2}\eta T(L_s - 1 - L_s/L_i). \tag{10}$$

In the past, measured values of $T_m(\text{SSB})$ tended to be higher than that given by Eq. (10) and have been attributed to noise contribution from diode series resistance and the parametric effect of the time-varying diode capacitance acting on the correlated components of the diode shot noise (Held and Kerr, 1978; Held, 1976). Recently, however, Mott junctions have been available, made on GaAs with very sharp epilayer/substrate doping transition (see Section V), and consequently with extremely little change in diode capacitance with bias voltage. When these diodes were used in mixers cooled to 18 K so that the noise contribution from the series resistance was negligible, it was possible to verify experimentally that Eq. (10) did, in fact, give the correct value of T_m, and thus that the measured mixer noise was consistent with a purely theoretical shot noise (Linke et al., 1978).

As materials with similarly sharp epilayer/substrate interfaces become more generally available (see Section V), it is reasonable to expect Eq. (10) to give generally correct values for T_m (except, of course, for the additional contribution of the diode series resistance in an uncooled mixer).

Examining Eq. (10), it is seen that in order to keep the mixer noise at a minimum, both T and η should be minimized. From the discussion in Section A, we recall that at room temperature this implies that N should be less than

$2 \times 10^{17}/\text{cm}^3$.* If the diode is cooled so that full field emission results, N should be reduced even further.

These considerations, important though they are, do not impose special constraints for Schottky diodes operating in the near-millimeter. They apply to all low-noise diodes, no matter what their wavelength of operation. The final term that determines $T_m(\text{SSB})$ in Eq. (10), namely the conversion loss, is, however, the parameter that changes dramatically at near-millimeter wavelengths. In Section IV this behavior is examined in detail.

IV.   Diode Conversion Loss

A.   DIODE STRUCTURE AND EQUIVALENT CIRCUIT

Figure 1 is a schematic illustration of a cross section of part of a Schottky-barrier diode chip. The semiconductor wafer is normally a GaAs epilayer structure, for reasons which will be explained below, consisting of an n-type epilayer of carrier concentration $2 \times 10^{16}/\text{cm}^3$–$2 \times 10^{17}/\text{cm}^3$ on an n^+ substrate of carrier concentration $\geq 2 \times 10^{18}/\text{cm}^3$, depending upon the donor species. An ohmic contact, consisting of a highly doped surface layer formed by alloying metallic layers into the semiconductor, is formed at the back of the wafer. The front of the wafer is covered with a passivation layer of SiO_2 in which holes have been produced either photolithographically, or using electron beam lithography. The Schottky barriers are formed by depositing Pt followed by Au through the holes onto the GaAs epilayer. The thickness of the SiO_2 is typically 0.3–0.4 μm, the epilayer is approximately 0.1–0.2 μm, and the substrate 100 μm.

The simplest lumped-equivalent circuit for the diode, which is likely to give realistic estimates for its behavior in a mixer, is illustrated in Fig. 2. In this diagram C_0 is the zero-bias parasitic capacitance of the space-charge region of the junction, R_o is the actual nonlinear resistance of the Schottky barrier, and R_s is the series resistance of the epilayer, substrate, and back contact. The parasitic inductance and fringing capacitance of the whisker are ignored (Kerr et al., 1978). Based upon this equivalent circuit, the figure of merit normally used to evaluate the high-frequency performance potential of the Schottky-barrier diode is the cutoff frequency $f_c = 1/2\pi R_s C_0$, where C_0 is the zero-bias junction capacitance. The cutoff frequency is thus seen to define the point at which the reactance of the shunting capacitance equals

---

* Another, and extremely important, reason for keeping N below this value is that avalanche noise has been observed during LO pumping on diodes made with $N = 3 \times 10^{17}/\text{cm}^3$ and with back-breakdown voltages of ~ 4 V. It is thus important to keep the back-breakdown voltage above this value by using lower doping because the presence of avalanche noise would result in considerable excess diode noise temperature.

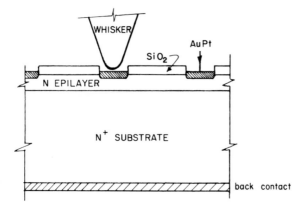

FIG. 1 Cross section of an epitaxial Schottky-barrier diode.

the series resistance. In view of the fact that R_s is frequency-dependent and consequently that f_c can actually be multivalued (Champlin and Eisenstein, 1977, 1978; Kelly and Wrixon, 1978a), the cutoff frequency is not a useful performance criterion at submillimeter wavelengths. It is necessary then to examine the diode parasitics in more detail so as to develop an improved model that will be useful in the frequency-dependent regime (see Section B). This improved model for a standard circular diode is used to predict the expected conversion losses in a typical submillimeter mixer configuration. Finally, the discussion is broadened to embrace alternative diode and chip geometries, some of which appear to offer significant advantages for submillimeter use.

B. SERIES RESISTANCE

The dc series resistance of a Schottky-barrier diode may be written

$$R_{dc} = R_{epi} + R_{sub} + R_{BC}. \tag{11}$$

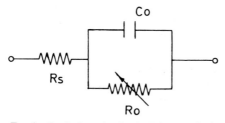

FIG. 2 Equivalent circuit of a Schottky diode.

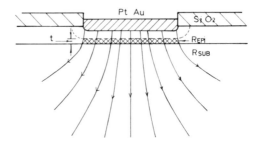

FIG. 3   Schematic representation of dc current flow.

In this expression R_{epi} is the contribution of the undepleted (cross-hatched) portion of the epilayer, as illustrated in Fig. 3, R_{sub} is the substrate contribution and R_{BC} is the back contact resistance. For typical Schottky diodes, the anode diameter is ≥ 1 μm, while the undepleted epilayer thickness is at most 0.1 μm. Therefore, the geometry is approximately as in Fig. 3, and the resistance of this undepleted epilayer is given to a good approximation by the expression

$$R_{epi} = t/\sigma^1 \pi a^2, \tag{12}$$

where t is the thickness of the undepleted epilayer, σ^1 is the epilayer conductivity, and a is the diode radius.

It has been shown by Dickens (1967) that the spreading resistance in the substrate R_{sub} is given by

$$R_{sub} = (2\pi\sigma a)^{-1} \arctan(b/a) \tag{13}$$

where σ is the substrate conductivity, and the chip is assumed to be cylindrical, with radius b and ohmic contacted sides. In practice Schottky diode chips are rectangular parallelepipeds with nonmetallized sides. However, since most of R_{sub} accumulates in the constricted region close to the anode, the shape of a large chip has negligible effect upon the calculated value of R_{sub}.

The back contact resistance R_{BC} can generally be ignored because contact resistivities of the order of 10^{-5} Ω cm^2 are easy to achieve [cf. for example, Kelly and Wrixon (1978b)] on n^+-type GaAs; therefore, for a normal chip area ($< 10^{-3}$ cm^2), R_{BC} makes a negligible contribution. Thus, the values obtained for R_s by means of the normal dc I–V curve-fitting procedure might be expected to be the sum of R_{epi} and R_{sub} as calculated from Eqs. (12) and (13). Measurements reported by Weinreb and Decker (1976) and Held (1976) indicate, however, that due to junction heating the dc value obtained for R_s is too low. A corrected 10 MHz measurement technique, i.e., faster than the shortest thermal time constant of the diode and its mounting structure, confirms that the correct value for R_s is approximately 25% higher than the dc-measured value. This is an important factor in correlating diode performance at rf frequencies with measured dc characteristics.

The most obvious effect of high-frequency operation upon the diode series resistance is an increase in R_s due to the restriction of current to a thin surface layer whose thickness is given approximately by the skin depth d_s, where

$$d_s = (2/\omega\mu\sigma)^{1/2}. \tag{14}$$

In this expression μ is the semiconductor's magnetic permeability. The inverse dependence of d_s upon frequency, as shown in Fig. 4, ensures that this effect will become important at short wavelengths. In this diagram the calculated magnitude of d_s is plotted as a function of frequency for various values of σ. The values used are 2000 ℧/cm corresponding to Te-doped GaAs substrates that have recently been used for high-quality epitaxial GaAs; 1000 ℧/cm, corresponding to more normal substrate carrier concentration levels; and 111 ℧/cm, corresponding to typical GaAs epilayer concentrations.

The series impedance Z_{skin} arising from the skin effect was shown by Dickens (1967) to be given by

$$Z_{skin} = [(1 + j)/2\pi\sigma d_s] \ln(b/a) \tag{15}$$

for a circular diode of radius a on a cylindrical chip of radius b. As was also true in the case of the expression for R_{sub}, the fact that real chips are rectangular does not appreciably alter Z_{skin}, since $\ln(b/a)$ is a slowly varying function for large values of b/a. Thus, assuming that the imaginary part of the impedance is detunable by the embedding circuit, the real part of Z must be included in the series resistance for high frequency use, giving

$$R_s = R_{epi} + R_{sub} + R_{skin}, \tag{16}$$

where R_{skin} is given by

$$R_{skin} = \ln(b/a)/2\pi\sigma d_s. \tag{17}$$

In Eq. (16) the expression used for R_{sub} is that given in Eq. (13). Although correct at dc, this becomes less accurate as frequency increases because, as depicted schematically in Figs. 3 and 5a, b for progressively higher frequencies, the current flow distorts as there is less penetration of rf current into the substrate. However, a calculation of R_{sub} from Eq. (13) shows that 80% of its complete dc value has been reached by a depth of $3a$, where a is the diode radius. Thus the approximation of Eq. (16) for R_s, using the dc value for R_{sub}, should be reasonably good up to frequencies where the skin depth d_s is $3a$, which corresponds to approximately 200 GHz for a 2-μm-diameter diode. At higher frequencies, R_{sub} is less accurate, but is also less important since R_{skin} is steadily increasing. Therefore, Eq. (16) should be a useful expression for small diodes at high frequencies. Experimental measurements

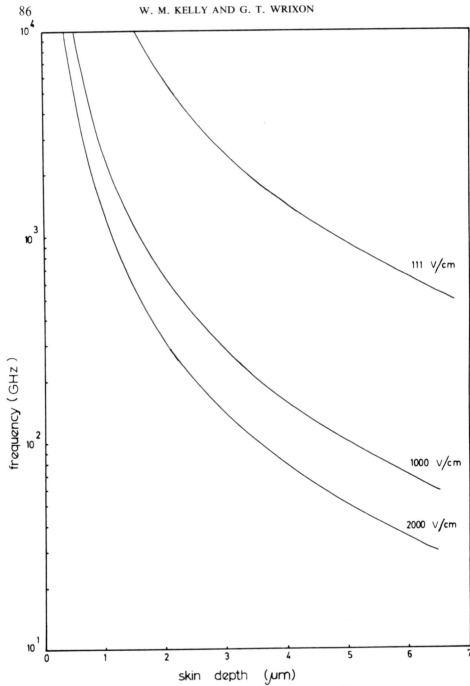

FIG. 4   Calculated skin depth for GaAs as a function of frequency.

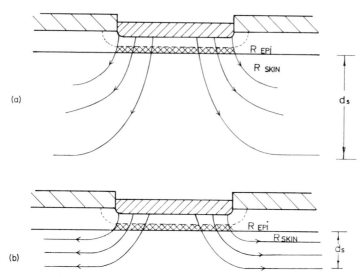

FIG. 5 Schematic of high-frequency current distribution. (a) $d_s \cong 3a$, (b) $d_s \cong 1a$ (a = diode radius).

by Held and Kerr (1978) of the skin resistance of a 2.5-μm-diameter GaAs diode at 115 GHz are in good agreement with the values calculated from Eq. (17). A more precise calculation of R_s based upon computer simulation by finite element techniques is in progress and will be published later (Campbell *et al.*, 1979).

The above treatment of series resistance ignores two potentially important high-frequency effects which were included in diode models by Hodges and McColl (1977), and in more detail by Champlin and Eisenstein (1977, 1978). First, above the dielectric relaxation frequency ω_d, where

$$\omega_d = \sigma/\varepsilon, \tag{18}$$

an appreciable fraction of carriers is carried via displacement currents. Second, carrier inertia becomes important above the scattering frequency ω_s given by

$$\omega_s = q/m^*\mu, \tag{19}$$

where m^* and μ are the carrier effective mass and mobility. Together these effects lead to carrier plasma oscillations that produce a considerable increase in the series impedance at frequencies close to the plasma frequency ω_p, where

$$\omega_p = (\omega_s \omega_d)^{1/2} = (nq^2/m^*\varepsilon)^{1/2} \tag{20}$$

and n is the carrier concentration.

Champlin and Eisenstein's model for bulk diodes has been extended to epitaxial diodes by Kelly and Wrixon (1978a, 1979a). By analogy with Eq. (16) the total, complex series impedance of the diode, Z_s, can be written

$$Z_s = Z_{epi} + Z_{sub} + Z_{skin} \tag{21}$$

Here, Z_{sub} and Z_{skin} are the substrate spreading resistance and the skin effect impedance discussed by Champlin and Eisenstein and are written as follows:

$$Z_{sub} = \frac{1}{2\pi\sigma a} \arctan(b/a) \left[\frac{1}{1 + j(\omega/\omega_s)} + j(\omega/\omega_d) \right]^{-1} \tag{22}$$

and

$$Z_{skin} = \frac{\ln(b/a)}{2\pi} \left(\frac{j\omega\mu}{\sigma} \right)^{1/2} \left[\frac{1}{1 + j(\omega/\omega_s)} + j(\omega/\omega_d) \right]^{-1/2}. \tag{23}$$

Note that in the low frequency limit, Z_{sub} and Z_{skin} reduce to the expressions in Eqs. (13) and (15). The term Z_{epi} in Eq. (21) is the contribution of the undepleted epilayer. It may be written in the thin epilayer approximation ($t \ll a$) as follows:

$$Z_{epi} = \frac{t}{\sigma'\pi a^2} \left[\frac{1}{1 + j(\omega/\omega_s)} + j(\omega/\omega_d) \right]^{-1}. \tag{24}$$

This is essentially the dc epilayer resistance $t/\sigma'\pi a^2$ modified by the plasma resonance term which increases strongly in the vicinity of f_p. Assuming that at any given operating frequency the imaginary parts of Z_s may be matched by careful embedding circuit design, then the diode parasitic conversion loss will be characterized by the real component of Z_s, $Re(Z_s)$, from which one could calculate a theoretical *minimum* value for conversion loss. The total series resistance is then given by

$$R_s = Re(Z_{epi}) + Re(Z_{sub}) + Re(Z_{skin}). \tag{25}$$

These quantities are shown in Table I for three operating frequencies. In the far infrared it is clear that the plasma resonance effects are very important, and as pointed out by Wrixon and Kelly (1978) have important implications for diode structures and materials. However, in the region of interest in this chapter, up to 1000 GHz, the dominant effect is the gradual increase ($\approx \omega^{1/2}$) in the skin impedance.

C. PARASITIC CONVERSION LOSS

The parasitic conversion loss of a Schottky diode is given by

$$L_p = \frac{R_s}{R_m} \left(1 + \frac{2R_s}{R_m} \right) \left(1 + \frac{R_m}{R_s} + \omega^2 C_0^2 R_m^2 \right). \tag{26}$$

TABLE I

R_s Constituents for a GaAs Schottky-Barrier Diode at
300 GHz, 1000 GHz, and 4100 GHz[a]

Frequency	Re(Z_{epi}) (Undepleted epilayer resistance)	Re(Z_{sub}) (Substrate spreading resistance)	Re(Z_{skin}) (Skin effect)	R_s (Total)
300 GHz	3.8	3.0	3.3	10.1
1000 GHz	4.3	3.0	6.8	16.1
4100 GHz ($\approx f_p$)	64.0	3.5	15.4	82.9

[a] GaAs diode series resistance (ohms). These figures assume a 2-μm-diameter n/n^+ Schottky diode with $n = 2 \times 10^{17}/cm^3$, $n^+ = 2 \times 10^{18}/cm^3$, with undepleted epilayer thickness $t = 0.125$ μm at a temperature of 295 K. The dc *measured* R_s of this diode would be $\sim 5\,\Omega$.

R_m is the barrier nonlinear resistance at the rf frequency and it is assumed, as pointed out by Burnues *et al.* (1976), that for a broadband mixer it is twice the IF resistance, R_{IF}. C_0 is the zero-bias junction capacitance. The capacitance of the junction as a function of voltage is given by

$$C_j = C_0\left(1 - \frac{V}{V_B}\right)^{-\gamma} \tag{27}$$

with

$$C_0 = \pi a^2[q\varepsilon N/2(\phi - V_B)]^{1/2}, \tag{28}$$

where ϕ is the barrier height, V_B the applied voltage, and γ is related to the geometry of the junction and the semiconductor doping profile. For a planar junction with uniform doping, γ is 0.5; but by making the depletion region extend to the epilayer/substrate interface, γ may be made close to 0.1 and then the use of C_0 in Eq. (26) is a better approximation. The parasitic conversion loss can be expected from Eq. (26) to be larger both for small radii, where R_s increases, and for large radii when the C_0^2 term dominates. Thus L_p is a function of diode size and minimizes for some optimum radius at a given frequency. In order to examine the variation in conversion loss as a function of diode diameter, Eqs. (22)–(25) are used to determine R_s, C_0 is assumed independent of frequency and applied voltage, and Eq. (26) gives the result as shown in Fig. 6.*

As expected, at a given frequency L_p has a minimum, which is sharper at shorter wavelengths. At 300 GHz the lowest conversion losses are obtained

* We have chosen $n^+ = 2 \times 10^{18}/cm^3$ in Fig. 6. However, good epitaxial layers have recently been reported by Lacombe *et al.* (1977) on substrates with over $5 \times 10^{18}/cm^3$ concentration.

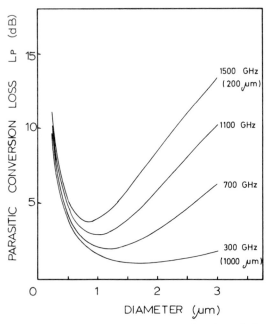

FIG. 6 Parasitic diode conversion loss plotted as a function of diameter for various frequencies assuming undepleted epilayer thickness $t = 0.125$ μm (GaAs: $n/n^+ = 2 \times 10^{17}/2 \times 10^{18}/\text{cm}^3$).

with a 2 μm diode, which fortuitously is about the limit obtainable with photolithographic techniques and is the size normally used for existing millimeter mixers. At higher frequencies the value of the minimum conversion loss is higher and corresponds to slightly smaller diameters. At a given frequency the capacitance dominates the parasitic losses for larger diameters, but when the diameter decreases below the optimum, R_s losses begin to dominate and L_p rises again. The steepness of the curve on the low-diameter side shows that it is better to have a diode whose diameter is greater rather than smaller than optimum. This is important since it indicates that it is not sufficient to simply reduce diode diameter below 1 μm, e.g., by using electron beam lithography, in order to obtain enhanced high frequency performance.

There are two separate effects adding together to give the results of Fig. 6:

(i) At any frequency, if the diameter is made small enough L_p increases steeply due to the R_s/R_m terms of Eq. (26).

(ii) The minimum L_p increases steadily with frequency, due to the skin effect. From Table I it is clear that for the diode configuration chosen, this is responsible for most of the increase up to about 1500 GHz.

D. INTRINSIC CONVERSION LOSS

Two models for a broadband mixer which might be useful are the Y mixer, in which all the out-of-band frequencies are short-circuited, and the Z mixer in which they are open circuited. The relative advantages of these models have been compared by Saleh (1971) who shows that, in theory, one needs infinite pump power and/or dc bias to optimize the performance of a Y mixer, while the Z mixer can be optimized with finite pump power and dc bias. In practice, however, it is difficult to obtain an ideal Z mixer because the diode capacitance will tend to short circuit the high-order out-of-band frequencies. In addition, the theoretical optimization of mixer performance sometimes demands very low dc bias, giving diode impedances that are far too high to be successfully matched in the embedding network. Because of these and similar considerations, Saleh concluded that no general statement could be made as to which type is better. In this section we describe McColl's (1977) treatment of the intrinsic conversion losses L_0 of a Y mixer.

When diode diameter decreases, intrinsic conversion losses can increase steeply due to the impedance requirements which the circuit places upon the device. In order for the diode to couple effectively to a circuit with a specific impedance it must pass approximately the same current independent of junction size ($R_o \propto 1/I$). Consequently, a physically smaller diode requires larger current densities and hence greater dc bias voltage V_o. This places a limitation on the useful amplitude of the local oscillator voltage V_1 because the junction I–V characteristic in the forward direction is nonlinear only for applied voltages less than the barrier height potential ϕ of the metal–semiconductor interface. Since $V_o + V_1 < \phi$, decreasing the area serves to limit V_1 and consequently will increase L_0. There is a further limitation upon the size of V_1 due to the increase in capacitance as the bias voltage approaches ϕ [see Eq. (28)].

The lower limit for intrinsic conversion loss L_0 is obtained when a thermionic emitting diode is optimally coupled in its embedding circuit. This lower limit L_{00} is given by*

$$L_{00} = \frac{2}{\eta}(1 + \sqrt{1 - \eta})^2, \tag{29}$$

where

$$\eta = 2I_1^2/\{I_0^2(1 + I_2/I_0)\}. \tag{30}$$

---

* McColl (1977) shows that by keeping the circuit impedances fixed and varying V_o and V_1 it is sometimes possible to reduce L_0 below L_{00}. For simplicity this is not included here.

I_0, I_1, and I_2 are modified Bessel functions of the first kind, with argument qV_1/kT. For a given diode size the magnitude of V_1 is obtained by solving

$$d/d_m = [(I_0 + I_2)\sqrt{1 - \eta}]^{-1/2} \exp(qV_1/2kT), \tag{31}$$

where d_m is the smallest possible diameter for which the impedance matching condition can be met, and is given by

$$d_m = (\pi q R_m A^* T/4k)^{-1/2}, \tag{32}$$

where A^* is the Richardson constant. The calculated optimum conversion loss L_{00} for a room-temperature GaAs diode is shown in Fig. 7 for a 100 Ω rf source impedance. Clearly, L_{00} increases strongly as d is reduced, so that materials with a large Richardson constant have a distinct advantage in that intrinsic conversion loss will not become significantly larger than 3 dB until the value of d nears that of d_m which, with $R_m = 100\ \Omega$ at 290 K, is $\sim 0.2\ \mu m$ for GaAs and $\sim 0.04\ \mu m$ for Si. In addition, cooling to 77 K causes a significant increase in d_m, thus increasing losses even more for small diameter diodes.

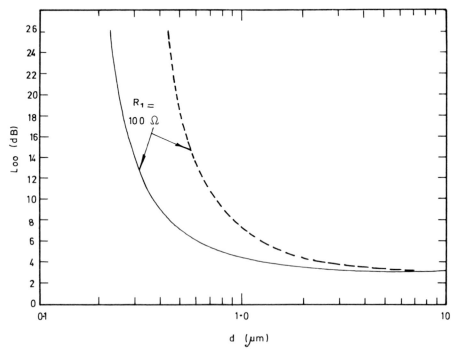

FIG. 7   Intrinsic diode conversion loss plotted as a function of diode diameter (———, $T = 290$ K; ———, $T = 77$ K).

By comparing Figs. 6 and 7 it is clear that for GaAs diodes the effect of this increase in intrinsic conversion loss will have little effect upon the results of the previous section. Since these indicated that optimum diode diameter is about 1 μm for operation at 1000 GHz, then the intrinsic conversion loss effect at room temperature adds about 1 dB (in addition to the minimum 3 dB contribution). It emphasizes, however, the necessity for not using unnecessarily small diodes, since this adds considerably to the already steep parasitic losses. For diodes used in cooled mixers this effect may be important.

E. TOTAL CONVERSION LOSS

The complete minimum conversion loss $L_p L_{00}$ can now be calculated for any diode from its diameter and epilayer geometry, knowing the doping concentration of epilayer and substrate. Thus, in Fig. 8 we show $L_p L_{00}$ for the same diode configuration as was used for Fig. 6. Comparing Figs. 8 and 6, L_{00} has added about 3 dB to the 300 GHz minimum, and about 4 dB to the 1500 GHz minimum. This is to be expected since L_{00} should be close to 3 dB, although it will rise somewhat at higher frequencies as the smaller optimum diode diameters approach d_m.

There are as yet few published data available to confirm the predicted diode behavior discussed above. Since submillimeter measurements vary widely depending on the precise details of the mixer structure, it is difficult to

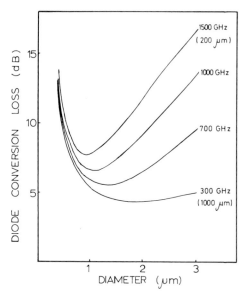

FIG. 8 Total optimum diode conversion loss plotted as a function of diode diameter at various frequencies for $t = 0.125$ μm (GaAs: $n/n^+ = 2 \times 10^{17}/2 \times 10^{18}/\text{cm}^3$).

extract reliable diode performance data from published mixer conversion loss. There are, however, results published by Fetterman *et al.* (1974) for 3 and 1.5 μm diodes operating at 1760 GHz in the same mixer, in which the 3 μm diode had a signal-to-noise ratio twenty times smaller than the 1.5 μm diode, although its cutoff frequency f_0 $(=1/2\pi R_s C_0)$ was larger, being 1193 GHz compared with 601 GHz. This experimental result is consistent with the guidelines developed here, which predict that an ideal 1.5 μm diode will have an optimum conversion loss at 1760 GHz that is 13.5 dB better than that of a 3 μm diode.

Erickson (1979) has also made a relatively complete breakdown of the performance of a waveguide mixer at 318 GHz (shown in Table II). The diode losses in this table total 6.3 dB, which is in good agreement with the 6.1 dB optimum losses calculated from the model.

Before concluding this discussion of conversion loss, it should be noted that the calculation of R_s based upon Eq. (25) can be rendered difficult because of the gradual epilayer/substrate transition in a real wafer (see Fig. 13 below). In this case an effective value for the undepleted epilayer thickness t can be chosen so as to make the calculated dc R_s match the measured R_s of the real diode, and the frequency dependent behavior of R_s then follows as above. Similarly, for small diodes the calculated value for C_0 can be in error due to edge effects, etc. Therefore, to provide improved accuracy, the measured value of C_0 can be used in Eq. (26). Some performance curves calculated in this manner are shown in Fig. 9 for C_0, R_s combinations measured on actual diodes. The three upper curves are for three typical sets of 2 μm Schottky diode characteristics and are contrasted with the lower curves, which are based upon "ideal" diodes. This diagram illustrates the importance of minimizing the sources of excess parasitic capacitance, e.g., fringing and overlay capacitance, and also illustrates the low-loss, wide

TABLE II

318 GHz MIXER LOSS BREAKDOWN

Waveguide impedance	1.8 dB
Diplexer insertion loss	0.2 dB
IF transformer, ohmic, and residual mismatch	0.2 dB
Intrinsic diode loss	3.9 dB
RF mismatch	0.1 dB
R_s at RF frequency	1.9 dB
R_s at IF frequency	0.4 dB

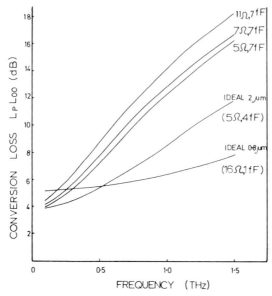

FIG. 9   Optimum conversion loss plotted for some Schottky diodes as a function of frequency.

bandwidth operation of the 0.8 μm ideal diode, this being the optimum size for frequencies near 1000 GHz.

The concepts discussed above are based upon relatively simple circuit considerations, in that the diode is assumed to be an ideal diode with exponential I–V characteristics at all frequencies, which interact with the remaining lumped elements in the circuit (parasitic capacitance, resistance, etc.). In order for the treatment to be complete, it ought to be complemented by a comprehensive treatment of transport mechanisms within the space-charge region itself, including transit time effects (van der Ziel 1976). There is a very interesting theory on Schottky-barrier diode rectification, developed by Tsang and Schwarz (1979) and based on first-principle Boltzmann calculations, which indicates anomalous high-frequency behavior *within the junction*. Of particular interest is the fact that small-signal mixing calculations indicate that the conversion efficiency is a function of IF frequency, and can be larger than the value one would estimate from the ideal diode equation. Unfortunately, the treatment used by Tsang and Schwarz does not give any information about mixer behavior with strong local oscillators. An extension of this very interesting model to include large-signal mixing would be of considerable interest to designers of submillimeter mixers.

V.   High-Frequency Optimization

A.   dc Diode Characteristics

The dc characteristics normally used to gauge Schottky diode performance are the series resistance R_s, the zero-bias capacitance C_0, the back-breakdown voltage V_{BB}, and the ideality factor η. R_s is normally deduced from a four-point I–V measurement, C_0 can be measured on a capacitance bridge, and the back-breakdown voltage is measured at 10 μA reverse current and should be as sharp as possible.

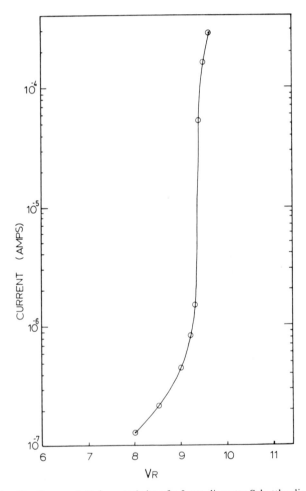

Fig. 10   Reverse I–V characteristics of a 2-μm-diameter Schottky diode.

The sharpness of the reverse characteristics is primarily a function of the final etching solution used prior to anode plating. Figure 10 contains an example of a satisfactorily sharp reverse breakdown measured on a 2 μm Schottky diode. This particular diode was etched with a 0.1% Br_2:MeOH solution at 20°C, but there are alternative techniques which give satisfactory results. For example, the anodization/etching sequence devised by Schneider et al. (1977) and by Cho and Schneider and Carlson (1977) and Cho and Schneider (1978) gives particularly good reverse breakdown characteristics.

The diode ideality factor η is defined by

$$I \cong I_0 \exp(qV/\eta kT), \tag{33}$$

which describes approximately the forward characteristic in the absence of series resistance. Figure 11 shows the forward I–V characteristics of a typical 1 μm Schottky diode. The I–V data are linear between 10^{-6} and 10^{-4} A, with a slope corresponding to $\eta = 1.145$, but there are deviations at both low and high currents. The nonlinearity in the 10^{-3} A region is due to a series resistance R_s of approximately 10 Ω. There is a slight leakage current, causing a shunt effect below 10^{-6} A. This is clearly undesirable, but does not seem to adversely affect diode noise performance when it is of the magnitude shown in Fig. 11. However, diodes demonstrating considerably larger amounts of leakage current have poor noise temperatures, and this effect seems associated with processing techniques rather than with any intrinsic property of the epitaxial GaAs. The particular diodes whose characteristics are plotted in Fig. 11 produced excellent (Vizard et al., 1979) conversion loss and noise temperature at 111 and 170 GHz (cf. Figs. 15–17).

Although the concepts discussed in Section IV indicate that the dc characteristics of a Schottky diode are not of themselves *quantitative* criteria for determining high-frequency performance, nevertheless they still serve a useful function. In fabricating diodes, the most crucial fabrication step is the final epilayer etch, which prepares the GaAs for anode deposition and adjusts the undepleted epilayer thickness to the required size. By measuring η, R_s, C_0, and V_{BB} this etching rate can be established precisely.

B. SEMICONDUCTOR MATERIALS

Schottky diodes are usually made from either Si or GaAs epitaxial material, GaAs being preferred for high-frequency use because of its greater mobility (five times greater than Si). However, high mobility is in itself insufficient to guarantee better high-frequency performance, since it must be combined with a sufficiently high solid solubility of the donor impurities in order to produce satisfactory resistivity. With arsenic-doped Si, for example, doping densities of up to 10^{20}/cm^3 are available (Burnues et al., 1976), with corresponding resistivities of 0.001 Ω cm, but this value for resistivity can now

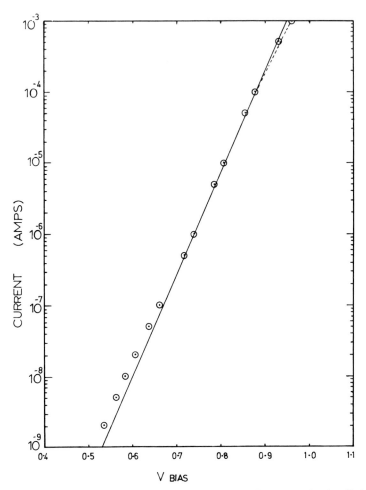

Fig. 11   Forward *I–V* characteristics (circles) of a 2-μm-diameter Schottky diode. The straight line corresponds to an ideal exponential diode with $\eta = 1.145$; the dotted line $R_s = 10\,\Omega$.

be routinely achieved with $2 \times 10^{18}/\text{cm}^3$ Si-doped GaAs. Thus comparing a GaAs and an Si Schottky diode with the same epilayer carrier concentration [seen from Eq. (28) to be necessary in order to give similar parasitic capacitances], the epilayer series resistance will be a factor of five times lower for GaAs. In addition, recently reported (Lacombe *et al.*, 1977) low-pressure organometallic VPE techniques have allowed the use of Te-doped GaAs substrates with a resistivity of less than $0.0005\,\Omega$ cm. Thus, there is little doubt that GaAs is the better material for high-frequency diode fabrication. There has been considerable interest in using some of the other, even higher mobility III–V compounds, but there have been practical difficulties. One of the most promising is InSb with a mobility some nine times higher than GaAs. Schottky-barrier diodes have been fabricated on InSb by Korwin-Pawlowski and Heasell (1975), Kelly and Wrixon (1978c) and McColl and Millea (1976), but have exhibited anomalously high series resistance in addition to relatively low shunt resistances. In addition to these practical difficulties, it has been pointed out by Wrixon and Kelly (1978), that attempts to benefit from the high mobility of InSb by using a low epilayer doping to reduce shot noise, will lead to a degradation in mixer performance due to carrier plasma oscillations in the epilayer. Another drawback of InSb is its low bandgap, which requires that it be cooled to produce a rectifying barrier. In summary then, because of its high mobility, ready availability, and room-temperature wide-bandwidth operation, GaAs is the most suitable material for the fabrication of submillimeter Schottky diodes.

C. EPILAYER STRUCTURE

The essential requirements for the epitaxial structure of a Schottky diode are that the Schottky barrier and its associated depletion region be situated on an epilayer of suitably low carrier concentration so that junction parasitic capacitance is low [cf. Eq. (28)], shot noise in the junction is reduced, and back-breakdown voltage is sufficiently high to avoid drawing reverse current during hard local oscillator pumping, which would result in noise. On the other hand, the concentration should be high enough so as to prevent carrier freezeout in a cooled diode, and to avoid making the undepleted epilayer series resistance unacceptably high. For room-temperature GaAs Schottky diodes the epilayer carrier concentration that is normally used to optimize these conflicting requirements is about $1 \sim 2 \times 10^{17}/\text{cm}^3$. For diodes that will be used in cooled mixers, it is better to choose a lower epilayer carrier concentration in order to reduce the cooled-mixer noise temperature. It has been found with University College Cork (UCC) diodes that $2 \times 10^{16}/\text{cm}^3$ epilayers produced a measured mixer noise temperature at 295 K, seven times larger than at 15 K, while the same ratio for diodes fabricated on $2 \times 10^{17}/\text{cm}^3$ epilayers was approximately 2.5. Thus, the lower doped epilayers increased the noise improvement by cooling, to yield a value of 90 K for T_m (SSB) at 115 GHz in a cryogenic receiver (Vizard et al., 1979; Keen, 1979).

Of crucial importance is the availability of epilayer material with sufficiently good control, both of the thickness of the epilayer and of the sharpness of the epilayer/substrate transition. Figure 12 shows typical carrier concentration profiles measured on material grown by three different techniques, vapor phase epitaxy, molecular beam epitaxy, and low-pressure organometallic vapor phase epitaxy (Lacombe et al., 1977). It is important for submillimeter

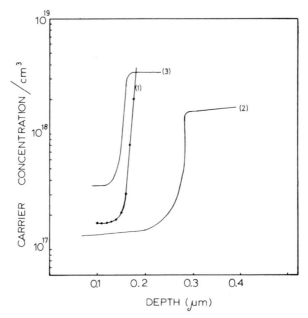

FIG. 12   Measured carrier concentration profiles of epitaxial GaAs. Curve number 1 was obtained from the C–V characteristics of a 1 mm Hg contact; the remaining curves were drawn with a POP profiler (Ambridge and Factor 1975). (1) Organometallic VPE, (2) standard VPE, (3) MBE.

diodes to use epitaxial material with sharp interfaces, like curves 1 and 3 in Fig. 12, for the following reasons. When a Schottky diode is fabricated on a relatively thick epilayer, as in Fig. 13a, the depletion depth changes with voltage as indicated, V_F, V_0, and V_R indicating the position of the edge of the depletion region under forward, zero, and reverse bias, respectively. The zero-bias barrier capacitance C_0 remains low, since the space charge region is within the epilayer and the back-breakdown voltage is relatively high (of the order of 12 V for an epilayer concentration of $2 \times 10^{17}/\text{cm}^3$). The shape of the interface is of little consequence in this instance. However, it is clear that the high resistivity epilayer would, in this case, cause the series resistance to be unnecessarily high, and it would be better to choose a thinner epilayer as in Fig. 11b. Although this lowers R_s, since the undepleted epilayer contributes less, the situation is still not ideal. With the illustrated doping profile, R_s could be reduced by making the epilayer thinner, but this would move the depletion region further up the interface and cause an increase in capacitance and reduction in back-breakdown voltage. However, with a very sharp profile, as in Fig. 11c, it is, in principle, possible to arrange for R_s to be minimized without degrading C_0 or V_{BB}. Ideally, the epilayer should be no more than 0.12-μm

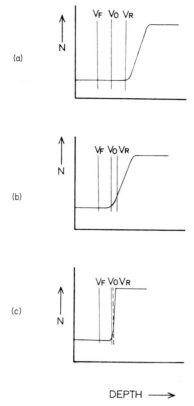

FIG. 13 Schematic illustration of the effect of epilayer structure upon the width of the space charge region.

thick so that a few hundred angstrom etch is sufficient to provide the final epilayer thickness required.

D. OPTIMUM CHIP TOPOGRAPHY

One of the important parameters for high-frequency optimization is the diode diameter. It has been known for some time that at short wavelengths use of a large-diameter diode leads to reduced operating efficiency because of its large parasitic capacitance. However, it was shown in Section IV that the use of a diode with an unnecessarily small diameter can also lead to very high conversion loss, so it is important to use the optimum diameter for a particular operating frequency. In this respect it should be noted that the diode diameter determines the conversion loss via the series resistance and capacitance terms in Eqs. (26) and (29). As pointed out by Wrixon and Pease (1975), the possibility exists of reducing the spreading resistance, while keeping the capacitance constant, by increasing the perimeter-to-area ratio of the diode. Wrixon and Pease fabricated cross-shaped epitaxial diodes of equal area with 1-μm-diameter circular diodes and confirmed that the spreading resistance was reduced by 30 % while the capacitance remained unchanged. This technique, which requires the use of electron beam lithography to attain the required resolution, can reduce optimum conversion loss below that shown in Fig. 8.

There has been some recent analytical investigation into the most effective diode shape for reducing the skin-effect component of series resistance, which is of importance for diodes operating in the submillimeter region.

TABLE III

COMPUTED SKIN-EFFECT CONTRIBUTION TO SPREADING RESISTANCE R_s
FOR DIODES OF AREA $\pi(\mu m)^2$ AT 400 GHz[a]

No.	Anode shape	Ohmic contact configuration	Skin-effect contribution to spreading resistance R_s (Ω)
1		B	4.4
2	2 μm ϕ	S	3.9
3		T	2.0
4		B	3.9
5	0.3 μm wide	S	3.3
6		T	1.4
7		B	4.2
8	0.5 μm wide	S	3.6
9		T	1.7
10		B	4.2
11	0.3 μm wide	S	3.6
12		T	1.7

[a] See text for further details.

The preliminary results of this work, which used the numerical finite element technique by Campbell *et al.* (1979), are summarized here in Tables III and IV. In Table III the diode shapes compared are of equal area to a 2-μm-diameter circle (1-μm-diameter circle in Table IV) and are assumed to be fabricated upon $2.10^{18}/cm^3$ GaAs since, as mentioned above, at submillimeter wavelengths the skin-effect resistance arises almost entirely within the substrate material. The anode is assumed to be located in the middle of the top surface of a square chip of 250 μm sides, which is 100 μm thick; three ohmic contact configurations are used, labeled B, S, and T. B indicates that the ohmic contact is on the back of the chip only, S indicates that the back and sides are metallized, and T indicates that back, sides, and top are metallized to within 12.5 μm of the diode. Thus, for example, configuration number 9 would consist of a cross-shaped area in the center of a 25 μm circle of unmetallized GaAs on a 250 μm square chip, the rest of whose surfaces have had an ohmic contact fabricated on them. These interesting results, which are calculated for $f = 400$ GHz but are scalable as $(f)^{1/2}$, show that a sizable reduction in high-frequency series resistance is obtainable, first, by using a metallized front surface, and second, by using a cross-shaped diode.

TABLE IV

COMPUTED SKIN-EFFECT CONTRIBUTION TO SPREADING RESISTANCE R_s FOR DIODES OF AREA $\frac{1}{4}\pi(\mu m)^2$ AT 400 GHz[a]

No.	Anode shape	Ohmic contact configuration	Skin-effect contribution to spreading resistance R_s (Ω)
13	1 μm ϕ	B	4.9
14		S	4.4
15		T	2.5
16	0.3 μm wide	B	4.7
17		S	4.2
18		T	2.3
19	0.5 μm wide	B	4.8
20		T	2.1

[a] See text for further details.

The importance of metallizing the epilayer surface to lower skin effect has been pointed out previously by Calviello and Wallace (1976) for large-diameter varactor diodes, and the above calculations confirm the improvements to be expected. The calculations of Tables III and IV have ignored the spreading resistance and the epilayer resistance, whose sum gives another shape-dependent, but only slowly frequency-dependent, resistance. Interestingly, one of the best shapes appears to be the stripe geometry, which is a relatively simple configuration and analogous to the bathtub-shape used by Schneider et al. (1977) to fabricate Schottky diodes for use at 100 GHz. The lower limit upon the dimensions of complex shapes will probably be determined by edge effects, which could actually cause the capacitance to increase for a shape with excessive perimeter/area ratio, thus offsetting the reduction in R_s.

It is likely that metallizing the sides of the diodes will lead to a larger improvement than is indicated in the tables, as a result of damage induced during scribing or sawing of the diode chip. This will cause a damage-induced mobility decrease near the surface of the chip, so that in practice the B values in Tables III and IV will be higher and, consequently, the actual improvements will be greater than those tabulated. Experimental work is under way in UCC at present (Langley et al., 1979; Kelly and Wrixon, 1979b) to make circular, crossed and striped diodes on chips with both metallized and non-metallized epilayers in order to confirm that the expected improvements are achieved in practice.

One additional point, which can be important at short wavelengths but is often overlooked, is that the physical dimensions of the chip should, if possible, be short compared with a wavelength, otherwise generation inefficiency and harmonic dropout can result. It has recently been suggested by Keen (1979) that a possible explanation is the relative phases of various rf currents flowing from the diode over the surface of the chip. The amount of out-of-phase attenuation is a function of chip geometry, but Fig. 14 shows the result for four different diode positions on a 250 μm square chip 100 μm thick. One additional advantage of the metallized front surface diode is that the maximum physical path lengths are reduced to about 20 μm (effectively 65 μm in the GaAs with $\varepsilon \approx 10.9$), so that this phase problem should not be important until well over 1000 GHz.

One of the inherent drawbacks of the Schottky diode–whisker structure is that the whisker-diode structure has a certain amount of inbuilt mechanical instability. This is not particularly important for a laboratory environment, but for space and airborne applications, stability is important to ensure high reliability. Thus, there is considerable interest in developing planar diodes with a photolithographically fabricated contact on the chip surface. A number of groups have developed techniques for reducing the associated

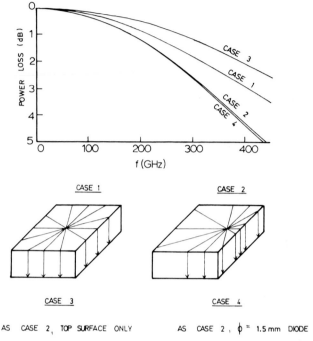

FIG. 14 A plot of the calculated increase in conversion loss as a function of frequency for a 2 μm diode on a 250-μm-square chip positioned as illustrated (due to Keen, 1979).

parasitic capacitance which is the major drawback of this technique, and the efforts in this direction have been comprehensively reviewed by Clifton (1979). Considerable progress has been made and the photolithographic technique suggests the possibility of integrating complete circuits on the GaAs surface. At present, however, the performance of planar diode structures in the submillimeter region falls short of normal Schottky diode–whisker combination.

E. EXPERIMENTAL RESULTS

The best results achieved with state of the art Schottky diode mixers above 100 GHz are presented in Figs. 15 and 16. The results are shown in terms of both mixer SSB noise temperature T_m and conversion loss L_m. For the purpose of facilitating comparison in these diagrams, results that were originally reported in terms of DSB operation have been converted to SSB, by the relationships

$$T_m(\text{SSB}) = 2T_m(\text{DSB}) \tag{34}$$

$$L_m(\text{SSB}) = 2L_m(\text{DSB}) \tag{35}$$

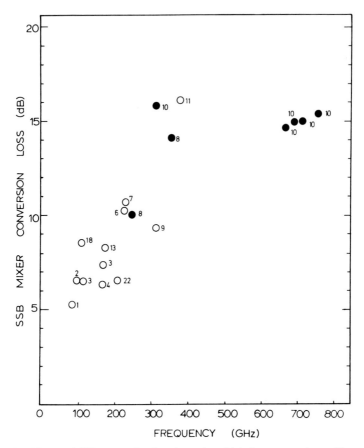

FIG. 15   Measured SSB conversion loss in various high frequency mixers. The numbers refer to the following references: (1) Held and Kerr (1978), (2) Carlson *et al.* (1978), (3) Vizard *et al.* (1979), (4) Kerr *et al.* (1977b), (6) Decker (1978), (7) Goldsmith and Plambeck (1976), (8) Gustincic *et al.* (1977), (9) Erickson (1978), (10) Fetterman *et al.* (1978), (11) Wrixon (1978), (13) Wrixon (1974), (14) Kerr (1975), (15) Keen *et al.* (1979), (16) Vizard (1979), (18) Kerr *et al.* (1977a), (19) Schneider and Carlson (1977), (20) Schneider *et al.* (1977), (21) Keen *et al.* (1978), (22) Zimmerman and Mattauch (1979). (●) Quasi-optical, (○) waveguide.

These equations assume that the conversion losses in the image and signal channels of the DSB mixer are equal, which should be a reasonable approximation for small IF frequencies.

The results plotted have several interesting features. First, it is clear that T_m increases very rapidly with frequencies over 100 GHz, but at the upper end of the frequency scale the quasi-optical mixer results of Fetterman *et al.* (1978) are very promising. The fact that the same mixer produced a relatively

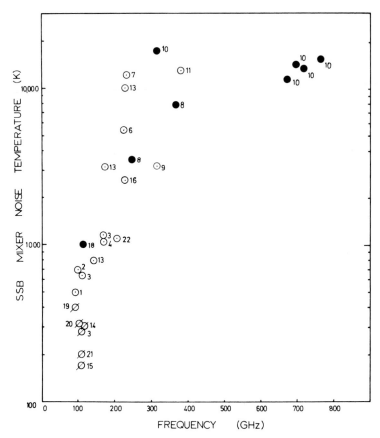

FIG. 16   Measured SSB mixer noise temperature in high frequency mixers. References numbered as in Figure 15. (●) Quasi-optical, (☉) waveguide, (∅) cooled waveguide.

poor result at about 320 GHz is due not to any intrinsic problem with this mixer type at longer wavelengths, but to the fact that the mixer was designed for 760 GHz. Similar mixers are under construction for the 0.55 to 0.75 mm region and should give an interesting comparison with the best waveguide-based results in this region. Most of the diodes used in these mixers were standard 1- to 2-μm-diameter Schottky-barrier diodes, with the exception of data point numbers 3, 15, 16, 19, 20, 21. The points numbered 19 were obtained with the notch-front diodes described by Schneider and Carlson (1977). These are relatively small chips (75 × 75 μm) whose side surfaces have ohmic contacts and were mounted in a stripline mixer. Relatively large, bathtub-shaped diodes were used for points 20 and 21 with a low-doped, thin epilayer, giving characteristics approaching those of a Mott

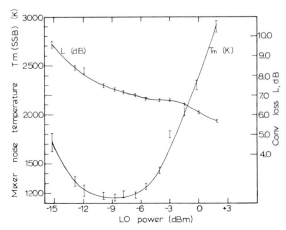

FIG. 17   Conversion loss and mixer noise temperature for a Mott diode in a 170 GHz superheterodyne receiver (f_{IF} = 4 GHz) (from Vizard et al., 1979).

diode. The diodes giving the 280 K T_m at 111 GHz and 1150 K at 170 GHz were also Mott diodes (Vizard et al., 1979), 1 μm in diameter, with epilayer concentration of 1.5 × 10^{17}/cm^3. These diodes gave a T_m of about 620 K in the same mixer at room temperature. The reverse-bias C–V characteristics of these diodes indicated relatively small capacitance variation corresponding to $\gamma = 0.2$, compared with $\gamma = 0.5$ for a typical Schottky diode. In addition, the LO power requirements were remarkably little, -6 dBm at 111 GHz and -9 dBm at 170 GHz. Figure 17 shows the variation of mixer noise temperature and conversion loss as a function of LO power at 170 GHz for a room-temperature mixer containing one of these diodes. The reasons for the exceptionally low LO power requirement are not yet completely clear, but it may be coupled with the slow capacitance–bias variation. Further variations on these diodes are being fabricated in an attempt to shed light on this question, which is clearly important, since diodes requiring little LO power will be a significant development in improving high-frequency performance at short wavelengths where LO power is difficult and expensive to obtain.

The results of Keen et al. (1979) were obtained with 2 μm Mott diodes made on similar epitaxial material to the diodes of Vizard et al. (1979), except the epilayer concentration was ~2 × 10^{16}/cm^3. The diodes produced approximately the same T_m at room temperature as the 2 × 10^{17}/cm^3 epilayer Schottky diodes but, when cooled, T_m was much lower. More recent measurements on these same diodes have resulted in an even lower mixer noise temperature than that shown in Fig. 16. A T_m of less than 90 K was obtained in a 115 GHz mixer cooled to 15 K, indicating once again the importance of using a lightly doped epilayer in cooled receivers in order to benefit from the attendant reduction in shot noise.

REFERENCES

Ambridge, T., and Faktor, M. M. (1975). *Inst. Phys. Conf. Ser. No. 24* Chapter 6, pp. 320–330.
Burnues, F., Kuno, H. G., and Crandall, P. A. (1976). *Microwaves* March, p. 46 ff.
Calviello, J. A., and Wallace, J. L. (1976). *Electron. Lett.* **12**, 648–650.
Campbell, J., Langley, J., and Wrixon, G. T. (1979). To be presented at the *Int. Submillimeter Conf., IR and Near MM waves, 4th, Florida, December.*
Carlson, E. R., Schneider, M. V., and McMaster, T. F. (1978). *IEEE Trans. Microwave Theory Tech.* **MTT-26** (10), 705–706.
Champlin, K. S., and Eisenstein, G. (1977). *Appl. Phys. Lett.* **31** (3), 221 ff.
Champlin, K. S., and Eisenstein, G. (1978). *IEEE Trans. Microwave Theory Tech.* **MTT-26** (1), 31–34.
Cho, A. Y., and Schneider, M. V. (1978). U.S. Patent 4,108,738 (Cl. 204-15; C25DS/02), August 22, 1978. Appl. 770,014, February 18, 1977.
Clifton, B. G. (1979). *The Radio and Electronic Engineer*, January.
Decker, D. R. (1978). Internal Report No. 195, Electronics Div., NRAO, Charlotteville, Virginia.
Dickens, L. E. (1967). *IEEE Trans. Microwave Theory Tech.* **MTT-15**, 101–109.
Dragone, C. (1968). *Bell Syst. Tech. J.* **47**, 1883–1902.
Erickson, N. R. (1978). *IEEE Int. Microwave Symp. Digest MTT-S* pp. 438–439. Ottawa.
Erickson, N. R. (1979). Private communication.
Fetterman, H. R., Clifton, B. J., Tannewald, P. E., and Parker, C. D. (1974). *Appl. Phys. Lett.* **24** (2), 70–72.
Fetterman, H. R., Tannewald, P. E., Clifton, B. G., Parker, C. D., Fitzgerald, W. D., and Erickson, N. R. (1978). *Appl. Phys. Lett.* **33**(2), 151–154.
Goldsmith, P. F., and Plambeck, R. L. (1976). *IEEE Trans. Microwave Theory Tech.* **MTT-24**, 859–861.
Gustincic, G. G., de Graauw, Th., Hodges, D. T., and Luhmann, N. C., Jr. (1977). Unpublished report.
Held, D. N. (1976). Sc. D. Dissertation, Columbia Univ., New York.
Held, D. N., and Kerr, A. R. (1978). *IEEE Trans. Microwave Theory Tech.* **MTT-26**, 49–61.
Hodges, D. T., and McColl, M. (1977). *Appl. Phys. Lett.* **30**, 5–7.
Keen, N. J. (1979). Private communication.
Keen, N. G., Haas, R., and Perchtold, E. (1978). *Electron. Lett.* **14** (25), 825–826.
Keen, N. G., Wrixon, G. T., and Kelly, W. M. (1979). *Electron. Lett.* **15** (21), 689–690.
Kelly, W. M., and Wrixon, G. T. (1978a). *Appl. Phys. Lett.* **32** (9), 525–527.
Kelly, W. M., and Wrixon, G. T. (1978b). *Electron. Lett.* **14** (4), 80–81.
Kelly, W. M., and Wrixon, G. T. (1978c). *Proc. Int. Conf. Submillimeter Waves Appl., 3rd, Guildford.*
Kelly, W. M., and Wrixon, G. T. (1979a). *IEEE Trans. Microwave Theory Tech.* **MTT-27** (7), 665–672.
Kelly, W. M., and Wrixon, G. T. (1979b). To be presented at *Int. Conf. Infrared Near Millimeter Waves, 4th, Florida* December.
Kerr, A. R. (1975). *IEEE Trans. Microwave Theory Tech.* **MTT-23** (10) 787.
Kerr, A. R. (1979). *IEEE Trans. Microwave Theory Tech.* Vol. *MTT-27*, (2), 135–140.

Kerr, A. R., Siegel, P. H., and Mattauch, R. G. (1977a). *IEEE MTT-S Int. Microwave Symp. Digest* IEEE Cat. No. 77CH 1219-5 MTT, pp. 96–98.

Kerr, A. R., Grange, G. A., and Lichtenberger, G. A. (1977b). *IEEE Trans. Microwave Theory Tech.* **MTT-25** (5), 399–401.

Kerr, A. R., Grange, G. A., and Lichtenberger, G. A. (1978). NASA Tech. Memo. 79616, Goddard Space Flight Centre, Greenbelt, Maryland.

Korwin-Pawlowski, M. L., and Heasell, E. L. (1975). *Solid State Electron.* **18**, 849 ff.

Lacombe, G., Duchemin, J. P., Bonnet, M., and Huyge, D. (1977). *Electron. Lett.* **13** (16), 472–473.

Langley, J. B., Kelly, W. M., and Wrixon, G. T. (1979). To be presented at the *Int. Conf. Infrared Near Millimeter Waves, 4th, Florida*.

Linke, R. A., Schneider, A. Y., Cho, A. Y. (1978). *IEEE Trans. Microwave Theory Tech.* **MTT-26**, No. 12.

McColl, M. (1977). *IEEE Trans. Microwave Theory Tech.* **MTT-25**, 54–59.

McColl, M., and Millea, M. F. (1976). *J. Electron. Mater.* **5**, 191–208.

Padovani, F. A., and Stratton, R. (1966). *Solid State Electron.* **9**, 695–703.

Rideout, V. L. (1975). *Solid State Electron.* **18**, 541–550.

Saleh, A. A. M. (1971). "Theory of Resistive Mixers." MIT Press, Cambridge, Massachusetts.

Schneider, M. V., and Carlson, E. R. (1977). *Electron. Lett.* **13** (24), 745–746.

Schneider, M. V., Linke, R. A., and Cho, A. Y. (1977). *Appl. Phys. Lett.* **31** (3), 219–221.

Tsang, D. W., and Schwarz, S. E. (1979). To be published.

Van der Ziel, A. (1976). *J. Appl. Phys.* **47**, 2059–2068.

Vizard, D. R. (1979). Private communication.

Vizard, D. R., Keen, N. G., Kelly, W. M., and Wrixon, G. T. (1979). *IEEE Microwave Tech. Int. Symp., Orlando, Florida* pp. 81–83.

Weinreb, S., and Decker, D. R. (1976). Unpublished result quoted by Held (1976).

Wrixon, G. T. (1974). *IEEE Trans. Microwave Theory Tech.* **MTT-22** (12), 1159–1165.

Wrixon, G. T. (1978). *Proc. Eur. Microwave Conf., Paris*.

Wrixon, G. T., and Kelly, W. M. (1978). *Infrared Phys.* **18**, 413–428.

Wrixon, G. T., and Pease, R. F. W. (1975). *Inst. Phys. Conf. Series No. 24* Chapter 2, pp. 55–60.

Zimmermann, P., and Mattauch, R. G. (1979). Presented at *IEEE MTT-S Int. Microwave Symp., Orlando, Florida*, April.

CHAPTER 3

Pyroelectricity and Pyroelectric Detectors

A. Hadni

Laboratoire Optique IR et du Solide
University of Nancy I
Nancy, France

Copyright © 1980 by Academic Press, Inc.
All rights of reproduction in any form reserved.
ISBN 0-12-147703-7

I. Introduction

A. PYROELECTRICITY

There is some confusion in the definition of pyroelectric materials. Let us define them as presenting spontaneous electric polarization P_s in a thermodynamic equilibrium state. We eliminate electrets because their electric polarization is obtained by applying an electric field to heated material, which is then cooled at a constant electric field. At room temperature the field is suppressed and polarization is held for a long time (weeks or months). However, because the polarization is unstable, we shall not consider electrets as pyroelectrics.

Pyroelectricity results from an electric dipole moment in the unit cell of the crystal in thermodynamic equilibrium. There are two subgroups of pyroelectrics (Zheludev, 1971):

(i) Linear polar dielectrics. This group is composed of the linear dielectric crystals belonging to polar classes such as the wurtzite-type crystals (ZnO, CdS, CdSe, etc.), tourmaline, lithium sulfate, saccharose, etc. For these pyroelectrics the spontaneous polarization cannot be reversed by application of an electric field. Only the induced polarization is field sensitive and this modification is reversible. The linear pyroelectrics belong to crystal classes 1, 2, 3, 4, 6, m, mm2, 3 m, 4 mm, 6 mm, (Nye, 1960) and to textures ∞, ∞ m (Zheludev, 1971).

(ii) Ferroelectrics. The second group is composed of ferroelectric crystals. Spontaneous polarization generally appears at some temperature at which a phase transition occurs from a nonpolar or polar neutral phase to a polar phase (1, 2, 3, 4, 6, m, mm2, 3 m, 4 mm, 6 mm). At this temperature the crystal splits into regions of different orientation of spontaneous polarization because there is no preferential orientation for the polar axis when the crystal undergoes the phase transition from the nonpolar to the polar phase. Spontaneous polarization can be reversed by a small electric field when the temperature is close to the phase transition temperature. The ferroelectric crystals are nonlinear dielectrics; they exhibit hysteresis loops when they are submitted to an alternating electric field. Linear pyroelectrics do not have such phase transitions: they do not split into domains, and they do not show hysteresis loops.

Close to the transition temperature the spontaneous polarization P_s of ferroelectric crystals decreases either sharply (first-order transition) or smoothly (second-order transition). In both cases the pyroelectric coefficient $\pi = dP_s/dT$ takes large values (i.e., 10^{-2}C m^{-2} K^{-1} and 10^{-3}C m^{-2} K^{-1}, respectively, for first- and second-order transitions). For linear pyroelectrics, π is very small at room temperature, however, it can take noticeable values at very low temperatures (Zheludev, 1971).

In conclusion, for all pyroelectrics crystals there is some range of temperatures where spontaneous polarization is very sensitive to temperature variation. The high value of pyroelectric coefficients can be used to detect infrared radiation. The radiations have to increase the crystal temperature, the increase of temperature is transduced into a change of spontaneous polarization which is electrically detected.

B. PYROELECTRIC DETECTION

In a 1963 report (Hadni, 1963) we considered most of the possibilities for the detection of infrared radiation. Indium antimonide, carbon, and germanium bolometers were already working at liquid helium temperatures with a remarkable detectivity ($D^* \simeq 10^{12}$ W^{-1} cm Hz$^{1/2}$) thanks to Putley, Low, Sievers, Zwerdling, and many others. On the other hand, with two promising exceptions the possibilities of pyroelectric detection were hardly known.

Cooper (1962) chose a barium titanate ceramic, working around 100°C, and noticed the advantage of these special thermal detectors for high frequencies, especially when they were made of a very thin plate of pyroelectric material. We deliberately preferred high-quality pyroelectric crystals, which were readily available following discovery of triglycine sulfate (TGS) ferroelectricity. As early as 1963 we obtained high detectivities at room temperature; e.g., D^* (6.5 Hz; 300 K; 10 mm^2; 500 K) = 3.10^8 W^{-1} cm Hz$^{1/2}$, which was comparable to the best thermopiles or bolometers widely used at the time for the detection of infrared, although it was lower by two orders of magnitude than the remarkably optimized detectivity of pneumatic detectors obtained by Dr. Golay during World War II.

We thus have had at our disposal since 1963 an infrared detector as sensitive as bolometers and with a flexibility of design that has led to the proposal of a pyroelectric vidicon with a retina made of a pyroelectric slab. The infrared image is absorbed by the pyroelectric retina, giving a relief of temperature, which is transduced into a distribution of electric charges that are read by a slow-velocity electron beam.

In the sixteen years since then a number of pyroelectric materials have been considered in several laboratories. Single crystals are still giving the best results. Among them TGS leads to the largest detectivity: D^* (6.5 Hz; 300 K; 10 mm^2; 500 K) $\simeq 3.10^9$ W^{-1} cm Hz$^{1/2}$, an order of magnitude higher than the first results. On the other hand, the pyroelectric vidicon is now giving infrared images of a quality approaching the visible TV, and TGS (or its isomorphs) is, thus far, giving the best targets.

These remarks call our attention to the theory of pyroelectric detection, which suggests a look at other pyroelectric materials or better use of those still employed. There is, of course, a maximum detectivity limited by thermal fluctuations. At room temperature D^*_{ideal} (300 K) $= 2.10^{10}$ W^{-1} cm Hz$^{1/2}$, which is only an order of magnitude higher than the best pyroelectric detectors working at low frequencies (i.e., 6.5 Hz), but several orders higher at megahertz frequencies. At low temperature D^* can be very high. D^*_{ideal} is proportional to $T^{-5/2}$, i.e., D^*_{ideal} (0.3 K) $= 6.10^{17}$ W^{-1} cm Hz$^{1/2}$, which is much higher than the detectivities reached by germanium bolometers. The possibilities of pyroelectric detectors at low temperature have also to be considered.

II.   Detectivity

There have been several papers on the calculation of detectivity: Cooper (1962), Hadni (1967), Putley (1971, 1977a and b), Liu and Long (1978), Beerman (1969), Hadni et al. (1969a), Glass and Abrams, (1970b), Schwarz and Poole (1970), Oseki et al. (1972), Peterson et al. (1974), Beerman (1971), Blackburn and Wright (1971), Hatanaka et al. (1972), Holeman et al. (1972), Leiba et al. (1972), Ziel and Liu (1972). The most recent is the one by Liu and Long (1978), which we shall summarize.

A.   Perfect Pyroelectric Crystals

Liu and Long first considered a perfect pyroelectric crystal with dielectric constant reduced to its real part ε'. Heat losses are purely radiative. For a unit area, they are given by $g_R = 4\alpha\sigma T^3$, σ being the Planck constant ($\sigma = 5.67 \times 10^{-12}$ W cm^{-2} K^{-4}), and α the absorption coefficient. In that approximation, the noise arises from thermal fluctuations only and the normalized detectivity,

$$D^*_{\text{ideal}} = (16kT^5\alpha\sigma)^{-1/2}. \tag{1}$$

With $\alpha = 1$ and $T = 300$ K, we get

$$D^*_{\text{ideal}}(300 \text{ K}) = 1.8 \times 10^{10} \quad \text{W}^{-1} \text{ cm Hz}^{1/2}.$$

B. LOSSY PYROELECTRIC CRYSTALS

Liu and Long are now considering a complex dielectric constant $\varepsilon = \varepsilon' - j\varepsilon''$ with $\tan \delta = \varepsilon''/\varepsilon'$, δ being the loss angle. They get

$$D^* = \frac{\pi}{c'} \frac{1}{(d\varepsilon' \tan \delta)^{1/2}} \frac{1}{(4 \, kT\omega\varepsilon_o)^{1/2}} \frac{\alpha\omega\tau}{(1 + \omega^2\tau^2)^{1/2}}, \qquad (2)$$

where $\tau = c'/g_R$ is the thermal time constant, c' is the volume specific heat, d is the pyroelectric sample thickness, π the pyroelectric coefficient.

An abridged demonstration of Eq. (2) can be given, assuming a uniform temperature of the sample and a noise dominated by the dielectric losses of the crystal, as in Liu's paper.

Two types of detector are possible (Fig. 1). In both types radiations are received on an area $A = Ll$. In type I this surface is perpendicular to the pyroelectric axis; in type II it is parallel to it. We shall see that flux detectivity D^* can be the same for both types. However, for type II the resistance equivalent to the detector is much larger, as is both noise and signal. The problem of amplification may be easier.

C. RESPONSIVITY

1. *Temperature Variation*

We assume the detector is surrounded by a background at temperature T_o, except for the solid angle Ω (Fig. 1) where the radiations have a luminance variation calculated between L_o (blackbody brilliancy at temperature T_o), with solid angle closed by a reflecting shutter, and $L_o + \Delta L$, with shutter completely opened and the whole detector area viewing the source of radiations through the solid angle Ω. It is also assumed that ΔL variations are sinusoidal (Fig. 2):

$$\Delta L = \tfrac{1}{2}\Delta L_M(1 - \cos \omega t). \qquad (3)$$

Let us also assume that temperature is uniform in the whole pyroelectric material. This means that diffusion length $l = (2\chi/\omega)^{1/2}$ and thermal wavelength $\Lambda = 2\pi l$ (χ being the thermal diffusivity) are larger than thickness d. Then the increase ξ of temperature is written:

$$\xi = \xi_1 + \xi_M \sin(\omega t + \phi),$$

with

$$\xi_1 = \alpha \mathscr{T} \Omega \, \Delta L_M/2\mathscr{G}, \qquad (4)$$

$$\xi_M = \frac{\alpha \mathscr{T} \Omega \, \Delta L_M}{2\mathscr{G}(1 + \omega^2\tau^2)^{1/2}}, \qquad (5)$$

$$\tan \phi = \mathscr{G}/\mathscr{C}\omega \qquad \text{and} \qquad \tau = \mathscr{C}/\mathscr{G}.$$

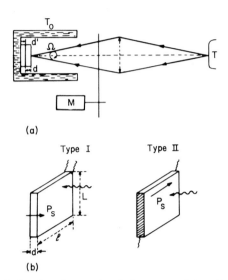

(a)

Type I   Type II

(b)

FIG. 1   (a) Schematic drawing of a pyroelectric detector receiving a beam of radiations from a blackbody source at temperature T, chopped by a synchronous motor M. (b) Two types of pyroelectric detectors (Hadni, 1970).

The thermal time constants τ generally range from 2 s for a TGS pyroelectric plate stuck on a thin mylar film suspended in vacuum, to 20 ms when it is directly glued on a copper block (heat sink). $\mathscr{C} = c'd$ is the heat capacity per unit area, c' being the specific heat per unit volume and d the sample thickness as said above.

$\mathscr{G} = (\lambda'/e') + 4\sigma T_o^3$ is the thermal loss flux per unit area, unit time, and 1 deg temperature difference between pyroelectric plate and background. λ' is the thermal conductivity, e' the thickness of the thermal insulator between heat sink and pyroelectric plate, σ is the Stefan constant ($\sigma = 5.67 \times 10^{-12}$ W cm^{-2} K^{-4}), α is the absorption coefficient, and $\mathscr{T}$ the transmission co-

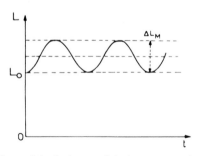

FIG. 2   Waveform of the liminance of the beam versus time t (Hadni, 1970).

efficient of the window generally used in front of the detector. For $T_o = 300\,\text{K}$, we get $4\sigma T_o^3 = 6 \times 10^{-4}\,\text{W cm}^{-2}\,\text{deg}^{-1}$.

It is seen that temperature bias ξ_1 is independent of chopping frequency ω, and temperature amplitude ξ_M decreases when ω increases. For angular frequencies ω large enough for $\omega\tau \gg 1$, we get

$$\xi_M \simeq \frac{\alpha \mathscr{T} \Omega\, \Delta L_M}{2\omega\mathscr{C}}, \tag{6}$$

where the temperature variation amplitude is inversely proportional to ω and independent of the thermal loss coefficient $\mathscr{G}$.

2. Flux Responsivity

The increase in temperature of the pyroelectric plate $\Delta T = \xi - \xi_1$ induces a polarization variation

$$\Delta P_s = \Delta T\, \frac{\partial P_s}{\partial T} + \frac{(\Delta T)^2}{2}\, \frac{\partial^2 P_s}{\partial T^2},$$

$$\Delta P_s = \pi_\sigma\, \Delta T + \tfrac{1}{2} - \mu_\sigma (\Delta T)^2,$$

where

$$\pi_\sigma = \left(\frac{\partial P_s}{\partial T}\right)_\sigma \quad \text{and} \quad \mu_\sigma = \left(\frac{\partial^2 P_s}{\partial T^2}\right)_\sigma$$

in the case of unclamped samples (constant stress σ), which is the general case (see Section VIII). We assume the pyroelectric plate to be covered by a conductive layer on each face. We have a pyroelectric capacitor that gets electrically charged when illuminated. This simple model (Fig. 3) contains a pure capacitor C in parallel with a pure resistor ρ', which takes into account all losses in the pyroelectric plate (ρ' is the internal resistance), and a load resistance ρ which can be chosen. In this model, the capacitor C receives a pyroelectric current

$$i(T) = \mathscr{A}\, \frac{dP_s}{dt}$$

with $\mathscr{A} = Ll$ for type I detectors, and $\mathscr{A} = Ld$ for type II detectors: We thus have

$$i = \mathscr{A}\, \frac{dP_s}{dT}\, \frac{dT}{dt}$$

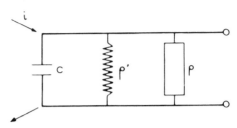

FIG. 3   Schematic diagram of a pyroelectric detector; ρ is the load resistance (Hadni, 1970).

or

$$i(t) = \frac{\alpha \mathcal{T} \mathscr{A}\Omega \, \Delta L_M \pi_\sigma \omega}{2\mathscr{G}(1 + \omega^2\tau^2)^{1/2}} \cos(\omega t + \phi) + \mathscr{A}\mu_\sigma \, \xi_M^2 \sin^2(\omega t + \phi). \qquad (7)$$

If we compare Eqs. (3) and (7), we see that $i(t)$ is in phase (or in phase opposition, depending on the spontaneous polarization orientation) with ΔL when $\phi = 0$ and where ϕ is the phase advance of the current over the illumination, with

$$\tan \phi = \mathscr{G}/\mathscr{C}\omega. \qquad (8)$$

Thus we get the potential difference s which appears across load resistor ρ: $s(t) = |Z| i(t)$, Z being the complex total impedance of the detector circuit, and $|Z| = R/(1 + \omega^2\tau'^2)$ with

$$1/R = (1/\rho) + (1/\rho'), \text{ and } \tau' = RC.$$

We obtain the amplitude S of this potential difference in the assumption that ξ_M^2 is small enough to allow neglect of the second term of $i(t)$:

$$S = \frac{\alpha \mathcal{T} \mathscr{A}\Omega \, \Delta L_M \pi_\sigma \omega}{2\mathscr{G}(1 + \omega^2\tau^2)^{1/2}} \frac{R}{(1 + \omega^2\tau'^2)^{1/2}}. \qquad (9)$$

The flux responsivity $\mathscr{R}$ is defined as $\mathscr{R} = S/\phi$, with $\phi = A\Omega \, \Delta L_M$ being the variation amplitude of the radiation flux: $\mathscr{R}$ is the signal normalized to a unit flux illuminating the whole detector area $A = Ll$, and is generally expressed in $\mu V/\mu W$.[*]

One equation for each type of detector is obtained:

$$\mathscr{R}_I = \frac{a.\mathcal{T} \pi_\sigma \omega}{2\mathscr{G}(1 + \omega^2\tau^2)^{1/2}} \frac{R}{(1 + \omega^2\tau'^2)^{1/2}}, \qquad (10)$$

$$\mathscr{R}_{II} = \frac{(\alpha.\mathcal{T} \pi_\sigma \omega \, d)/l}{2\mathscr{G}(1 + \omega^2\tau^2)^{1/2}} \frac{R}{(1 + \omega^2\tau'^2)^{1/2}}. \qquad (11)$$

[*] There is sometimes a need to consider a luminance responsivity $\mathscr{R}' = S/\Delta L_M$. Assuming that radiations cover the whole surface area A, $\mathscr{R}'$ is expressed in μV per μW cm^{-2} sd^{-1}. This responsivity is useful when an unlimited flux of brilliancy ΔL_M is at our disposal. This may occur in far infrared spectrometers; $\mathscr{R}$ is used in astronomy where the acceptance $E = A\Omega$ is strictly given.

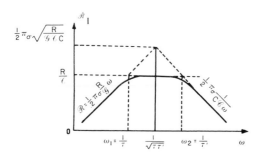

FIG. 4   Responsivity $\mathscr{R}$ versus angular frequency ω.

The flux responsivity is dependent on both ω and R. On the other hand, while $\tau_I = \tau_{II}$, we generally have $\tau'_I \neq \tau'_{II}$ and no immediate relations can be drawn for the ratio $(\mathscr{R}_I/\mathscr{R}_{II})$.

3.   Discussion on Responsivity $\mathscr{R}_I$

a.   *General Dependence on Frequency ω.*   The variation of $\mathscr{R}_I$ versus ω is shown in Fig. 4. It is seen that $\mathscr{R}_I$ is maximum between $\omega_1 = 1/\tau$ and $\omega_2 = 1/\tau'$; ω_1 is the thermal frequency; ω_2 is the electrical frequency.

For $\tau = 4 \times 10^{-2}$s, $\omega_1 = 25$ rad s^{-1}. If we assume that $C = 10^{-9}$F and $R = 10^4\Omega$, then $\tau' = 10^{-5}$s and $\omega_2 = 10^5$ rad s^{-1}. We thus have a wide frequency band ($\omega_2 - \omega_1 \simeq 10^5$ rad s^{-1}) where $\mathscr{R}_I$ is maximum and fairly constant. For higher frequencies $\mathscr{R}_I$ decreases as $1/\omega$.

b.   *Dependence on R.*   For a given value of ω,

$$\frac{d\mathscr{R}_I}{dR} = \frac{\alpha\tau\pi_\sigma\omega}{2\mathscr{G}(1 + \omega^2\tau^2)^{1/2}} \frac{1 - [\omega^2\tau'^2/2(1 + \omega^2\tau'^2)^2]}{(1 + \omega^2\tau'^2)^{1/2}}.$$

This derivative is always positive and thus responsivity is maximum for R maximum. We are led to make R maximum by choosing an infinite value of the bias resistance ρ (or at least higher than the internal resistance ρ'), then $R \simeq \rho'$ and

$$\mathscr{R}_I^{opt} = \frac{\chi \mathscr{T} \pi_\sigma\omega\rho'}{2\mathscr{G}(1 + \omega^2\tau^2)^{1/2}(1 + \omega^2\tau'^2)^{1/2}}, \tag{12}$$

with $\tau' = \rho'C$; $\tau' = r_o\varepsilon_o\varepsilon'$ being the electrical time constant characteristic of the given material and r_o the resistivity.

However it is seen that $d\mathscr{R}_I/dR \to 0$ for any value of R when $\omega \to \infty$: For frequencies much higher than ω_1 and ω_2, responsivity becomes independent of bias resistance ρ. This is easy to understand since $\rho' \to 0$ when $\omega \to \infty$ and any value of ρ is short circuited.

D. Optimized Responsivity

1. *Type II Detectors May Have a Higher Responsivity*

For both types of detectors, normalized thermal conductivity $\mathscr{G}$ and heat capacity $\mathscr{C}$ are the same. However, for type I, $\rho_I' = r_o d/Ll$, and thus from Eq. (10):

$$\mathscr{R}_I^{\mathrm{opt}} = \frac{\alpha \mathscr{T} \pi_\sigma \omega r_o\, d}{2\mathscr{G}Ll(1 + \omega^2\tau^2)^{1/2}(1 + \omega^2\tau'^2)^{1/2}}. \tag{13}$$

For type II, $\rho_{II}' = r_o l/Ld$, and from Eq. (11):

$$\mathscr{R}_{II}^{\mathrm{opt}} = \frac{\hat{\alpha} \mathscr{T} \pi_\sigma \omega}{2\mathscr{G}L(1 + \omega^2\tau^2)^{1/2}(1 + \omega^2\tau'^2)^{1/2}}. \tag{14}$$

Now $\tau_I' = \tau_{II}' = \varepsilon_o r_o \varepsilon'$, and we get

$$\mathscr{R}_{II}^{\mathrm{opt}}/\mathscr{R}_I^{\mathrm{opt}} = l/d. \tag{15}$$

It is seen that for very thin pyroelectric film ($d \ll l$), responsivity is higher for type II detectors (see also Simhony *et al.*, 1973).

2. *The Case of Insulating Materials*

The losses are purely dielectric and

$$r_o = 1/\varepsilon_o \varepsilon'' \omega. \tag{16}$$

It is known that at room temperature ε'' is independent of ω at least up to 100 MHz, except for some acoustic resonance (Fig. 5). The result is that r_o

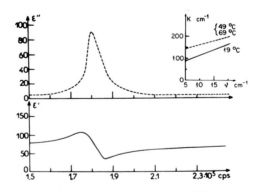

FIG. 5   Except for an acoustic resonance at 180 kHz, the dielectric constant does not change up to 250 kHz at 25°C. In the right it is seen that in the far infrared the absorption coefficient K increases with frequency $\dot{v}$, and with temperature up to the Curie point (TGS 25°C; 150 v/mm) (Hadni, 1970).

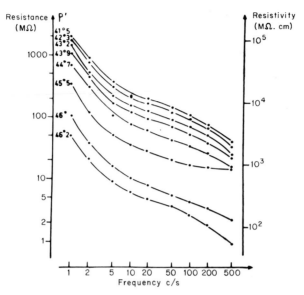

FIG. 6 Resistivity versus frequency for a TGS (40%)–TGSe (60%) mixed crystal at different temperatures (surface $= 1$ cm^2, $e = 0.17$ mm).

decreases as $1/\omega$ when ω increases. This is well checked by experiment (Fig. 6). Then

$$\mathscr{R}_I^{\text{opt}} = \frac{\alpha \mathscr{T} \pi_\sigma d}{2\mathscr{G} L l \varepsilon_o \varepsilon''(1 + \omega^2\tau^2)^{1/2}(1 + \omega^2\tau'^2)^{1/2}}, \tag{17}$$

$$\mathscr{R}_{II}^{\text{opt}} = \frac{\alpha \mathscr{T} \pi_\sigma}{2\mathscr{G} L \varepsilon_o \varepsilon''(1 + \omega^2\tau^2)^{1/2}(1 + \omega^2\tau'^2)^{1/2}}. \tag{18}$$

E. HIGH FREQUENCIES

When $\omega > \omega_2 > \omega_1$ Eqs. (10) and (11) can be simplified directly without any assumption on R:

$$\mathscr{R}_I^{\text{hf}} \simeq \frac{\alpha \mathscr{T} \pi_\sigma \omega R}{2\mathscr{G}\omega^2\tau\tau'}, \qquad \mathscr{R}_{II}^{\text{hf}} \simeq \frac{\alpha \mathscr{T} \pi_\sigma \omega R}{2\mathscr{G}\omega^2\tau\tau'} \frac{d}{l}.$$

Hence

$$\mathscr{R}_I^{\text{hf}} \simeq \frac{\alpha \mathscr{T} \pi_\sigma}{2\mathscr{C} C_I \omega}, \qquad \mathscr{R}_{II}^{\text{hf}} \simeq \frac{\alpha \mathscr{T} \pi_\sigma}{2\mathscr{C} C_{II} \omega} \frac{d}{l}.$$

Now

$$C_I = \frac{\varepsilon_o \varepsilon_R L l}{d}; \qquad C_{II} = \frac{\varepsilon_o \varepsilon_R L d}{l};$$

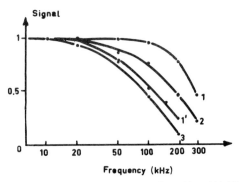

FIG. 7   Signal versus frequency for type I detectors from 10 to 300,000 Hz for several load resistors: curve 1, $\rho = 10\,k\Omega$; curve 2: $\rho = 22\,k\Omega$; curve 3, $\rho = 47\,k\Omega$ (detector area 7 mm^2); curve 1′, $\rho = 10\,k\Omega$ (detector area 80 mm^2).

then

$$\mathscr{R}_{\mathrm{I}}^{\mathrm{hf}} = \frac{\alpha \mathscr{T} \pi_{\mathrm{s}} d}{2 \mathscr{C} \varepsilon_o \varepsilon_{\mathrm{R}} L l \omega}, \tag{19}$$

$$\mathscr{R}_{\mathrm{II}}^{\mathrm{hf}} = \frac{\alpha \mathscr{T} \pi_{\mathrm{s}}}{2 \mathscr{C} \varepsilon_o \varepsilon_{\mathrm{R}} L \omega}. \tag{20}$$

In both cases it is seen again that responsivity is independent of R and $\mathscr{G}$, is proportional to $1/\omega$, and is generally larger for type II detectors: $\mathscr{R}_{\mathrm{II}}^{\mathrm{hf}}/\mathscr{R}_{\mathrm{I}}^{\mathrm{hf}} = l/d$. Figure 7 summarizes some measurements on two TGS type I detectors with different target area.

F.   NOISE*

Let us suppose that noise B is practically limited to Johnson noise. It occurs from the real part Z_{R} of the total impedance Z of the whole detector (crystal plus bias resistor):

$$Z_{\mathrm{R}} = R/(1 + \omega^2 \tau'^2)^{1/2},$$
$$B = 2(RkT\,\Delta f)^{1/2}/(1 + \omega^2 R^2 C^2)^{1/2}, \tag{21}$$

where Δf is the bandwidth, f the chopping frequency, $\omega = 2\pi f$, and k the Boltzmann constant ($k = 1.380 \times 10^{-23}$ J K^{-1}). For high frequencies such that $\omega \gg \omega_2$,

$$B^{\mathrm{hf}} \simeq (2kT\,\Delta f)^{1/2}/R^{1/2} C \omega. \tag{22}$$

* See Van der Ziel and Liu (1972) and Logan and Moore (1973).

1. *The Case in Which R Is a Constant*

   $B^{hf} \propto 1/\omega$: noise is proportional to the frequency reciprocal.

2. *The Case in Which ρ Is High Enough for $R \simeq \rho'$*

   We get

$$R_I = \frac{d}{\varepsilon_o \varepsilon'' \omega L l}, \qquad R_{II} = \frac{l}{\varepsilon_o \varepsilon'' \omega L d}.$$

Remembering that $C_I = \varepsilon_o \varepsilon' L l/d, \qquad C_{II} = \varepsilon_o \varepsilon' L d/l$, we get

$$B_I^{opt, \, hf} = \frac{(2kT \, \Delta f \varepsilon'')^{1/2}}{(\omega \varepsilon_o)^{1/2} \varepsilon'} \left(\frac{d}{Ll}\right)^{1/2},$$

$$B_{II}^{opt, \, hf} = \frac{(2kT \, \Delta f \varepsilon'')^{1/2}}{(\omega \varepsilon_o)^{1/2} \varepsilon'} \left(\frac{l}{Ld}\right)^{1/2}.$$

   (i) $B_{opt}^{hf} \propto 1/\omega^{1/2}l$,
   (ii) $B_{II}^{hf, \, opt}/B_I^{hf, \, opt} = l/d$.

For thin films ($d \ll l$), noise is higher for type II detectors, but we have seen that the responsivity is also higher, they show same detectivity.

In Fig. 8 it is seen that noise effectively decreases when frequency ω increases, but there is a limit for high frequencies (amplifier noise). For high frequencies, noise is inversely proportional to R.

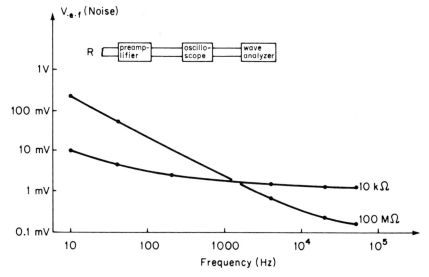

FIG. 8   The noise of the electronics versus frequency for two values of the input impedance R.

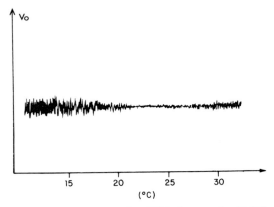

FIG. 9   Noise versus temperature for a TGSe detector, $f = 12.5$ Hz, $T = 5s$.

Figure 9 shows a minimum of noise at the Curie temperature where ε' is maximum.

G.   DETECTIVITY

1.   Signal to Noise Ratio, Detectivity D

By comparing signal [Eqs. (10) and (11)] and noise [Eq. (21)], in the most general case we get

$$D_{\mathrm{I}} = \frac{\mathscr{R}_{\mathrm{I}}}{B} = \frac{\alpha \mathscr{T} \pi_\sigma \omega R^{1/2}}{4\mathscr{G}(1 + \omega^2\tau^2)^{1/2}(kT\,\Delta f)^{1/2}}, \tag{23}$$

$$D_{\mathrm{II}} = \frac{\mathscr{R}_{\mathrm{II}}}{B} = \frac{\alpha \mathscr{T} \pi_\sigma \omega R^{1/2} d/l}{4\mathscr{G}(1 + \omega^2\tau^2)^{1/2}(kT\,\Delta f)^{1/2}}. \tag{24}$$

D_{I}, D_{II} are the responsivities normalized by taking the noise as a unit; they are called detectivities. For high enough bias resistance, $R \simeq \rho'$ and

$$D_{\mathrm{I,II}}^{\mathrm{opt}} = \frac{\alpha \mathscr{T} \pi_\sigma d^{1/2}\omega^{1/2}}{4\mathscr{G}(1 + \omega^2\tau^2)^{1/2}(kT\,\Delta f)^{1/2}(\varepsilon_o\varepsilon'')^{1/2}(Ll)^{1/2}}. \tag{25}$$

Both types of detectors have the same detectivity. Let us notice that parameters in Eq. (25) are not independent (i.e., τ depends on d, etc.) and we cannot discuss their importance.

2.   High Frequencies

Let us suppose $\omega \gg \omega_1$; then

$$D_{\mathrm{I,II}}^{\mathrm{opt,\,hf}} = \frac{\alpha \mathscr{T} \pi_\sigma}{4c'(kT\,\Delta f\,\varepsilon_o\varepsilon'')^{1/2}(Lld)^{1/2}\omega^{1/2}}. \tag{26}$$

Now all parameters are independent, and we can see how to increase detectivity:

(i) decrease Δf,
(ii) decrease thickness d,
(iii) decrease target area Ll (assuming that it still receives a unit flux),
(iv) decrease ω, and
(v) increase $M(T) = \pi_\sigma / c'(\varepsilon'' T)^{1/2}$. $\qquad(27)$

$M(T)$ is the factor of merit of the pyroelectric material. It will be discussed in Section III.

3. Noise Equivalent Power (NEP)

The NEP is the radiation flux that produces a pyroelectric signal, the value of which is the same as the noise. Hence

$$\text{NEP} = 1/D.$$

It is found from Eqs. (23) and (24) that

$$\text{NEP}_\text{I} = \frac{4\mathscr{G}(1 + \omega^2 \tau^2)^{1/2}(kT\,\Delta f)^{1/2}}{\alpha \mathscr{T} \pi_\sigma R^{1/2} \omega}, \qquad(28)$$

$$\text{NEP}_\text{II} = \frac{4\mathscr{G}(1 + \omega^2 \tau^2)^{1/2}(kT\,\Delta f)^{1/2}}{\alpha \mathscr{T} \pi_\sigma R^{1/2} \omega}\frac{l}{d}, \qquad(29)$$

and from Eq. (26) that

$$\text{NEP}_\text{I or II}^{\text{opt, hf}} = \frac{4(Lld)^{1/2}(k\varepsilon_o\,\Delta f)^{1/2}}{\alpha \mathscr{T} M(T)}\,\omega^{1/2}. \qquad(30)$$

4. Normalized Detectivity D*

We have seen $D_\text{I, II}^{\text{opt}} \propto (A\,\Delta f)^{-1/2}$; hence the suggestion to normalize D by putting

$$D^* = DA^{1/2}\,\Delta f^{1/2}.$$

We get

$$D_\text{I or II}^{*\,\text{opt}} = \frac{\alpha \mathscr{T} \pi_\sigma d^{1/2}\omega^{1/2}}{4\mathscr{G}(1 + \omega^2\tau^2)^{1/2}(kT)^{1/2}(\varepsilon_o\varepsilon'')^{1/2}}, \qquad(31)$$

$$D_\text{I or II}^{*\,\text{opt, hf}} = \frac{\alpha \mathscr{T} M(T)}{4(k\varepsilon_o)^{1/2}d^{1/2}\omega^{1/2}}. \qquad(32)$$

It is seen in Fig. 10 that detectivities D^* as high as $6 \times 10^9 \text{ W}^{-1} \text{ cm Hz}^{1/2}$ have been obtained. More commonly, detectivities observed are of the order of $10^8 \text{ W}^{-1} \text{ cm Hz}^{1/2}$ (i.e., for commercial pyroelectric detectors).

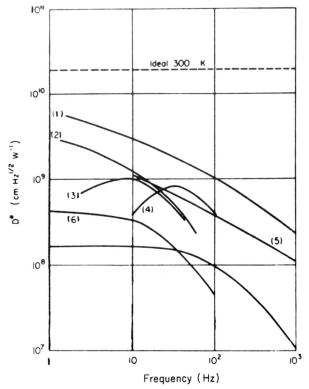

FIG. 10   Performance of some uncooled thermal detectors: (1) Mullard research sample using alanine-doped TGS 10-μm thick and area 1.5×1.5 mm^2, (2) spectroscopic thermopile, (3) Golay cell, (4) TRIAS cell, (5) Mullard production TGS detector in ruggedized encapsulation with area 0.5×0.5 mm^2, (6) evaporated film thermopile, (7) immersed thermistor (from Putley, 1977a).

III.   The Choice of Pyroelectric Materials

We have defined a factor of merit $M_1(T) = \pi_\sigma/c'(\varepsilon''T)^{1/2}$ [Eq. (27)] which is the same for type I and type II detectors when they are optimized by a bias resistor high enough, and when they work at a frequency higher than the thermal frequency.

1.   Pyroelectric Coefficient

Liu and Long (1978) have experimentally seen that π_σ is roughly proportional to $\varepsilon'^{1/2}$ (see Fig. 11):

$$\pi_\sigma = 3 \times 10^{-9} \varepsilon'^{1/2} \quad \text{C cm}^{-2} \text{ K}^{-1}. \qquad (33)$$

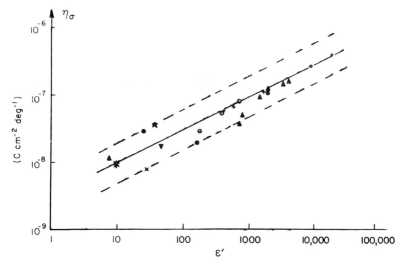

FIG. 11   Relationship between the pyroelectric coefficient and the dielectric constant at room temperature solid line: $P = \sqrt{10} \times 10^{-9}\,\varepsilon'^{1/2}$, $T = 300$ k, $\triangle$, NaNO$_2$; $*$, Li$_2$SO$_4$H$_2$O; $\bullet$, ATGs; X, LiNbO$_3$; $\blacktriangledown$, LiTaO$_3$; $*$, TGS; $\bullet$, BaTiO$_3$; $\odot$, SBN; $+$, La SBN; $\blacktriangle$, PLZT (from Liu and Long, 1978. Reprinted from Proc. IEEE **66** (1), 14–26 (Jan. 1978). © 1978 IEEE).

They have given a demonstration in the case of displacive transitions with a soft mode. As seen in Fig. 11 it is also checked by ferroelectrics with an order–disorder transition. This equation is not fulfilled by linear pyro-electrics of the wurtzite type or by improper ferroelectrics (the primary-order parameter is not the spontaneous polarization) such as tourmaline and boracites. In any case, it is a rough approximation (see Fig. 11), satisfied within a factor of ± 2, which cannot be taken into account by the difficulty to get repetitive values of dielectric constants because of domains reversals. It has been shown that with some care it is possible to get a well-defined value of the dielectric constant ε'.

a.   *Application to Responsivity.*   From inspection of Eqs. (19) and (20) it is seen that $\mathscr{R}_{\mathrm{I,II}}^{\mathrm{hf}}$ is proportional to

$$\frac{\pi_\sigma}{c'\varepsilon'} = \frac{\pi_\sigma}{c'\varepsilon'^{1/2}}\,\frac{1}{\varepsilon'^{1/2}}.$$

From Liu and Long the first term being roughly a constant, responsivity is proportional to $\varepsilon'^{-1/2}$ and decreases when materials with increasing dielectric constant ε' are considered. This prediction is checked by experiment (see Fig. 12).

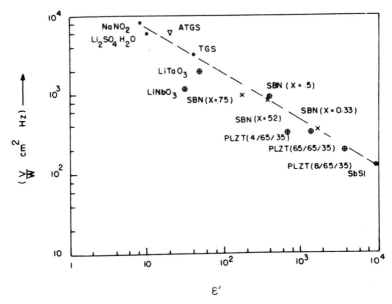

FIG. 12 $\pi_s/C'\varepsilon_0\varepsilon'$ versus ε' for various pyroelectric materials ($T = 300$ K). Data from: $\oplus$, Putley; $\times$, Glass; $\bullet$, Liu et al.; ∇, Pock (from Liu and Long, 1978. Reprinted from Proc. IEEE **66** (1), 14–26 (Jan. 1978). © 1978 IEEE).

b. *Applications to Detectivity.* $M_1(T)$ can be written, $M_1(T) = \pi_\sigma/(c'\varepsilon''^{1/2})\, 1/T^{1/2}$. Now π_σ is roughly proportional to $\varepsilon'^{1/2}$. Then

$$M_1(T) \propto \frac{1}{c'}\left(\frac{\varepsilon'}{\varepsilon''}\right)^{1/2}\frac{1}{T^{1/2}}.$$

It thus appears that $M_1(T) \propto (\tan\delta)^{-1/2}$. From Liu et al., all materials should be equivalent if they have the same loss factor.

2. *Loss Angle*

While the volume specific heat c' does not vary significantly for different materials and is easy to measure, as is the pyroelectric coefficient π_σ, which can be obtained from the slope of the curves $P_s(T)$ versus T, the loss angle is difficult to get with some confidence. Measurements have to be made after a high dc electric field E_{dc} has been applied for some time (i.e., 1 h). Then the crystal is in a state represented by point F of Fig. 13 (hysteresis cycle). All ferroelectric domains are parallel and the applied field is null. In these condition the values of ε'' (and ε') are repetitive for one sample and much smaller than without electrical hardening. The effect has been to remove all domain walls and even any hint of a nucleation center (Hadni et al., 1976).

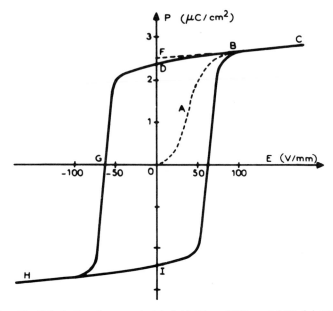

FIG. 13   Polarization P versus electric field E for a TGS crystal (Hadni, 1970).

Figure 14 gives the dielectric constant versus temperature in two cases: (i) $E_{dc} = 750$ V/cm and (ii) $E_{dc} = 150$ V/cm. In the second case, $\varepsilon'' = 3$ (instead of 1.1) and $\varepsilon' = 1900$ (instead of 600), and the ac measurement field (300 V/cm) is sufficient to switch some domains and increase both parts of the dielectric constant [see also Micheron and Godefroy (1971)].

Figure 15 is a pyroelectric map of a TGS crystal plate especially prepared to show a number of domain walls. The dielectric constant at room temperature is especially high ($\varepsilon' = 260$), while we have $\varepsilon' = 33$ for a single domain crystal plate. In any case the loss angle tangent δ depends on the care taken in the preparation of the pyroelectric crystal, as has been shown in the case of strontium barium niobate (SBN) (Liu and Maciolek, 1975; Maciolek *et al.*, 1976).

3. *Discussion*

Table I gives ε'', π_σ, c', and $M_1(T)$ for some pyroelectric materials. The data concerning TGFB are from Felix *et al.* (1978). It is seen that $M_1(T)$ ranges from 1.34×10^{-10} cm J^{-1} K$^{-1/2}$ for TGSe to 0.01×10^{-10} for BaTiO$_3$. For most materials, there is a lack of data concerning ε'', but there are many papers concerning methods and measurements of the pyroelectric coefficient and dc and ac dielectric constants:

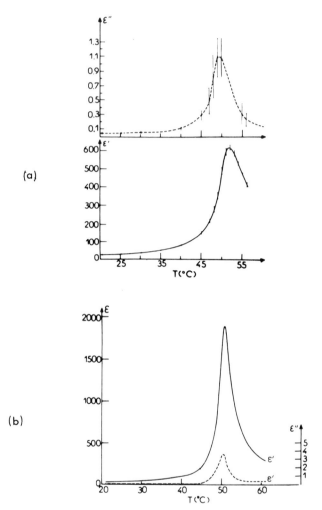

Fɪɢ. 14   (a) TGS dielectric constant in a 750 V/mm dc field measured with a 30 V/mm ac field at 1.592 Hz. (b) TGS dielectric constant in a 150 V/mm dc field measured with a 30 V/mm ac field at 1.592 Hz (Hadni, 1970).

"Méthode et résultats concernant la pyroélectricité de BaTiO$_3$ et autres cristaux" (Gladkii and Zheludev, 1965); "Etude des propriétés pyro-électriques de quelques cristaux et de leur utilisation à la détection du rayonnement" (Hadni et al., 1965); "Method for the measurement of the pyroelectric coefficient, dc dielectric constant and volume resistivity of a polar material" (Lang and Steckel, 1965a); "Study of the ultrasensitive pyroelectric thermometer" (Lang and Steckel, 1965b); "The pyroelectric

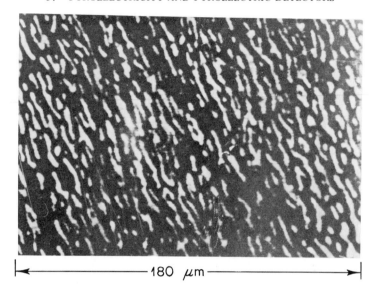

FIG. 15 Pyroelectric map of a small TGS plate area 180 μm long ($e = 7$ μm) after applying a 1600 cps alternative voltage, the amplitude of which has been progressively reduced from 2 MV^{-1} m to zero while the hysteresis loop area is reduced to the origin of coordinates in the P–E plane. The net polarization $P \simeq 0$. The dielectric constant $\varepsilon_R = 260$ (Hadni and Thomas, 1976b).

detector of infrared radiation" (Beerman, 1969); "Pyroelectric thermometer for use at low temperatures" (Lang *et al.*, 1969); "Constantes optiques du sulfate de glycocolle de l'infrarouge proche à l'infrarouge lointain. Application à la pyroélectricité" (Hadni *et al.*, 1969a); "Constantes optiques du Séléniate de glycocolle de l'Infrarouge proche à l'infrarouge lointain et dans le domaine radio-électrique" (Grandjean *et al.*, 1970); "Dispersion de la

TABLE I

PYROELECTRIC PROPERTIES OF A FEW SINGLE CRYSTALS

			ε''	π_σ (10^{-4} C m^{-2} K^{-1})	c' (10^6 J m^{-3})	$M_1(T)$ (10^{-10} cm J^{-1} K$^{-1/2}$)
TGSe	295 K	(22°C)	5.6	13	2.6	1.34
TGFB	338 K	(65°C)	1.26	10	2.5	0.19
DTGS	323 K	(50°C)	1.44	10	2.7	0.17
TGS	299 K	(26°C)	0.18	3	2.5	0.16
SBN	333 K	(60°C)	25	11	2.3	0.04
SBN:La	300 K	(27°C)	9.8	12	2.3	0.10
GASH	300 K	(27°C)	0.02	0.1	1.2	0.03
BaTiO$_3$	333 K	(60°C)	300	7	3.0	0.01

constante diélectrique du Sulfate de glycocolle dans l'infrarouge très lointain " (Hadni *et al.*, 1970); " $PbTiO_3$ Pyroelectric Infrared detector " (Yamaka
et al., 1971); "High performance pyroelectric detectors" (Baker *et al.*, 1972);
"Measurement of the pyroelectric coefficient and permittivity from the
pyroelectric response to step radiation signals in ferroelectrics" (Simhony
and Shaulov, 1972), etc.

Others factors of merit M_2 and M_3 are useful for pyroelectric vidicons (see
Section VII).

It is worthwhile to note that most of the first materials we started with are
still the best. This is probably due to their good crystalline quality. It is seen,
however, that for proper ferroelectrics the best detectivity is not obtained
at the temperature (Curie point) where π_s is maximum, since at this same
temperature the dielectric constant ε' is also maximum. For improper ferroelectrics (i.e., boracites, rare earth molybdates, or propionates) the anomaly
of the dielectric constant at the transition temperature may be negligible
(Smith *et al.*, 1979). Shaulov (1979) has shown that the factor of merit $M_3 =$
$\Pi_\sigma/c'\varepsilon'$ (see Section VII) can be maximum at the Curie point for some
boracites.

IV. Ferroelectric Domains

A. LINEAR PYROELECTRIC CRYSTALS

In the calculation of responsivity, noise, and detectivity, we have assumed
that spontaneous polarization has one direction in the whole material, i.e.,
there is no domain texture. It is the case for all linear pyroelectrics, such as
wurtzite-type crystals considered (ZnS, ZnO, CdS, CdSe, etc.), where the
hexagonal axis is oriented as can be checked by cutting a perpendicular plate
and looking at the etching figures. With brominated methyl alcohol in the
case of CdSe, for instance, a cadmium face and a selenium face can be distinguished. Of course, the observation can be complicated by the appearance
of twins, i.e., interpenetration of two single crystals of different orientation.
These defects can be avoided by appropriate growth methods. In the case of
linear pyroelectrics we have already cited boracites such as tourmaline,
lithium sulfate, saccharose, etc. All these crystals are devoid of phase transition and belong to polar crystallographic classes: 1, 2, 3, 4, 6, m, 2m, 3m, 4m,
6m, or to textures ∞, ∞m, over the entire temperature range in which they
are not decomposed.

B. FERROELECTRIC CRYSTALS

Ferroelectric crystals generally present a phase transition at some temperature, and the low-temperature phase is often the polar one (ferroelectric

phase). These crystals have a domain texture because the spontaneous polarization has no preferential direction, and macroscopic polarization for any large sample is generally null or much weaker than P_s. However, by application of an electric field, it is possible to get a single domain, to study its spontaneous polarization, pyroelectric coefficient, dielectric constants, etc., and to use it as a pyroelectric detector.

The specific problem met in using ferroelectrics as pyroelectric detectors is to control their domain texture.

C. Visualization of Domains

Visualization of domains is relatively recent. Up to World War II they had been seen only on Rochelle salt by using the powders method. When lead oxide (red one) and sulfur powders are mixed, they take, respectively, positive and negative charges. The blend is sprayed on the pyroelectric crystal and the grains are attracted selectively by domains of opposite charges.

Since this pioneering work, the etch method has been widely used. For instance, in the case of TGS, positive domains are more rapidly soluble in water or acetone. Many other methods have also been employed (Zheludev, 1971). One of the most sensitive in spatial and time resolution is probably the laser scan pyroelectric microscopy in real time.

Figure 16 is a schematic diagram of the laser beam pyroelectric microscope. This technique is now capable of inspecting a $100 \times 100 \ \mu m$ TGS surface in a few seconds with a 2-μm limit of resolution. The method is described by Hadni and Thomas (1976b), with calculation of the pyroelectric signal in different situations. With the 200 kHz chopping frequency generally used, only a thin surface layer is inspected, corresponding to the thermal diffusion length l at this frequency, $l \simeq 0.64 \ \mu m$. The sideways limit of resolution is larger and depends on the laser spot diameter ($2R_1 \simeq 2 \ \mu m$). It might be possible to reduce this value by a maximum factor of 4–5, but then the chopping frequency would have to be increased. Figures 17, 18, and 19 show some applications (see also Fig. 15).

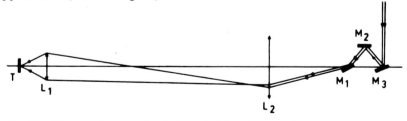

FIG. 16   Diagram of the sweeping device and of the microscope focusing the laser beam on a small point of the pyroelectric target; T, target; L_1, objective; L_2, eyepiece; M_1, fixed mirror; M_2, horizontal axis oscillating mirror; M_3, vertical axis oscillating mirror.

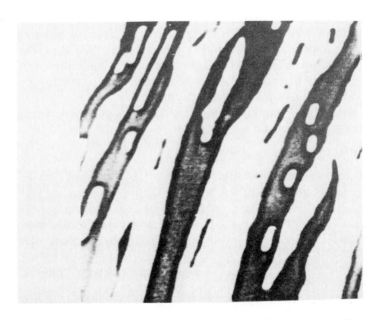

FIG. 17   Pyroelectric map of a TGS virgin crystal plate, 130-μm long and 10-μm thick; chopping frequency = 170 kHz. Small details at lower left (white) and right (black) are smaller than 2 μm and are seen repeatedly (Hadni *et al.*, 1976c).

D.   SELECTION OF FERROELECTRIC TARGETS

We have prepared a number of TGS pyroelectric targets with a 3-mm diameter. When they are observed with the laser beam pyroelectric probe, most of them show a number of tiny domains with polarization opposed to the dominant one. They can be switched into the right orientation, but they will more or less rapidly return to the initial polarization. Moreover, domain walls correspond to macroscopic heterogeneities, which are fragile (Fig. 20) and are also sources of dielectric losses. In some way, domains are similar to twins. It is better to select targets that appear as single domains, and repeat the test after some time. Such crystals generally keep their uniform polarization.

E.   MICROSCOPIC POLARIZATION SWITCHING BY LOCALIZED ILLUMINATION

It has been shown by Hadni *et al.* (1971) that microscopic polarization switchings are observed when even a low-power laser beam is focused on a TGS crystal plate cut perpendicular to the monoclinic axis.

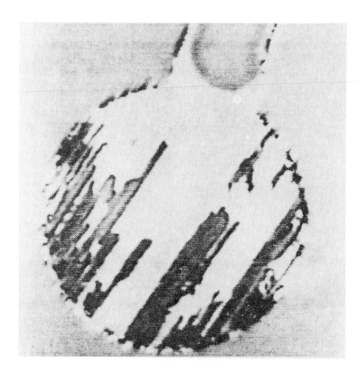

Fɪɢ. 18   Texture of a BaTiO₃ crystal plate with a small magnifying power (G = 10).

1. *Reversible Switching*

The localized polarization switching is reversible (i.e., P_s switches back to its original state when illumination has ceased) for illuminations lower than a rough limit of about 40 W cm^{-2}. It starts at much lower values, depending on which part of the crystal is illuminated (Fig. 21). The parts that are free of defects due to impurities or ionizing irradiation are easy to switch, as are the parts that have been switched recently, either by intense illumination or ac applied field. What is somewhat surprising is that such reversible switching can be repeated at high frequencies for hours without any hint of fatigue (Fig. 22).

Such switching has been explained by the gradient of temperature due to illumination. This leads to a gradient of spontaneous polarization, a density of bound charges, and a pyroelectric field that can switch the polarization. The switching starts from the nonilluminated face of the plate and propagates into the direction of the illuminated face. It is reversible as long as this face is not attained (Hadni *et al.*, 1971).

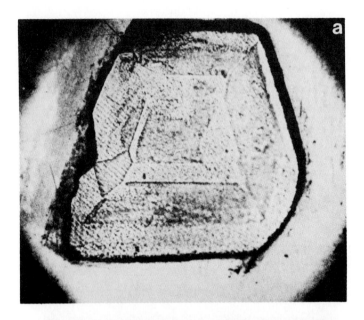

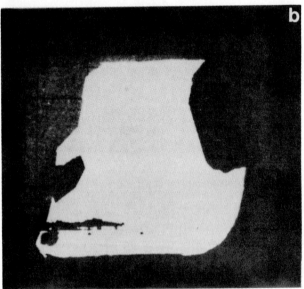

Fig. 19 Comparison of the rub etch technique and the pyroelectric probe technique: (a) Microphotograph of a crystal that has been rub etched to bring out the domain structure. The pebbled areas are reported to be negative domains, the gray smoother areas to be positive domains (the circle diameter is 20 mm, the crystal thickness 0.1 mm). (b) Pyroelectric map of the same crystal after a thin layer of gold has been deposited on each face. The polygonal domain on the left side matches exactly a gray area in (a). The same correspondence can be found on the right side.

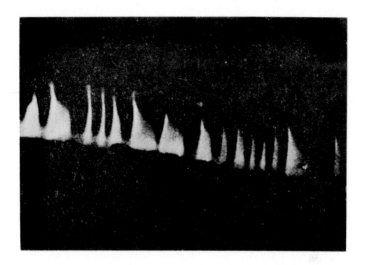

FIG. 20 Domains naturally seen on a side of the crystal parallel to the ferroelectric axis after breaking the plate. There is no need to use either the powder deposition method or the dip etch method to reveal the domain structure. The crystal thickness is 47 μm (Hadni and Thomas, 1974a).

This reversible switching is important for pyroelectric detection because it adds an extra contribution $2AP_s$ to the classical $A\pi_\sigma\, dT$ pyroelectric signal, which is much smaller. However this extra signal cannot improve the low-frequency NEP of pyroelectric detectors. Because of their high sensitivity at low frequency (close to the background-limited sensitivity), the involved illuminations are too low to switch a domain. For high frequencies (i.e., MHz) we have to use much higher illuminations and they lead to extra signals. It is certainly thanks to these extra signals that it has been possible to read ferroelectric memories at such high frequencies as 1 M bits/s.

2. Irreversible Switching

For higher fluxes Φ falling on a given area, or longer illumination time t, switching becomes nonreversible. It is thus possible to write domains on a pyroelectric detector. Figure 23 gives such an example. It has been shown by Hadni and Thomas (1974b) that the minimum time t^+ of exposure for irreversible switching is proportional to $e^{A/\phi}$, A being a constant.

F. BLOCKING SPONTANEOUS POLARIZATION

We have seen that spontaneous polarization is easy to reverse locally by illumination or by heat transport leading to some gradient of temperature.

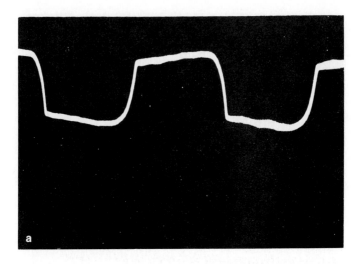

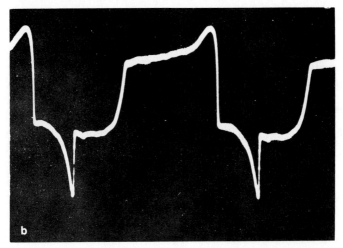

FIG. 21   (a) Response of a TGS detector to a He–Ne laser beam chopped at 16.6 cps, $T = 42°C$, and focused on a spot where polarization is clamped. (b) Response of a TGS detector to a He–Ne laser beam chopped at 16.6 cps, $T = 42°C$, and focused on a spot close to a nucleation site: The laser beam switches periodically to a small domain, giving an extra signal (sharp peak) (Hadni and Thomas, 1972a).

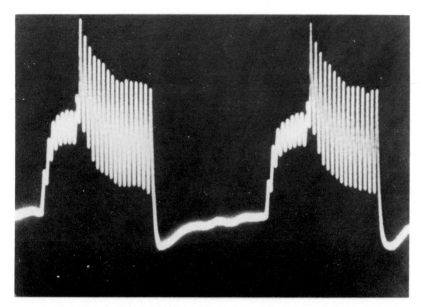

FIG. 22 Double modulation of the laser beam: $f_1 = 16.6$ cps; $f_2 = 800$ cps; $T = 42°C$; $\rho = 470\ k\Omega$: The laser beam switches the domain without any sign of fatigue.

FIG. 23 The word NANCY is written on a 100-μm-long rectangle.

Such domain switching can be permanent, transforming a monodomain crystal into a mosaic of tiny domains, with the net detectivity of the corresponding pyroelectric detector decreased to zero. To keep a reliable detectivity, we have to check that there is no domain texture and also find a way of blocking this single domain state. On the other hand, this is beneficial for the detectivity, since it reduces the loss angle and thus improves the factor of merit $M_1(T)$. Fortunately, there are several ways to block spontaneous polarization.

1. Electric DC Field

As we have seen polarization switching is often made by an internal electric field (i.e., pyroelectric field, or electric field due either to free charges localized at defects or piezoelectric bound charges produced by local stresses). These electric fields can be cancelled by applying a permanent dc field to the whole crystal.

The method is an efficient one. However, it implies an additional dc voltage source and also it leads to a decrease of detectivity. The pyroelectric current is written:

$$i = A\pi_\sigma \frac{dT}{dt} + \frac{V\varepsilon_o A}{d} \frac{\partial \varepsilon'}{\partial T} \frac{dT}{dt}.$$

Now generally $\pi_\sigma < 0$ and $\partial \varepsilon'/\partial T > 0$ when $T < T_c$. Voltage V brings a contribution of opposite sign to the pyroelectric contribution. From various experiments the result is a 10% decrease of responsivity.

2. Irradiation by X or γ Rays

Hadni et al. (1971) and others have shown that ionizing radiation prevents polarization switching because of induced electric charges that give rise to an electric field parallel to polarization. There is also a decrease of the Curie temperature (6°C for 2 MR irradiation).

3. Impurities

The introduction of an impurity in an asymmetric site (Lines and Glass, 1977) produces a dielectric dipole $\Delta\vec{\mu}$ which is the sum of its proper value and of the dipoles it induces around it. Let us assume impurities diluted enough (N per unit volume) to let us neglect their interaction:

$$\Delta\vec{P} = N \Delta\vec{\mu} \qquad \text{and} \qquad E_i = \Delta P/\varepsilon'\varepsilon_o.$$

This internal electrical field is uniform if N is uniform and $\Delta\vec{\mu}$ is a constant (generally $\Delta\vec{\mu}$ is parallel to $\vec{P_s}$ when the crystal growth is operated in the ferroelectric phase).

In the case of l- or d-alanine doped TGS, $\Delta\vec{\mu}$ keeps its direction when $T > T_c$. For lower temperatures, the crystal is single domain with internal fields up to 100 kV/cm. The Curie temperature may be increased up to 60°C. With such crystals Keve (1975) has obtained one of the highest detectivities $[D^*(6.5 \text{ Hz}) \simeq 6.10^9 \text{ W}^{-1} \text{ cm Hz}^{1/2}]$ as can be seen in Fig. 15 (see also Lock, 1971).

V.   Thin, Oriented Pyroelectric Films

A.   INTRODUCTION

1.   Single Crystals

Since D^* is proportional to $d^{-1/2}$, pyroelectric detectors have to be made as thin as possible. If we start from bulk single crystals, careful polishing can lead to plates as thin as 10 μm. Ion milling and chemical etching have also been used with success in a few cases, but they cannot lead to much thinner plates. Ion milling is difficult in some cases (i.e., DTGFB).

2.   Evaporated Films, Polymers

ZnO films 0.1–10-μm thick have been prepared by sputtering. Thin films of polyvinylidene fluoride of the same thickness (Glass et al., 1971; Phelan et al., 1971) can also be prepared, but they have to be stressed and polarized after some heating. Spontaneous polarization can be as high as 3 μC cm^2 and detectivity D^* reaches 10^8 W^{-1} Hz$^{1/2}$ cm. It is nearly constant between 80 and 300 K.

3.   Powders

Polycrystalline TGS thin films have been made by pressing the powder (Jansson, 1974). At IBM, a spinning disk, called *tournette* in French, has been used to flatten a drop of TGS solution in water. Rapid evaporation leads to a mosaic of flat and thin ($d \simeq 1$ μm) single crystals of any orientation.

In both these cases the film is not oriented and is not well suited for pyroelectric applications, although the method of the *tournette* might be somewhat better because it gives flat single crystals with sizable dimensions.

B.   POSSIBILITY OF GROWING ORIENTED, THIN FILMS

The pyroelectric application is a typical case where there is a need for thin, oriented crystals. Sizable lateral dimensions are also required to decrease the number of defects, brought by grain boundaries, that can increase the dielectric losses. Moreover, for type I detectors the pyroelectric film must be coated on each side by a conductive layer.

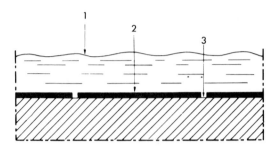

FIG. 24   The metal layer (2) shows pin holes (3) where the TGS solution (1) is in close contact with the single crystal.

We have described a method that gives an oriented, thin film of TGS on any thin support. The thin, oriented film is made of a mosaic of flat oriented, single crystals with lateral dimensions up to 400 μm. This is the ENSH method, i.e., epitaxial nucleation in submicroscopic holes of an intermediate thin film of either a metal, any salt, or an organic plastic compound closely applied on a bulk, single crystal.

For instance on a polished plane face of a TGS single crystal, we first evaporate a thin layer of silver with $R_\square \simeq 1\ \Omega$. The TGS plate is fixed on a rotating disk (*tournette*) with the silver layer on the upper side. It is heated to 75°C and a drop of TGS solution in water at 80°C is deposited (Fig. 24).

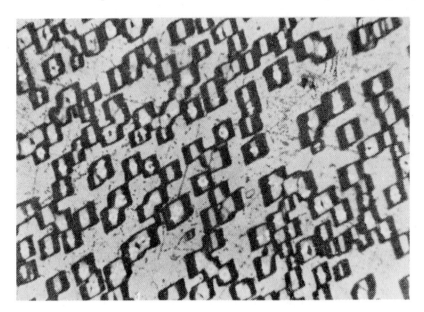

FIG. 25   Oriented TGS single crystals on a polystyrene film.

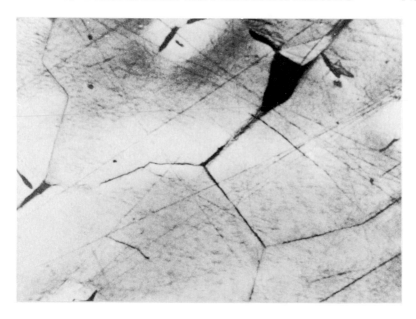

FIG. 26   Mosaic, on flat, oriented, TGSe single crystals closely joined together on a silver layer $R_c = 0.05\ \Omega$ (Hadni *et al.*, 1977a).

There is a rapid evaporation leading to a number of oriented, single crystals, with dimensions ranging from 10 to 100 μm (Fig. 25).

For a square resistance around 0.05 Ω, the oriented crystal's growth is especially flat (Fig. 26) and leads to a mosaic closely covering the whole metal layer. For many applications this mosaic is equivalent to an oriented, single crystal film. The thickness may be lower than 1 μm; no method up to now could give such thin crystals.

To make a pyroelectric detector, we have only to put a second metal layer on top of the thin TGS crystal (Fig. 27). As is seen in Fig. 28, in such a detector responsivity is larger than for a 20-μm-thick bulk detector as soon as frequency $f > 1000$ Hz. Noise is roughly equivalent. The smaller responsivity for low frequencies is due to the thermal contact between detector and holder. By separating the thin detector from the single crystal holder, the photon-noise-limited detectivity could be achieved at 10 Hz frequency with a 1-μm-thick TGSe pyroelectric detector (Hadni, 1969). At higher frequencies the detectivity is farther and farther away, but still much higher than with a bulk TGS detector. For such frequencies it is not necessary to separate the detector from the holder. This is a promising application that should be considered, although there is still some technology to be developed for avoiding grain boundaries, thickness differences, etc.

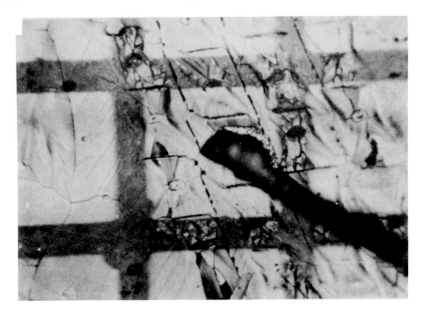

FIG. 27   TGS pyroelectric detector 0.8-μm thick.

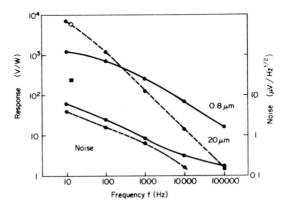

FIG. 28   The noise of two detectors (thickness 0.8 μm and 20 μm) has the same order of magnitude over the whole frequency range. However, for $f = 10^4$ Hz the thin receptor has a detectivity that surpasses that of the thick receptor ($D \cong 10^8$) by a ratio of the order of 2.3 for the same frequency. One notes that the ratio of the thickness is of the order 5^2, and a detectivity about five times greater is expected for high frequencies (Claudel and Hadni, 1976).

C. CONCLUSION

It is possible to make thin, oriented pyroelectric films. We hope to achieve in the near future thin, single crystals of appreciable dimensions (i.e., $150 \times 200 \, \mu m$) with a small thickness ($d < 1 \, \mu m$), which may lead to improved pyroelectric detectors, either type I or II, since $D^* \propto d^{-1/2}$; also, responsivity is expected to increase, especially for type II detectors.

However, we must be conscious that up to now we have nearly no data on such thin crystals. We have seen that spontaneous polarization P_s is not modified for $d > 1 \, \mu m$, but coercive field E_c increases dramatically. We have to look at the dielectric constant and the Curie temperature for a d as thin as possible. From the literature (Burfoot, 1978) we can expect the following:

(i) pyroelectric coefficients that depend more on the geometry of the crystal than on the lattice structure.

(ii) large shifts of Curie temperature, i.e., pyroelectricity in KNO_3 films has been observed at room temperature while $T_c = 110°C$ (Nolta et al., 1962).

(iii) lattice cell parameters that can be drastically modified at the surface layers. For $BaTiO_3$, 3% variations have been observed for thin powders (grain dimensions smaller than $0.1 \, \mu m$).

(iv) a depolarizing field that is not completely cancelled by surface charges, as Wurfel and Batra (1973) have shown in the case of semiconductor electrodes.

A fundamental study of pyroelectric, thin, single crystals seems to be needed.

The first results are concerning the infrared trichroism of some ferroelectric crystals recently studied on oriented films as thin as $0.8 \, \mu m$ [Gerbaux et al. (1977); (1978a and b)].

VI. Miscellaneous Problems

A. MICROPHONICS

Pyroelectric crystals are always piezoelectric (the reciprocal is not true). The result is that any mechanical vibration leads to an unwanted electric signal (Liu et al., 1978). It is shown that the piezoelectric voltage V is proportional to the acceleration value a of the crystal:

$$V = \tfrac{1}{2}(\delta/\varepsilon_o \varepsilon')\rho a d^2, \tag{34}$$

δ and ρ being the piezoelectric coefficient and the density. The result is that V decreases as d^2, hence another benefit to use thin crystals plates. It is also useful to put the crystal in vacuum and use a window. An electric filter can reduce the effect of mechanical resonances.

B. AMPLIFIER NOISE

Low-noise amplifiers are required. The problem is difficult because of the high values ($10^{12}\,\Omega$) of the impedance of the best pyroelectric crystals. Amplifier noise, thus far, is the limitation for pyroelectric detectors working at high frequencies or at low temperatures because, in both cases, the pyroelectric crystal Johnson noise can be negligible.

C. CLAMPING

Up to now we have considered an unclamped crystal, i.e., a sample free to expand, and have used a constant stress pyroelectric coefficient π_σ.

The second possibility is the case of a crystal that is completely clamped (fixed volume and size), and we can define a constant strain pyroelectric coefficient, which is also called primary coefficient π_I. It is shown that $\pi_\sigma = \pi_I + \pi_{II}$. π_{II} is the secondary pyroelectric coefficient due to piezoelectric contributions arising from the deformation of the crystal (see Section VIII).

The third case is that of a thin pyroelectric slab fixed on a bulk holder at constant temperature. Since the slab is free to expand only in the perpendicular direction, the corresponding pyroelectric coefficient is written π_{pc} (partially clamped).

Table II [from Zook and Liu (1978)] gives π_I, π_σ, and π_{pc} for three ferroelectrics and five linear pyroelectrics. It is shown that $\pi_I \simeq \pi_\sigma \simeq \pi_{pc}$ for ferroelectrics, but for linear pyroelectrics, π_σ (unclamped) is slightly higher than π_I (clamped) and π_{pc} is negligible.

We can conclude that the way the pyroelectric plate is held is not very stringent for ferroelectrics (at least for the ones considered in Table II, to be proper ferroelectrics), but for linear pyroelectrics it is fundamental to avoid any clamping and especially partial clamping. To avoid clamping, the sample may be held between spring-loaded, gold-plated electrodes (Lang, 1971), or

TABLE II[a]

	π_I	π_σ	$\pi_{pc}\ (10^{-6}\ cm^{-2}\ K)$
$LiTaO_3$	-180	-176	-161
$Pb_5Ge_3O_{11}$	-110	-95	-87
$Sr_{0.5}Ba_{0.5}Nb_2O_6$	-500	-550	-430
CdS	-3	-4	-0.13
CdSe	-2.9	-3.5	-0.67
ZnO	-6.9	-9.4	-0.35
$Li_2SO_4 \cdot H_2O$	60	86.3	
Tourmaline	0.48	4	

[a] From Zook and Liu (1978).

glued on one point of the heat sink, or suspended by two wires. In the last case, exchanges with the heat sink can be increased by introducing some He gas in the sample room.

D.   ABSORPTION

It has been shown above that detectivity is proportional to the absorption coefficient α; thus the problem of absorbing the infrared radiations is of prime importance. The case of type I and type II pyroelectric detectors have to be considered separately.

1.   Type I Detectors

For type I detectors the infrared radiations strike the outer conductive electrode of the pyroelectric capacitor. The conductive electrode must have a specific thickness to absorb around 75 % of the incident flux. It is shown that the optimized square resistance $R_c = (\mu_o/\varepsilon_o)^{1/2}$. About 50 % of the flux enters the pyroelectric material, and the absorption coefficient is increased selectively at some wavelengths. The absorption is more uniform versus frequency by using a very thin, gold-black layer. Gold is evaporated in a bell jar containing about 30 Torr of argon. The gold vapor condenses on the cold gas, giving a number of tiny crystals with dimensions much smaller than 1 μm. The theory is not simple but confirms experiments showing a high absorption in the near- and mid-infrared.

The problem of absorption becomes more difficult for long wavelengths when $\lambda \gg d$. It has been shown (Fig. 29) that the absorption coefficient of a thin plate stuck on a bulk conductive holder becomes lower than 10 % for

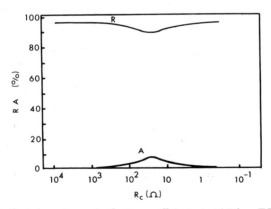

FIG. 29   Calculated absorption and reflection coefficients A and R for a TGS film: $d = 2\ \mu$m, semimetallized with gold on the front surface (square resistance R_c) and with the back surface stuck on a copper block versus R_c, for $\lambda = 337\ \mu$ (Claudel and Hadni, 1976).

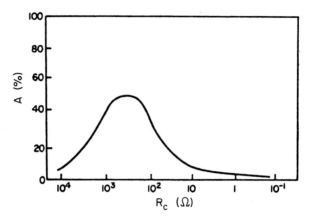

FIG. 30   Absorption coefficient A for a TGS film of thickness 2 μm, semimetallized with gold on two faces as a function of the square resistance R_c of the gold film ($\lambda = 337 \mu$m).

$\lambda/d > 100$. Such a highly conductive electrode has to be avoided and there is a need to put very thin electrodes on both side of the pyroelectric plate. It is seen in Fig. 30 that for $R_c \simeq 1000 \, \Omega$, the absorption coefficient can reach a 50% value.

2. Type II Detectors

a. *Intrinsic Lattice Absorption.*   In the case of type II detectors (Fig. 1) the infrared radiations are directly striking the crystal, and can excite a lattice vibration if their frequency is adjusted to match an infrared active vibration, and if the electric field is parallel to the crystal electric dipole variation. We can expect a quasi-instantaneous response when two conditions are fulfilled:

(1)   The excited dipole has a component on the pyroelectric axis.
(2)   There is enough anharmonicity to give a nonzero time-average value.

During thermalization, this instantaneous response should be progressively relayed by a much smaller thermal pyroelectric response.

b. *Extrinsic Electronic Absorption.*   In some cases the pyroelectric crystal can be doped with transition metals (i.e., Cr, Cu, etc.) or rare earth elements. When these impurities are excited, the electric dipole moment is modified by an amount $\Delta\mu$. In a pyroelectric crystal all these vectors have the same orientation; hence a change of polarization $\Delta P = n \, \Delta\mu$, n being the concentration of impurities in the excited state. This has been shown by Auston and Glass (1972) in the case of transition metals (for rare earths the 4 d shell shields the 3 f electrons from the crystalline field, and the above

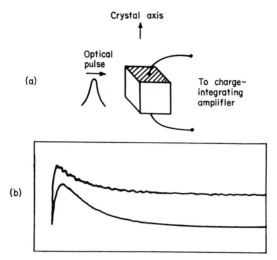

FIG. 31 (a) Sketch of the experimental geometry used for measurement of excited-state polarization, and (b) the experimental results (oscilloscope traces) showing optical emission (lower trace) and polarization change (upper trace) of $LiNbO_3$: Cr at 173 K. The vertical scale in (b) is arbitrary, but the horizontal scale is determined as 5 μs per division. The noise on the upper trace is piezoelectric ringing due to thickness resonances of the crystal (from Lines and Glass, 1977).

considerations are not applicable). Figure 31 shows the scheme of the experiment. The type II pyroelectric detector is made of lithium niobate doped with chromium. The excitation is made with visible light to excite an electronic transition in the chromium ions. The optical pulse is very short (10^{-9} s). There is excitation of electrons in the 4 d shell from state $^4A_2$ to $^4T_2$, followed by relaxation with emission of light ($\tau_{relax} \simeq 15\ \mu$s at 173 K). The lower curve in Fig. 31 gives the emission of light after the laser pulse. It lasts for about 15 μs. The upper curve shows polarization change, which increases rapidly after the laser pulse as a result of the change of electronic dipole moment and then decreases towards a constant value corresponding to the new thermal equilibrium after relaxation of the electronic excitation via nonradiative transition. This value stays on a much longer time (thermal relaxation time $\simeq$ ms).

c. *Black Coatings.*[*] In the case where type II pyroelectric crystal thickness is too small to give enough intrinsic or extrinsic absorption, a black coating can be used, either selectively (i.e., alkali halides) or nonselectively (i.e., gold black). In this case there is no direct excitation of dipoles in the

[*] See Annis and Simpson (1974) and Blevin and Geist (1974a).

crystal, and the response is purely thermal and is delayed until thermal equilibrium is reached between black coating and crystal. The delay is of the μs order for gold black.

In any case, with type II detectors there is no boundary problem at the limit of the electrodes if the electric field of the radiation is oriented perpendicular to the electrodes.

E.   RESPONSIVITY TO VERY SHORT PULSES

1.   *Piezoelectric Ringing*

When pyroelectric detectors are illuminated with short pulses of energy, the rapid heating of the crystal generates an acoustic wave (Glass and Abrams, 1970a) that oscillates at a frequency determined by the crystal dimensions (Fig. 32). This acoustic wave is seen as a spurious peizoelectric signal. It can be suppressed by cementing the crystal to a suitable, heavier holder (Roundy and Byer, 1972).

It has been shown by Simonhy and Bass (1979) that the onset of piezoelectric oscillations in PZLT detectors occurs 40 to 60 ns after the start of the pyroelectric response and always with the same sign.

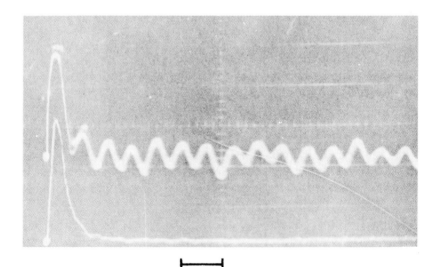

1 μs / div

FIG. 32   Response of a LiTaO$_3$ detector, showing piezoelectric ringing when mounted on brass (upper trace) and on silver (lower trace). Horizontal scale, 1 μs per division (Roundy and Byer, 1972).

2. Pyroelectric Voltage Response to Short Signals

When the time constants τ and τ' differ appreciably, the rise of the response and its decay are exponentials whose time constants are, respectively, the smaller and the larger ones from among τ and τ'. The peak voltage values are proportional to the load resistance ρ when $\tau' \ll \tau$; they become independent of ρ and proportional to τ when $\tau' \gg \tau$. The initial step of the response is found to be independent of ρ for a given illumination. A figure of merit (Simhony and Shaulov, 1971) is

$$M_3 = \pi_\sigma/c'\rho\varepsilon'.$$

Pulses as short as 1 ns from a mode-locked CO_2 TEA laser at 10.6 μ have been detected (Roundy et al., 1972) and show a 500 ps rise time. [See also Lawton (1972), Shaulov and Simhony (1972a), Shaulov and Simhony (1972b), Lavi and Simhony (1973), Roundy et al. (1974).]

It seems (Roundy et al., 1972) that the time of heat transfer from the black coating to the pyroelectric detector differs widely according to its nature (gold black: 1 μs; thin metal layer: 1 ns).

F. HETERODYNE DETECTION

Let us assume a coherent signal at frequency ω_S

$$\hat{E}_S = \hat{E}_{OS} \cos \omega_S t,$$

and a local oscillator at frequency ω_L close to ω

$$\hat{E}_L = \hat{E}_{OL} \cos(\omega_L t + \phi).$$

If the electric field directions are parallel and if they are mixed into a pyroelectric detector, the pyroelectric current is proportional to the total flux, i.e., to the square of the net electric field at the surface of the detector:

$$
\begin{aligned}
i &= K(E_S + E_L)^2 \\
&= KE_{OS}^2 \cos^2 \omega_S t + KE_{OL}^2 \cos^2(\omega_L t + \phi) \\
&\quad + KE_{OS}E_{OL} \cos[(\omega_S + \omega_L)t + \phi] + KE_{OS}E_{OL} \cos[(\omega_S - \omega_L)t - \phi].
\end{aligned}
$$

The last term has a frequency $\omega_S - \omega_L$ that can be lower than 100 MHz, i.e., located in the possible bandwidth of a pyroelectric detector. It is clear that the amplitude S of this low-frequency term is proportional to the product of local and signal electric field:

$$S_H = KE_{OS}E_{OL}.$$

In direct detection the amplitude of the current is therefore written:

$$S_D = KE_{OS}^2.$$

Hence

$$S_H/S_D = (\varnothing_L/\varnothing_S)^{1/2}.$$

Now assuming the same noise (detector noise in both cases), we could also write

$$D_H^*/D_D^* = (\varnothing_L/\varnothing_S)^{1/2}.$$

For instance, with $D_D = 10^9 \text{ W}^{-1} \text{ cm Hz}^{1/2}$, $\varnothing_S = 10^{-10} \text{ W}$, and $\varnothing_L = 10^{-2} \text{ W}$, $D_H^*/D_D^* = 10^4$, and thus $D_H^* \simeq 10^{13} \text{ W}^{-1} \text{ cm Hz}^{1/2}$. In fact, at least in the far infrared, the laser power fluctuations are up to 10% and it is difficult to reach this high value. We have been able to detect milliwatt powers with a local HCN oscillator ($\lambda_L = 337 \mu m$) and difference frequencies ($\omega_S - \omega_L$) up to 20 MHz. We must also cite similar work by Gebbie *et al.* (1967) at $337 \mu m$, by Leiba (1969) at $10 \mu m$, and by Abrams and Glass (1969).

With most stabilized lasers the limitation would be photon noise from the local oscillator. It seems that the primary advantage is to get the frequency ω_S of the signal with great accuracy since ω_L and $\omega_S - \omega_L$ can be known precisely. Pyroelectric detectors might be as good candidates around $50 \mu m$ for such high-accuracy emission spectroscopy as quantum detectors (i.e., HgCdTe) are at $10 \mu m$, where they have been used with much success by Townes for infrared astronomy.

VII.   Infrared Imaging with Pyroelectrics

A.   ELEMENTAL DETECTOR AND MECHANICAL SWEEPING

1.   *Introduction*

The first cameras for the mid-IR $(\lambda > 3 \mu m)$ used an elemental detector on which all points of the object were scanned by two mirrors oscillating around two perpendicular axes. The first mid-IR images were obtained by Golay after World War II with a pneumatic detector and by Barnes Engineering Co. with a thermistor. Several minutes were needed for 100×100 picture elements.

More recently, quantum detectors have been used. They have to be cooled down to same temperature T_o and their sensitivity is limited to a maximum wavelength λ_o. Ge:Hg ($T_o = 20 \text{ K}$; $\lambda_o = 12 \mu m$); HgCdTe ($T_o = 80 \text{ K}$; $\lambda_o = 12 \mu m$); InSb ($T_o = 80 \text{ K}$; $\lambda_o = 7 \mu m$). To get a picture with same resolution as above, a fraction of second is enough.

2.   *A Rough Calculation of the Maximum Sweeping Frequency*

Let us assume a flux $\varnothing = (\sigma T^4/\pi)A_o\Omega$ arriving on an elemental detector, A_o being the object area, T its average temperature, and Ω the solid angle

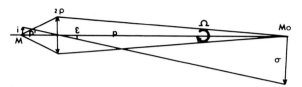

FIG. 33   Optical scheme of an infrared camera.

from which the camera objective is viewed at the object location. The minimum temperature variation dT that an elemental detector of detectivity D can see with a signal to noise ratio S/B is written

$$(4\sigma T^3/\pi)A_o\Omega\, dT = (S/B)/D. \tag{35}$$

Figure 33 shows a zoom objective, diameter 2ρ, conjugating any point M_o of the axis, at a distance p, with point M at the center of the detector, which is at a constant distance p' from the objective. However the farthest is M_o, the largest is the area A_o corresponding to the constant detector area $A = 4i^2$. Now $A_o\Omega = 4i^2(\pi\rho^2/p'^2)$ and $D(f, \Delta f) = D^*/A^{1/2}\,\Delta f^{1/2}$, f being the chopping frequency and Δf the bandwidth of the amplifier, and Eq. (35) can be written as

$$(8\sigma T^3 i\rho^2/p'^2)\, dT = (S/B)\,\Delta f^{1/2}/D^*(f). \tag{36}$$

Then T, dT, i, ρ, p', and S/B being given, we look for the maximum value we could give to f. It depends on the time τ we want to spend on each element of the object. As a rough approximation we can assume the following:

(i) $\tau \geq 1/f$, i.e., the time of observation contains at least one period of modulation. Then $f \geq 1/\tau$ and we shall take the smaller value of f in order to keep the highest detectivity $D(f, \Delta f): f = 1/\tau$.

(ii) $\tau_o = 1/\Delta f$ being the amplifier time constant, $\tau \geq 1/\Delta f$, i.e., the time of observation is at least equal to the time constant in order to get more than $1 - 1/e \simeq 63\%$ of the responsivity. Then $\Delta f \geq 1/\tau$ and we shall take the smallest value in order to again keep the highest detectivity: $\Delta f = 1/\tau$.

With these assumptions $f = \Delta f = 1/\tau$.

Now in the case of a pyroelectric detector

$$D^*(f) = D^*(f_o)f_o^{1/2}/f^{1/2}.$$

Hence

$$\frac{8\sigma T^3 i\rho^2}{p'^2}\, dT = \frac{(S/B)f}{D^*(f_o)f_o^{1/2}},$$

or

$$f = \frac{D^*(f_o)f_o^{1/2}8\sigma T^3 i \rho^2}{p'^2(S/B)} dT. \tag{37}$$

Applications: $S/B = 1$, $T = 300$ K, $dT = 1$ K, $i = 0.1$ cm, $\rho = 5$ cm, $p' = 10$ cm, $\sigma = 5.7 \times 10^{-12}$ W cm^{-2} K^{-4}. With D^* (10 Hz) $= 10^9$ W^{-1} cm Hz$^{1/2}$, we get $f = 10^5$ Hz.

Hence $\tau \simeq 10^{-5}$ s and within the relaxation time of the human eye retina (0.1 s), 10^4 spatial elements can be inspected.

Now let us take $S/B = 10$ to increase the image quality; we get $f = 10^4$ Hz and $\tau = 10^{-4}$ s, 1 s is needed now to inspect 10^4 elements. It is also possible to keep $f = 10^5$ Hz and $\tau = 10^{-5}$ s, and add 10 images with $S/B = 1$ in a 1 s observation.

Let us remark that with most thermal detectors which are not pyroelectric,

$$D^*(f) < D^*(f_o) \times \frac{f_o}{f},$$

and for the same detectivity $D^*(10$ Hz) $= 10^9$ W^{-1} cm Hz$^{1/2}$, we should get $f < 2.10^3$ Hz. This explains the time needed with thermal detector cameras to get an infrared image. The advantage of pyroelectric detectors is their better detectivity at high frequencies, with special emphasis for thin epitaxial films.

B. LINEAR ARRAY*

Linear arrays bring two obvious advantages:

(i)   Mechanical sweeping is reduced to one direction only, which leads to a simpler device. For qualitative observation, mechanical sweeping can even be canceled out; panning the camera is enough.

(ii)   Let us assume a linear array of N detectors and an image made of N^2 elements. With the same D^* for each array detector as for the elemental detector considered above, the time needed to get an image is $N\tau$ instead of $N^2\tau$. The gain of time is N—the number of detectors in the array.

There are commercially available (from Plessey) linear arrays of 64 pyroelectric detectors 0.5 mm apart; 64 detectors, however, is not enough for such application and they are much too separated.

Each pyroelectric detector may also be associated with a field effect transistor (FET), with each signal applied to an AsGa light emitting diode. The use of charge coupled devices (CCD) is also promising. Not every device

---

* See Blackburn *et al.* (1972), McIntosh and Sypek (1972), and Watton (1976).

that gives IR images with an array of pyroelectric detectors has been described thus far to our knowledge. However, at the International Symposium on Applications of Ferroelectrics, held June 13–15, 1979, in Minneapolis, Minnesota, U.S.A., N.E. Byer and A. Van der Jagt of the Martin Marietta Laboratories, Baltimore, Maryland described monolithic pyroelectric arrays. It is clear that with pyroelectric vidicons the gain of time is N^2 instead of N for a linear array. However, vidicons are not well adapted to use very thin epitaxial films: detectivity is lower for one element, as is the spatial resolution (see Section C).

C. PYROELECTRIC VIDICONS

1. *Introduction*

That an infrared camera equipped with a linear array of pyroelectric detectors has not been developed is probably due to the success of the pyroelectric vidicon, which is the best way today to get thermal images in the mid and far Infrared. The principle is simple (Hadni, 1963). As is seen in Fig. 34, the pyroelectric plate that is cut perpendicular to the polar axis is covered with a thin conductive layer on the side receiving the IR image. Such a layer, with $R_c \simeq (\mu_o/\varepsilon_o)^{1/2}$, is capable of absorbing the radiations, giving a thermal image transduced into a bound charge image on each side of the plate. The inner side is inspected by a beam of low-velocity electrons. If there is any local change of temperature between two inspections, the corresponding element will exhibit a change of potential. For instance, a decrease will lead to more electrons repulsed towards the electron photomultiplier. This gives the video signal, which can be used to modulate the spot of a cathode ray oscillograph (CRO) synchronized with the pyroelectric vidicon spot.

The greatest difficulty arises as a result of the high impedance of the pyroelectrics used up to now (i.e., TGS). The negative charges deposited by the electron beam are not evacuated between two inspections and will repulse definitively all inspection electrons. This problem has been considered by several investigators: Holeman and Wreathall (1971), Putley (1971),

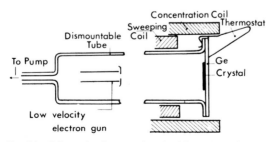

FIG. 34   Schematic of a pyroelectric vidicon (Hadni, 1963).

Steinhage et al. (1971), Tompsett (1971), Wreathall and Savage (1971), Logan and McLean (1972), Thiboumery (1972), Singer et al. (1974), Harmer (1976), Singer and Lalek (1976), Garn and Petito (1977), Hoeksma (1978).

2. Anode Potential Stabilization

The anode potential can be stabilized by three means:

(i) Pyroelectric semiconductive materials. As was said above, the pyroelectrics used up to now (i.e., TGS, PVF_2, etc.) have a high impedance. Conductive pyroelectrics seem to involve a high value of ε'', which should reduce detectivity.

(ii) Residual gases. In the first experiments to make a pyroelectric vidicon at the French Compagnie générale de télégraphie Sans Fil (CSF) (Le Carvennec, 1971; Charles and Le Carvennec, 1972) the vidicon tube was not baked and, therefore, was only partially evacuated. This was fortunate since the residual gas is ionized by the electron's impact. Positive ions are attracted by the negatively charged retina which is thus kept at a small positive value (pedestal).

(iii) Secondary emission. Between two lines of inspection with slow velocity electrons, there is a rapid fly-back with high velocity electrons. It is well known that with a high enough speed, the secondary emission coefficient is > 1, and thus the pyroelectric target surface loses electrons after every line inspection. This solution was first proposed by Le Carvennec and is now used by many others (Conklin and Stupp, 1974; Nelson, 1976; etc.).

There is another way to use secondary emission, and that is to keep the target at some positive potential (the one of the grid, 200 V for instance). The electrons strike the target with a high velocity, giving a secondary emission that depends on the small variations of the local potential due to illumination. There is some risk of damage to the target.

Singer has shown (1979) that the pedestal current (20 nA/cm^2) is the main source of noise. By a digital treatment of the signal, this noise has been substantially reduced. A pyroelectric vidicon system has been presented which has elemental $D^* = 10^8 \text{ W}^{-1} \text{ cm Hz}^{1/2}$. This is said to represent a factor of 5 improvement in signal to noise as compared to standard systems.

3. Performance

a. Reticulation, Factor of Merit. The best results have been obtained with circular TGS plates, 100-μm thick and 20 mm in diameter. Thermal conductivity is the limiting factor for the spatial resolution (Logan and McLean, 1972; Harmer and Wreathall, 1976b). With a 15 Hz chopping frequency and an optics opened to $f/1$, the spatial resolution is limited to three

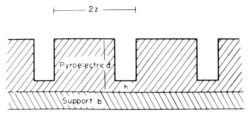

FIG. 35 Schematic section of a reticulated target (Watton, 1978).

pairs of lines per millimeter for a 1 K temperature variation sensitivity on the object.

Reticulation (Fig. 35) increases the spatial resolution up to five pairs of lines per millimeter (Fig. 36). It is made by lithography, laser, or ion milling (Yamaka *et al.*, 1976).

Different factors of merit have been proposed for the pyroelectrics used in pyroelectric vidicons. For the current responsivity it is $\pi_\sigma/c' = M_2$. For the potential, it depends either on $\pi_\sigma/c'\varepsilon' = M_3$, or $\pi_\sigma/c'\varepsilon''^{1/2}T^{1/2} = M_1$. Table I shows that TGS is among the best materials. Deuterated triglycine fluoberylate might be still better. B. Singer, *et al.* (1979) described the "Performance characteristics of pyroelectric vidicons with a reticulated DTGFB target" at the International Symposium on Applications of Ferroelectrics, June 13–15, 1979, in Minneapolis, Minnesota, U.S.A. The main advantage is a higher Curie temperature.

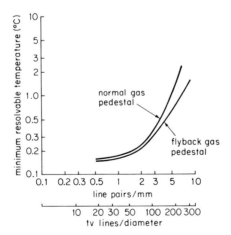

FIG. 36 Minimum resolvable temperature for panning mode. Pan rate = 4.5 mm/s^{-1} (Nelson, 1976).

4.   *Other Possibilities*

(i)   Chopping the infrared radiation is best for quantitative observation because the video signal can be treated in a computer. For instance, target local sensivity differences can be corrected, as well as remanence and blurring due to high temperature variations. Fluctuations due to chopping are also completely eliminated (Helmick, 1976; Nelson, 1976). TV color display is easy where different colors correspond to different temperatures.

(ii)   However, for qualitative observation chopping can be canceled and replaced by panning or orbiting the camera.

(iii)   Reading the pyroelectric plate inner surface might be possible without electron beam scanning (Alting-Mees and Koda, 1976). The inner surface could be covered by a photoemissive layer, the emission coefficient of which is often very sensitive to electric field. The target is kept in vacuum and is illuminated by visible light. Photoelectrons are emitted and focused on a phosphor target where a visible image, conjugated of the thermal image produced on the pyroelectric plate, is made.

(iv)   pyroelectric two-dimensional arrays have also been considered. The limiting noise sources are caused by thermal fluctuations due to the contact between the pyroelectric and silicon array, and $1/f$ noise in CCD multiplexed arrays (Singer, 1979).

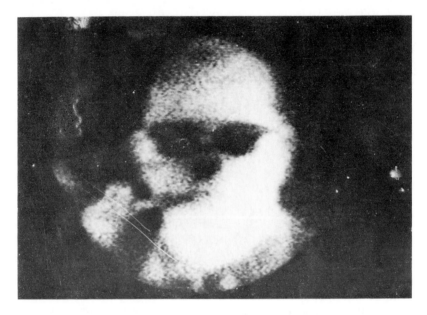

FIG. 37   Image of a face taken with a first version of the French Pyricon (Mangin *et al.*, 1974).

D. APPLICATIONS

The advent of the infrared Pyricon (Pyricon is a trademark from Thomson CSF, France) has led to a new vision of the world. It is now possible to see in the dark by the infrared light emitted from the objects. The most striking observation is that nearly everything in a dark room can be seen. This is evidence that thermodynamic equilibrium is an extrapolation that is far from common. The smallest heat capacity objects take new temperatures faster and are easier to see in a background of heavy heat capacity. It is also possible to see a difference between areas at the same temperature, but which show different values of the emission coefficient.

For instance, Fig. 37 shows a man's head. The glasses are black because they are much cooler than the skin. There are many applications to far infrared optics. Most classical textbook experiments in the visible can be repeated in the far infrared with the help of the Pyricon. For instance, it is possible to see far infrared fringes (Fig. 38) or to look at the cross section of a far infrared laser mode (Fig. 39).

With some improvements in sensitivity, the Pyricon could be a good candidate to replace, in the infrared, the photographic plate so useful for visible and UV spectroscopy, since it provides the Felgett advantage of seeing all spectral elements at one time. For more details on pyroelectric vidicon and their applications, see Wreathall and Chaplin (1972), Helmick (1973),

FIG. 38   Fresnel diffraction by a circular aperture ($\phi = 8$ mm) at 20 cm from the camera ($\lambda = 10 \ \mu$m) (Felix *et al.*, 1976).

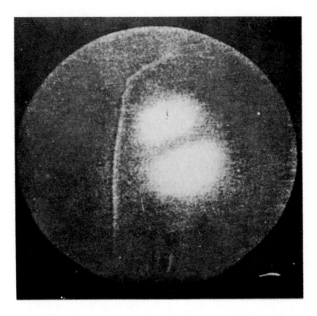

FIG. 39   TEM$_{10}$ mode of a HCN laser ($\lambda = 337 \, \mu$m) (Felix *et al.*, 1976).

Taylor and Boot (1973), Watton *et al.* (1973), Boot and Castled (1974), Conklin (1976), Garn and Sharp (1974), Watton (1974), Watton *et al.* (1974a, b), Helmick (1975), Klozenbe (1975), Conklin and Stupp (1976), Weimer (1976), Watton *et al.* (1977), Hatanaka *et al.* (1978), Shepard (1978), Talmi (1978), Felix *et al.* (1976).

VIII.   Low-Temperature Pyroelectricity

A.   INTRODUCTION

1.   *Low-Temperature Pyroelectricity*

For most pyroelectric materials used as room temperature pyroelectric detectors, the pyroelectric coefficient π'_σ decreases with temperature.

However, there are examples where pyroelectricity is observable only at low temperature, as is the case of linear pyroelectrics such as saccharose, where pyroelectricity is negligible at room temperature. This is also the case for ferroelectrics where Curie temperature is low. For KDP (K_2HPO_4): $T_c = 80$ K, and pyroelectricity occurs only for $T < T_c$.

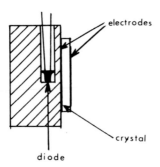

FIG. 40 The crystal plate is mounted with a dot of silver glue on a copper finger and is assumed to be unclamped.

2. Primary and Secondary Pyroelectric Coefficients

In most experiments considered up to now the pyroelectric plate is unclamped, i.e., free to expand (Fig. 40). The corresponding pyroelectric coefficient is written as

$$\pi_\sigma = (\partial P'_s / \partial T)_\sigma,$$

with σ representing the stresses that are assumed to be kept constant during temperature change.

On the other hand, in most theories a clamped crystal is considered; i.e., stresses are applied to keep a constant shape and volume. The corresponding pyroelectric coefficient is written as

$$\pi_s = (\partial P'_s / \partial T)_s,$$

s representing the strains that are kept constant, equal to zero.

It is shown that

$$\pi_\sigma = \pi_I + \pi_{II}, \tag{38}$$

where $\pi_I = \pi_s$ (primary pyroelectric coefficient) is independent of the unit cell deformation and, π_{II} is the pyroelectric contribution due to its deformation via the expansion coefficients and the piezoelectric terms. π_{II} is the secondary pyroelectric coefficient; it is either positive or negative. At low temperature it is shown that π_{II} is proportional to T^3.

B. EXPERIMENTAL METHODS

Two methods are used to get the pyroelectric coefficient versus temperature.

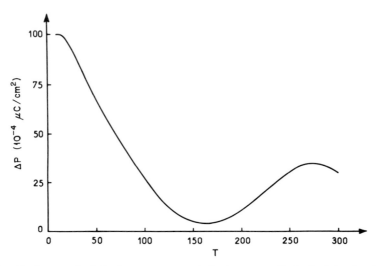

FIG. 41 Increase in $\Delta P'_s$ of spontaneous polarization versus T obtained by the charge integration technique (Mangin and Hadni, 1979).

1. Charge Integration Technique

The pyroelectric plate is cut perpendicular to the polar axis, both faces are covered with a conductive layer, and the pyroelectric condenser is put into a closed thermostat, the temperature of which is raised slowly from liquid helium temperature to room temperature, for instance.

With an operational amplifier, charges appearing on one electrode of the pyroelectric condensor are integrated and the crystal plate is kept short circuited. With a voltmeter we thus get $\Delta P'_s$ versus T. Figure 41 shows such a curve in the case of a saccharose single crystal, and Fig. 42 shows $\pi_\sigma = (\partial \Delta p'_s / \partial T)_\sigma$ deduced by derivation of the $\Delta P'_s(T)$ function represented in Fig. 41. It is seen that π_σ passes twice through a null value and exhibits a broad maximum at 14 K.

The charge integration technique is useful; however, it is not easy because it is a dc method affected by drifts that are difficult to eliminate.

2. Responsivity Technique

With some improvements, this is mainly the Chynoweth method (Chynoweth, 1956). The pyroelectric detector is put into a thermostat fitted with a window to illuminate one face of the pyroelectric condenser. We have seen that responsivity $\mathcal{R}$ is written

$$\mathcal{R} = \frac{\alpha \mathcal{T} \pi_\sigma \omega R}{2\mathcal{G}(1 + \omega^2 \tau^2)^{1/2}(1 + \omega^2 \tau'^1)^{1/2}},$$

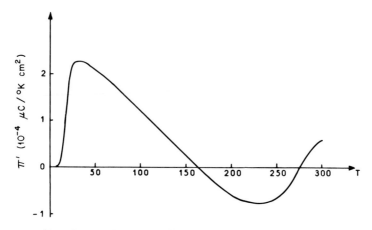

FIG. 42 Pyroelectric coefficient π' versus T derived from Fig. 41.

and for ω high enough to allow $\omega^2\tau^2 > 1$ and $\omega^2\tau'^2 > 1$, we have

$$\mathcal{R} \simeq \alpha\mathcal{T}\pi_\sigma/2\omega\mathcal{C}C.$$

From $\mathcal{R}$ versus T we can thus get $\pi_\sigma/\mathcal{C}$ versus T. Figure 43 gives such a curve in the case of saccharose.

a. *Heat Capacities at Low Temperatures.* To get $\pi_\sigma(T)$, $\mathcal{C}(T)$ must be known. Unfortunately, heat capacity data are rare for ferroelectrics at low

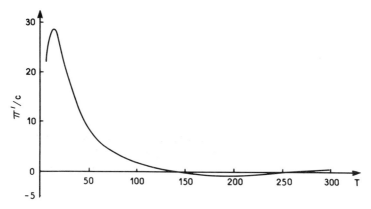

FIG. 43 Pyroelectric response $\pi'/\mathcal{C}$ versus T (experimental). The unit is 10^{-6} μC/J d° (Mangin and Hadni, 1979).

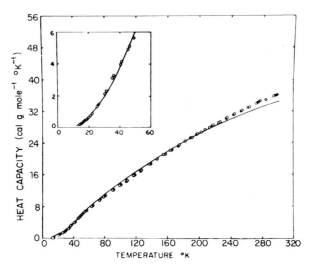

FIG. 44   Heat capacity (at constant pressure) of lithium sulfate monohydrate. The theoretical curve consists of the sum of the one Debye function and five Einstein functions, and includes a factor to correct from constant volume heat capacity to constant pressure. The characteristic temperatures were selected so as to minimize the sum of squares of the relative differences between the experimental points, and the theoretical curve rather than the sum of squares of the absolute differences. This improved the data fit at low temperatures, as shown in the inset. The average standard deviation per point was 4.5 % (from Lang, 1971).

temperatures (Fig. 44). According to Lawless (1976), they should have a characteristic behavior, with

$$\mathscr{C}_V(T) = aT^3 + bT^{3/2}. \tag{39}$$

There is a simple way to get information directly on $\mathscr{C}$ by the responsivity technique. Let us look back at Eq. (8), which gives the phase advance ϕ of the pyroelectric current on the sine wave illumination:

$$\tan \phi = \mathscr{G}/\mathscr{C}\omega, \tag{8}$$

$\mathscr{G}$ being the thermal loss per unit area and 1 K temperature difference between detector and heat sink, and $\mathscr{C}$ the heat capacity per unit area. At low temperatures, $\mathscr{C}$ decreases, and in most cases we have observed that $\tan \phi$ increases rapidly. Assuming $\mathscr{G}$ to be a constant, any discontinuity on $\mathscr{C}$ will give a corresponding one on $\tan \phi$.

b.   *Clamping*

(i)   Clamping at low frequencies.   The frequencies used for such measurements are of kilohertz order, much lower than mechanical resonance fre-

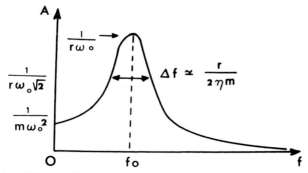

FIG. 45 Amplifying coefficient A for mechanical forced oscillation of an oscillator of natural frequency f_o versus f.

quencies of the crystal plate. According to Fig. 45, which gives the amplifying coefficient versus frequency for forced oscillations of a mechanical system with a resonance frequency at f_o, we shall assume that the crystal plate is unclamped, i.e., free to expand at the illumination frequency. In order to reduce crystal heating, the chopper is generally made of a blade with openings such that the duty cycle is 1 : 4. For a 1 kHz chopping frequency and a 1 : 4 duty cycle, the time of illumination is 125 μs, much shorter, even at low temperature, than $\tau = \mathscr{C}/\mathscr{G}$, the relaxation time of the crystal ($\tau \simeq 10^{-4}$ s at 20 K), and the pyroelectric signal is generally kept as a square pulse.

(ii) *Clamping at high frequencies.* When the chopping frequency is much higher than all the mechanical resonance frequencies, we get a thermal wave at this frequency. We shall take the case of a crystal thickness smaller than the diffusion length $[l = (2\chi/\omega)^{1/2}]$ in order to assume a uniform temperature in the pyroelectric plate. However, it is seen on Fig. 45 that the mechanical response of the system is negligible for such high frequencies. The mechanical forced vibration at a frequency much above the resonance cannot be excited; this is a case where the crystal is inertially clamped. The pyroelectricity which is then observed does not keep any contribution from π_{II}, and we get directly $\pi_{I} = \pi_{s}$.

For practical purposes, a dye laser pulse can be used (Lines *et al.*, 1978) for such measurements. Its time length is shorter than 150 ns and Fig. 46 shows its Fourier analysis. The high frequencies give a clamped pyroelectric signal. It is followed by oscillations at the resonance frequency.

C. PRESENT STATUS

1. *Experimental Data up to 1976*

As far back as 1914 Boguslawski had studied $Li_2SO_4 \cdot H_2O$ from 23–300 K and claimed that at low temperature π_σ was proportional to T. In 1966, ZnO

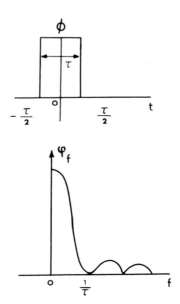

FIG. 46   Square pulse $\phi(t)$ and its Fourier components $\phi_f(f)$.

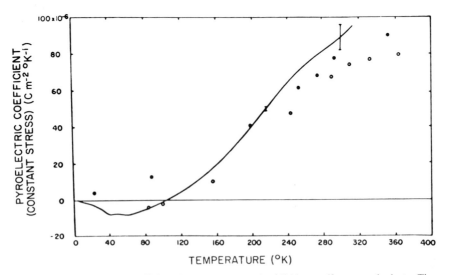

FIG. 47   Pyroelectric coefficient (at constant stress) of lithium sulfate monohydrate. The larger errors near 300 K were caused by difficulties in controlling the rate temperature change. ———, this work; ●, Ackerman (1915); ○, Gladkii and Zheludev (1965) (from Lang, 1971).

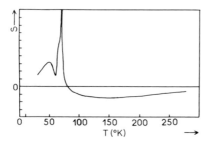

FIG. 48 Variation of the gadolinium bromate pyroelectric signal versus temperature (Poulet *et al.*, 1975).

was studied by Heiland *et al.* down to 9 K, and π_I was found proportional to T^3. In 1971, Lang looked again at lithium sulfate from 300 K–4 K. He did not find the proportionality of π_σ to T at low temperature but found instead a change of sign at 110 K (Fig. 47). The contribution of π_{II} to π_σ is calculated from the piezoelectric tensor and expansion coefficients. It is an important one. Thus π_I is obtained versus T. It is shown that π_I, like π_σ, is not proportional to T at low temperature and changes its sign at 110 K. In 1979, neodymium bromate nonahydrate was studied (Poulet *et al.*, 1975). The pyroelectric coefficient is very small at room temperature, changes its sign at 80 K, and shows a clear maximum at 70 K (Fig. 48).

2. Most Recent Data

A study of tourmaline (Fig. 49) was made by Donnay (1977) who confirms most of Boguslawski's results. Glass (1976) finds a maximum responsivity for both lithium niobate and lithium tantalate at 30 K (Fig. 50). In 1977 the pyroelectricity of saccharose was studied down to 2 K and showed a maximum responsivity at 14 K, and also a maximum pyroelectric coefficient (see Fig. 42). In 1978, several pyroelectrics were considered down to 2 K, and most of them showed a maximum responsivity at low temperatures: TGS, Fig. 51, gives a maximum at 14 K; TGSe, Fig. 52, at 13 K; KDP, Fig. 53, at 33 K; guanidine aluminum sulfate hexahydrate (GASH), Fig. 54, at 10 K; thiourea, Fig. 55, at 7 K. Barium strontium niobate shows a maximum responsivity at 50 K and barium sodium niobate at 5.6 (Gerbaux, unpublished). For lithium sulfate the results of Lang (1971) have been perfectly confirmed. However, the maximum responsivity at low temperatures now shows two components at 20 and 30.5 K (Fig. 56), and at very low temperatures the responsivity seems to diverge. The measurements have to be carried down to 0.3 K to clarify this point. Such a divergence could be explained by the contribution of the acoustical phonons to the pyroelectric coefficient. According to Born (1945), as explained below, it should be proportional to

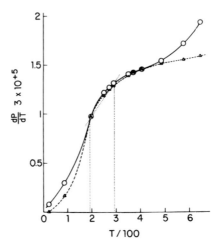

FIG. 49   Observed (circles) and calculated (triangles) values of $(3 = 10^{+5})\, dP/dT$ for pinkish-red tourmaline (Donnay, 1977).

T. Now if temperature is low enough, it is the only contribution; responsivity, being proportional to π_σ/c', should vary as $1/T^2$ and should diverge.

D.   TENTATIVE EXPLANATION

1.   *The Need for a Dynamic Model for Pyroelectric Material*

For ferroelectrics with an order–disorder transition, spontaneous polarization P_s is an order parameter that increases rapidly toward a constant value

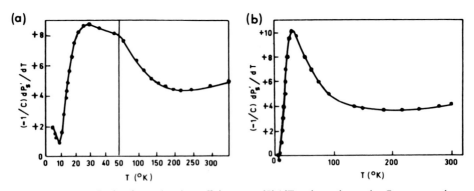

FIG. 50   (a) Ratio of pyroelectric coefficient $\pi = dP'_s/dT$ to thermal capacity C as measured for LiTaO$_3$ by the infrared detection technique. Also shown (crosses) are equivalent measurements for an effectively clamped crystal). (b) Ratio of pyroelectric coefficient $\pi = dP'_s/dT$ to thermal capacity C as measured for LiNbO$_3$ by the infrared detection technique (from Glass and Lines, 1976).

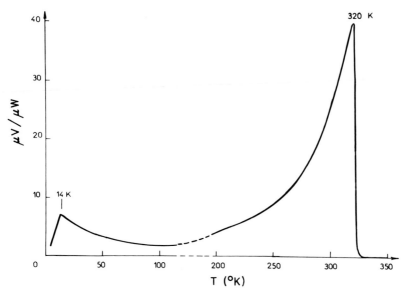

FIG. 51   Responsivity of a TGS pyroelectric detector versus T, $e = 0.5$ mm and $F = 800$ Hz (Hadni *et al.*, 1978).

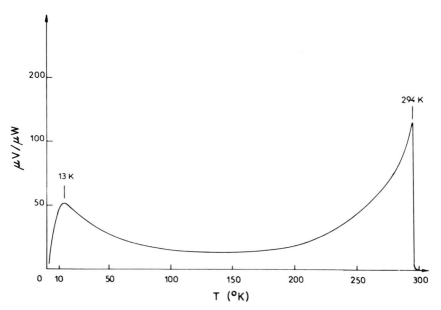

FIG. 52   Responsivity of a TGSe pyroelectric detector versus T, $e = 0.25$ mm and $F = 800$ Hz (Mangin and Hadni, 1978).

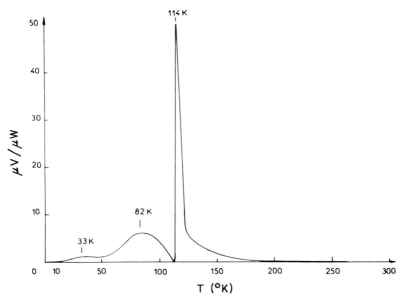

FIG. 53   Responsivity of a KDP pyroelectric detector versus T, $e = 0.3$ mm and $F = 800$ Hz (Mangin and Hadni, 1978).

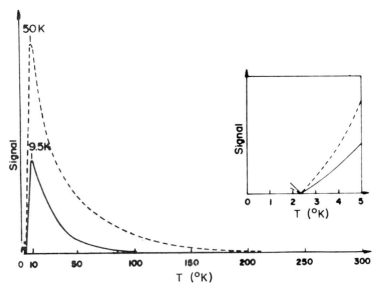

FIG. 54   Responsivity of a guanidine aluminium sulfate hexahydrate pyroelectric detector $e = 0.15$ mm; ———, $F = 2000$ Hz; ---, $F = 800$ Hz (Hadni et al., 1978).

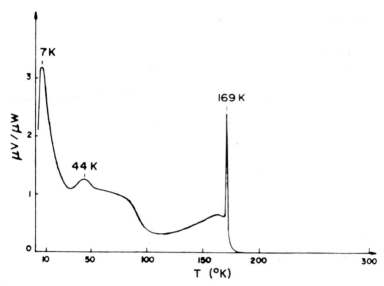

FIG. 55 Responsivity of a thiourea $[SC(NH_2)_2]$ pyroelectric detector, $e = 0.3$ mm and $F = 800$ Hz (Mangin and Hadni, 1978).

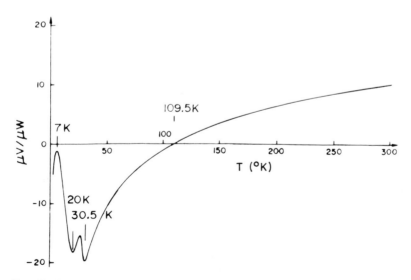

FIG. 56 Responsivity of a pyroelectric detector made of lithium sulfate monohydrate (LSM), $e = 0.3$ mm and $F = 800$ Hz (Hadni et al., 1978).

when temperature is decreased at constant volume from the Curie value T_c. At low temperature P_s should be a constant and the primary pyroelectric coefficient $\pi_I = \pi_s$ (constant strain, i.e., constant volume and shape, i.e., clamped sample) should be negligible. The same situation occurs for displacive ferroelectrics (i.e., $BaTiO_3$).

For linear pyroelectrics of the wurtzite type, for instance, spontaneous polarization is the sum of the dipole moments of all unit cells contained in the unit volume. The unit cell dipole moment is caused by the atomic structure (eclipsed in the case of wurtzite). It should not change when temperature is modified at constant strain, and π_I should be equal to zero.

In conclusion, the static dipole moment of the unit cell cannot explain the pyroelectric coefficient $\pi_I = \pi_s$ of a clamped sample either for a linear pyroelectric at any temperature or for a ferroelectric at temperatures much lower than the Curie point.

Born has considered a dynamic polarization that is temperature sensitive even at constant strain.

2. Born's Model

Born (1945) has shown that an average electric dipole moment component $\langle p_i \rangle$ may occur from any anharmonic oscillator of v_i frequency. Its value is not proportional to the oscillator energy but to the mean square amplitude:

$$\langle p_i \rangle = \frac{A_i}{v_i^2} \frac{h v_i}{e^{h v_i / kT} - 1}. \tag{40}$$

It is assumed that $\langle p_i \rangle$ is the electric dipole component measured on the polar axis; thus, A_i is either positive or negative according to the oscillator considered.

In a crystal, oscillation frequencies are distributed on three acoustical and $3n - 3$ optical branches, n being the number of atoms in the primary unit cell. Every branch brings a dynamic contribution to the spontaneous polarization. From Eq. (40) it is temperature sensitive and also gives a contribution to the primary pyroelectric coefficient π_I. Hence, at low enough temperature, where ferroelectricity does not bring any contribution to π_I (i.e., perfect order), we have

$$\pi_I = \pi_I^a + \pi_I^o. \tag{41}$$

a. *Acoustical Contribution.* Born has considered only the acoustical spectrum. Assuming a Debye frequency distribution, he did find a contribution

$$\pi_I^a = B(\theta_D / T),$$

where B is the Born function, θ_D the crystal Debye temperature, and T the crystal temperature. At very low temperatures, $B(\theta_D/T)$ should be proportional to T (Born, 1945). The problem is discussed by several authors, i.e., Grout and March (1975) and (1976a and b), Grout et al. (1973), Grout et al. (1978), Szigetti (1975 and 1976), Glass and Lines (1976).

b. *Contribution of Optical Branches to Pyroelectric Coefficient.* The contribution of optical phonons to the pyroelectric coefficient was first considered by Lang (1971) in his important paper on lithium sulfate pyroelectricity. Dispersion is neglected on optical branches in a first approach. Now the contribution of frequency ν_i is written

$$\Delta P_i = N \langle p_i \rangle,$$

where N is the number of primitive unit cells in the unit volume (or the number of modes in an optical branch). Then

$$\Delta P_i = \frac{NhA_i}{\nu_i(e^{h\nu_i/kT} - 1)},$$

$$\pi_i = \left(\frac{\partial \Delta P_i}{\partial T}\right), \qquad \pi_i = \frac{Nh^2A_i}{k}\frac{e^{\theta_i/T}}{T^2(e^{\theta_i/T} - 1)^2}, \tag{42}$$

and

$$\pi_I = B(\theta_D/T) + \sum \pi_i + \pi_I^o. \tag{43}$$

The primary pyroelectric coefficient π_I is the sum of the Born function $B(\theta_D/T)$, plus Einstein functions π_i, plus eventually a term π_I^o corresponding to the classical pyroelectricity observed in ferroelectrics close to the Curie point. At low temperature, in any case, $\pi_I^o = 0$.

We must recall that as early as 1914 von Boguslawski had empirically used one Einstein function to express the variations of π_I versus T for a tourmaline crystal.

The Einstein function representation is well known, with a positive slope decreasing toward zero for high temperature and a point of inflexion when the slope is maximum. There is, of course, an analogy between the contribution of a lattice oscillation to the specific heat and to the pyroelectric coefficient. However, the total specific heat of one gram-atom is written:

$$C = \sum_i 3R\Theta_i^2 e^{\theta_i/T}/T^2(e^{\theta_i/T} - 1)^2. \tag{44}$$

Every term of the sum is positive in Eq. (44), while in Eq. (42) the term π_i is either positive or negative according to the sign of A_i, which depends on the dipole orientation. The absolute value of A_i, depends both on the oscillator strength and anharmonicity. The result is that the curve $\sum \pi_i$ versus T can be much more complicated than the classical specific heat curves C versus T. Figure 42 shows such an example in the case of saccharose.

c. *Contribution of Optical Branches to Pyroelectric Detector Flux Responsivity.* Flux responsivity is proportional to $\pi_\sigma/\mathscr{C}$. Now

$$\pi_\sigma = \pi_I + KT^3 = B(\theta_D/T) + \sum \pi_i + \pi_I^o + KT^3.$$

Let us assume that T is small enough so that $\pi_I^o = 0$ and all π_i (Einstein function) are negligible, except π which corresponds to the lowest optical frequency (Einstein temperature θ). Then

$$\pi_\sigma = B(\theta_D/T) + KT^3 + \pi.$$

Flux responsivity is written as $\mathscr{R} = \pi_\sigma/\mathscr{C}$, and assuming $\mathscr{C} = K'T^3$, we have

$$\mathscr{R} = \frac{B(\theta_D/T)}{K'T^3} + \frac{K}{K'} + \frac{Nh^2A}{kK'T^5}\frac{e^{\theta/T}}{(e^{\theta/T} - 1)^2},$$

and in the case for which the first two terms should be negligible, we have

$$\mathscr{R} \simeq \frac{Nh^2A}{kK'T^5}\frac{e^{\theta/T}}{(e^{\theta/T} - 1)^2}. \tag{45}$$

Such a function shows a maximum for $T = \theta/5$ and goes to zero with T. It might give a possible interpretation of the experimental data concerning saccharose (Fig. 43). The farthest infrared absorption band is located at 45 cm^{-1} ($\theta = 60$ K; $\theta/5 = 12$ K), polarized parallel to the polar axis, while the maximum of $\pi_\sigma/\mathscr{C}$ is located around 14 K. However since π_σ itself shows a maximum (Fig. 42, with the assumption that $\pi' \simeq \pi_\sigma$), it is suggested that at least two Einstein functions are needed to explain the pyroelectricity of saccharose at temperatures around 14 K.*

On the other hand, while the Einstein functions give a negligible contribution to the responsivity at very low temperatures, the other terms:

$$\frac{B(\theta_D/T)}{K'T^3} + \frac{K}{K'}$$

should be revealed. In the case of saccharose, TGS, TGSe, KDP, and thiourea, there is no hint of such a contribution down to 2 K. For GASH (Fig. 54) and lithium sulfate monohydrate (Fig. 56) it seems there is some divergence of responsivity for $T < 2$ K.

---

* In fact, there is an absorption line at 64 cm^{-1} with the same polarization as the one at 45 cm^{-1}.

E. THE POSSIBILITIES OF PYROELECTRIC DETECTION AT LOW TEMPERATURES

1. Very Low Temperatures ($T < 20$ K)

We just have seen that pyroelectric responsivity proportional to π_σ/c' is as high at very low temperature as at room temperature for some pyroelectric materials.

If we look at the factor of merit:

$$M_1(T) = \frac{\pi_\sigma}{c'(\varepsilon'' T)^{1/2}}, \tag{46}$$

this shows that detectivity D^* then depends only on $(\varepsilon'' T)^{1/2}$. When temperature is reduced from 300 K to 0.3 K, there is first a $(10^3)^{1/2} \simeq 32$ increase ratio due to T. With ε'' probably going to zero at low temperature, Johnson noise should be negligible and $M(T)$ should increase dramatically. This means that detectivity should be limited by photon noise. We have seen in that case:

$$D^*_{\text{ideal}}(0.3 \text{ K}) \simeq 6 \times 10^{17} \text{ W}^{-1} \text{ cm Hz}^{1/2}.$$

Most probably in fact, with the present technology D^* should be limited to a smaller value by amplifier noise. There is some hope of improvements with the use of AsGa FETs cooled down to 4 K.

2. Low Temperatures ($T \simeq 80$ K)

Paul Richards at the University of California (Berkeley) has recently called our attention to the need for pyroelectric detectors cooled down to around 80 K. They should be useful in satellites (Brown, 1975) where it is easy to cool a detector to this temperature without any energy consumption (passive cooling). Many of these satellites use a TGS pyroelectric detector temperature controlled to around 300 K. If the same responsivity could be gotten at around 80 K, Johnson noise should be much smaller (because of smaller T and, more important, smaller ε''). At least one order of magnitude could be gained and the pyroelectric detector that generally looks at Earth should become photon noise limited.

Such detectors are especially needed for the wavelength range 15–80 μm, where there is a lack of high detectivity quantum detectors. There are several ferroelectric crystals with Curie temperature close to 80 K that might be good candidates.

IX. Conclusion

This brief survey has shown that pyroelectric detectors are now widely used for infrared laser pulse detection, in satellites, and in a number of devices where high sensitivity, low cost, and ruggedness must be combined. For infrared imaging, the pyroelectric vidicon is one of the best solutions. Commercially available instruments now have an elementary $D^* = 10^8 \, \text{W}^{-1}$ cm $\text{Hz}^{1/2}$; i.e., the minimum detectable temperature is 0.2 K on the object, with 16 frames per second, and a spatial resolution of five line-pairs per millimeter on the target.

However, several new ways of using pyroelectricity have been discussed and improvements of prime importance seem to be possible. The potentials of using new materials, new epitaxial films, or new devices (i.e., CCD) have also been considered. Arrays of very sensitive pyroelectric detectors can become competitive with the pyroelectric vidicon.

ACKNOWLEDGMENTS

We are grateful to the many colleagues who helped us during friendly discussions, visits, or conferences. There are too many to be cited. The names of some of them will be found in the references. Thanks also must be given to Miss Christine Bozzoli for the care taken in typing the manuscript and preparing the figures.

REFERENCES

Abrams, P. L., and Glåss, A. M. (1969). *Appl. Phys. Lett.* **15**, 251.
Alting-Mees, N. R., and Koda, N. J. (1976). *Ferroelectrics* **11**, 323.
Annis, A. D., and Simpson, G. (1974). *Infrared Phys.* **14**, 199–205.
Auston, D. H., and Glass, A. M. (1972). *Appl. Phys. Lett.* **20**, 398.
Baker, G., Charlton, D. E., and Lock, P. J. (1972). *Radio Electron. Eng.* **42**, 260.
Beerman, H. P. (1969). *IEEE Trans. Electron. Devices* **ED-16**, 554.
Beerman, H. P. (1971). *Ferroelectrics* **2**, 123.
Beerman, H. P. (1974). *Am. Ceram. Soc., Bull.* **53**, 606.
Beerman, H. P. (1975). *Infrared Phys.* **15**, 225–231.
Blackburn, H., and Wright, H. C. (1971). *Infrared Phys.* **10**, 191.
Blackburn, H., Wright, H. C., Eddington, R., and King, R. S. (1972). *Radio Electron. Eng.* **42**, 369.
Blevin, W. R., and Geist, J. (1974a). *Appl. Opt.* **13**, 1171–1178.
Blevin, W. R., and Geist, J. (1974b). *Appl. Opt.* **13**, 2212–2217.
Boot, H. A. H., and Castledi, J. G. (1974). *Electron. Lett.* **10**, 452.
Born, M. (1945). *Rev. Mod. Phys.* **17**, 245.
Brown, F. G. (1975). *SPIE Infrared Technol.* **62**, 201.
Burfoot, J. C. (1978). Pyroelectric and ferroelectric thin film devices. "Active and Passive Thin Film Devices" (T. J. Coutts, ed.), pp. 687–741. Academic Press, New York.
Bye, K. L., Whipps, P. W., Keve, E. T., and Josey, M. R. (1976). *Ferroelectrics* **11**, 525–534.
Charles, D. R., and Le Carvennec, F. (1972). *Adv. Electron. Electron Phys.* **33**, 279.
Chynoweth, A. G. (1956). *J. Appl. Phys.* **27**, 78.
Claudel, J., and Hadni, A. (1976). *Proc. Conf. Electro-Opt./Laser Int., United Kingdom* (H. G. Jerrard, ed.), p. 149. IPC Science and Technology Press.
Conklin, T., and Stupp, E. H. (1976). *Opt. Eng.* **15**, 510–515.

Cooper, J. (1962). *Rev. Sci. Instrum.* **33**, 92.
Day, G. W., Hamilton, C. A., Gruzensk, P. M., and Phelan, R. J. (1976). *Ferroelectrics* **10**, 99–102.
Donnay, G. (1977). *Acta Crystallogr. Sect. A* **33**, 927.
Felix, P., *et al.* (1976). *Opt. Laser Technol.* **8**, 75–80.
Felix, P., Gamot, P., Lacheau, P., and Raverdy, Y. (1978). *Ferroelectrics* **17**, 543.
Garn, L. E., and Petito, F. C. (1977). *IEEE Device* **24**, 1221–1228.
Garn, L. E., and Sharp, E. J. (1974). *IEEE Parts Ph.* **10**, 208–221.
Gebbie, H. A., Stone, N. W. B., Putley, E. H., and Shaw, N. (1967). *Nature (London)* **214**, 165.
Gerbaux, X., and Hadni, A. (1978). *J. Opt.* **9**, 57.
Gerbaux, X., Hadni, A., and Waldschmidt, J. M. (1977). *Opt. Commun.* **21**, 425.
Gerbaux, X., Waldschmidt, J. M., and Hadni, A. (1978). *Appl. Opt.* **17**, 1616.
Gladkii, V. V., and Zheludev, I. S. (1965). *Kristallografiya* **10**, 50.
Glass, A. M., and Abrams, R. L. (1970a). *J. Appl. Phys.* **41**, 4455.
Glass, A. M., and Abrams, R. L. (1970b). *J. Appl. Phys.* **41**, 281.
Glass, A. M., and Auston, D. H. (1974). *Ferroelectrics* **7**, 187.
Glass, A. M., and Lines, M. E. (1976b). *Phys. Rev. B* **13**, 180.
Glass, A. M., and Vonderli, D. (1976a). *Ferroelectrics* **10**, 163–166.
Glass, A. M., McFee, J. H., and Bergman, J. G. (1971). *J. Appl. Phys.* **42**, 5219.
Grandjean, D., Claudel, J., Brehat, F., Hadni, A., Strimer, P., and Thomas, R. (1970). *J. Phys.* **31**, 471.
Grout, P. J., and March, N. H. (1975). *J. Phys. C* **8**, L594–L595.
Grout, P. J., and March, N. H. (1976a). *Phys. Rev. B* **14**, 4027.
Grout, P. J., and March, N. H. (1976b). *Phys. Rev. Lett.* **37**, 791.
Grout, P. J., March, N. H., and Thorp, T. L. (1975). *Solid State Phys. C* **8**, 2167.
Grout, P. J., March, N. H., and Ohmura, Y. (1978). *Appl. Phys. Lett.* **32**, 453–454.
Hadni, A. (1963). *J. Phys.* **24**, 694.
Hadni, A., (1967). "Essentials of Modern Physics Applied to the Study of the Infrared." Pergamon, Oxford.
Hadni, A. (1969). *Opt. Commun.* **1**, 251.
Hadni, A. (1970). Thermal far infrared detectors. *Proc. Symp. Submillimeter Waves*, p. 251. Polytechnic Press, Polytechnic Institute of Brooklyn, Brooklyn, New York.
Hadni, A., and Thomas, R. (1972a). *Ferroelectrics* **4**, 39.
Hadni, A., and Thomas, R. (1972c). *Opt. Commun.* **6**, 314.
Hadni, A., and Thomas, R. (1974a). *Opt. Commun.* **10**, 366.
Hadni, A., and Thomas, R. (1974b). *Ferroelectrics* **6**, 241.
Hadni, A., and Thomas, R. (1975b). *Phys. Status Solidi (a)* **31**, 71.
Hadni, A., and Thomas, R. (1976a). *Appl. Phys.* **10**, 91.
Hadni, A., and Thomas, R. (1976b). *Ferroelectrics* **11**, 493.
Hadni, A., Henninger, Y., Thomas, R., Vergnat, P., and Wyncke, B. (1965a). *C. R. Acad. Sci. Paris* **260**, 4186.
Hadni, A., Henninger, Y., Thomas, R., Vergnat, P., and Wyncke, B. (1965). *J. Phys.* **26**, 345.
Hadni, A., *et al.* (1969a). *J. Phys.* **30**, 377.
Hadni, A., Thomas, R., and Perrin, J. (1969b). *J. Appl. Phys.* **40**, 2740.
Hadni, A., Grandjean, D., Claudel, J., and Gerbaux, X. (1970). *J. Phys.* **31**, 899.
Hadni, A., Perrin, J., Thomas, R., and Schoumacher, P. (1971). *C. R. Acad. Sci. Paris* **273**, 537.
Hadni, A., Gerbaux, X., Chanal, D., Thomas, R., and Lambert, J. P. (1973a). *Ferroelectrics* **5**, 259.
Hadni, A., Lambert, J. P., Pradhan, M. M., and Thomas, R. (1973b). *Infrared Phys.* **13**, 305.
Hadni, A., Bassia, J. M., Gerbaux, X., and Thomas, R. (1976c). *Appl. Opt.* **15**, 2150.

Hadni, A., Thomas, R., and Erhard, C. (1977a). *Phys. Status Solidi* **39**, 419.
Hadni, A., Thomas, R., Mangin, J., and Bagard, M. (1978). *Infrared Phys.* **18**, 663.
Harmer, A. L. (1976). *IEEE Trans. Electron. Devices* **23**, 1320–1325.
Harmer, A. L., and Wreathall, W. M. (1976). *Adv. Electron.* **A40**, 313–322.
Hatanaka, Y., Kamiryo, K., and Kano, T. (1972). *Appl. Phys.* **11**, 1788.
Hatanaka, Y., Okamoto, S., and Nishida, R. (1978). *Ferroelectrics* **19**, 171.
Heiland, G., and Ibach, H. (1966). *Solid State Commun.* **4**, 353.
Helmick, C. N. (1973). *Proc. Electro-Opt. Syst. Design Conf.* p. 195.
Helmick, C. N. (1975). *Proc. Soc. Photo-Opt. Instrum. Eng., August 19–20.* San Diego, California.
Helmick, C. N. (1976). *Ferroelectrics* **11**, 309–313.
Hoeksma, G. S. (1978). *Electron. Lett.* **14**, 146–148.
Holeman, B. R., and Wreathall, W. M. (1971). *J. Phys. D* **4**, 1848.
Holeman, B. R., Baldock, S. E. R. L., and Hertz, Q. K. (1972). *Infrared Phys.* **12**, 125.
Jansson, P. A. (1974). *Appl. Opt.* **13**, 1293–1294.
Keve, E. T. (1975). *Philips Tech. Rev.* **35**, 247.
Klozenbe, J. P. (1975). *Infrared Phys.* **15**, 87–93.
Lang, S. B. (1971). *Phys. Rev. B* **4**, 3603.
Lang, S. B. (1975). *Ferroelectrics* **9**, 65.
Lang, S. B. (1976). *Ferroelectrics* **11**, 315–319.
Lang, S. B. (1978a). *Ferroelectrics* **17**, 553.
Lang, S. B. (1978b). *Ferroelectrics* **19**, 25–60.
Lang, S. B. (1978c). *Ferroelectrics* **19**, 175–209.
Lang, S. B., and Athenstaedt, H. (1977). *Science* **196**, 985–986.
Lang, S. B., and Athenstaedt, H. (1978). *Ferroelectrics* **17**, 511–519.
Lang, S. B., and Steckel, F. (1965a). *Rev. Sci. Instrum.* **36**, 929.
Lang, S. B., and Steckel, F. (1965b). *Rev. Sci. Instrum.* **36**, 1817.
Lang, S. B., Shaw, S. A., Rice, L. H., and Timmerhavs, K. D. (1969). *Rev. Sci. Instrum.* **40**, 274.
Latham, R. V. (1976). *J. Phys. D* **9**, 2295–2304.
Lavi, S., and Simhony, M. (1973). *J. Appl. Phys.* **44**, 5187.
Lawless, W. N. (1976). *Phys. B* **14**, 134.
Lawton, R. A. (1972). *Electron. Lett.* **8**, 318.
Lawton, R. A., (1973). *IEEE Instrum.* **IM22**, 299–306.
Lax, M. (1976). *Phys. Rev. B* **13**, 1759–1769.
Le Carvennec, F. (1971). *Eur. Infrared Symp., 3rd, Malvern*, May, 17–21.
Leiba, E. (1969). *C.R. Acad. Sci. Paris* **268**, B31.
Leiba, E., Hadni, A., Thomas, R., and Blondel, M. (1972). *Nouv. Rev. Opt. Appl.* **3**, 263.
Lines, M. E. (1975). *J. Phys. C* **8**, L 589–L 593.
Lines, M. E., and Glass, A. M. (1977). "Principles and Applications of Ferroelectrics and Related Materials." Oxford Univ. Press (Clarendon), London and New York.
Liu, S. T., (1976). *Ferroelectrics* **10**, 83–89.
Liu, S. T., and Long, D. (1978). *Proc. IEEE* **66**, 14–26.
Liu, S. T., and Maciolek, R. B. (1975). *J. Electron. Mater.* **4**, 91–100.
Logan, R. M., (1973). *Infrared Phys.* **13**, 91.
Logan, R. M. (1975). *Infrared Phys.* **15**, 51–64.
Logan, R. M., and McLean, T. P. (1972). *Infrared Phys.* **3**, 15–24.
Logan, R. M., and Moore, K. (1973). *Infrared Phys.* **13**, 37.
Logan, R. M., and Watton, R. (1972). *Infrared Phys.* **12**, 17–28.
Maciolek, R. B., Schuller, T. L., and Liu, S. T. (1976). *J. Electron. Mater.* **5**, 415.
Mangin, J., and Hadni, A. (1978). *J. Phys. Lett.* **23**, L447.

Mangin, J., and Hadni, A. (1979). *Phys. Rev. B* **18**, 7139.
Mangin, J., *et al.* (1974). *Nouv. Rev. Opt.* **5**, 305.
Martens, U., and Kneubühl, F. (1974). *Appl. Opt.* **13**, 1455-1459.
Martens, U., Jeannet, P., and Kneubühl, F. (1975). *Appl. Opt.* **14**, 1177.
Micheron, F., and Godefroy, L. (1971). *C.R. Acad. Sci. Paris* **T273**, 143.
Nelson, D. F., and Lax, M. (1976). *Phys. B* **13**, 1785-1796.
Nelson, D. F., (1978). *J. Acoust. Soc. Am.* **63**, 1738-1748.
Nelson, P. D., (1976). *Electron. Lett.* **12**, 652.
Nolta, J. P., Schubring, N. W., and Dork, R. A. (1962). *Phys. Rev. Lett.* **9**, 285.
Nye, J. F., (1960). "Physical Properties of Crystals." Oxford Univ. Press (Clarendon), London and New York.
Phelan, R. J., and Cook, A. R. (1973). *Appl. Opt.* **12**, 2494.
Phelan, R. J., Malher, R. J., and Cook, A. R. (1971). *Appl. Phys. Lett.* **19**, 337.
Poulet, H., Mathieu, J. P., Vergnat, D., Vergnat, P., Hadni, A., and Gerbaux, X. (1975). *Phys. Status Solidi (a)* **32**, 509.
Putley, E. H. (1971). *Eur. Infrared Symp., 3rd, Malvern, United Kingdom*, May, 17-21.
Putley, E. H. (1977a). *In* "Semiconductors and Semimetals" (R. K. Willardson and A. C. Beer, eds.), Vol. 12, Chapter 7, p. 441. Academic Press, New York.
Putley, E. H. (1977b). Optical and infrared detectors. "Topics in Applied Physics," Vol. 19. Springer, Berlin and New York.
Roundy, C. B., and Bryer, R. L. (1972). *Appl. Phys. Lett.* **21**, 512.
Roundy, C. B., and Byer, R. L. (1976). *Ferroelectrics* **10**, 215.
Roundy, C.B., Byer, R. L., Phillion, D. W., and Kuizenga, D. J. (1974). *Opt. Commun.* **10**, 374-377.
Royer, M., and Micheron, F. (1978). *C. R. Acad. Sci. Paris* **287**, B-145.
Salomon, R. E., Oh, B. K., and Labes, M. M. (1976). *J. Appl. Phys.* **47**, 1710-1711.
Schmid, H., Genequand, P., Tippmann, H., Pouilly, G., and Guedu, H. (1978). *J. Mater. Sci.* **13**, 2257.
Schwarz, F., and Poole, R. R. (1970). *Appl. Opt.* **9**, 240.
Shaulov, A. (1979). *Commun. Int. Symp. Appl. Ferroelectrics, Minneapolis, Minnesota.*
Shaulov, A., and Simhony, M. (1972a). *J. Appl. Phys.* **43**, No. 4, 1440.
Shaulov, A., and Simhony, M. (1972b). *Appl. Phys. Lett.* **20**, 6.
Shaulov, A., and Simhony, M. (1976). *J. Appl. Phys.* **47**, 1-5.
Shaulov, A., Rosenthal, A., and Simhony, M. (1972). *J. Appl. Phys.* **43**, 4518.
Shepard, A. G., (1978). *Opt. Spectra* **12**, 52.
Simhony, M., and Bass, M. (1979). *Appl. Phys. Lett.* **34**, 426.
Simhony, M., and Shaulov, A. (1971). *J. Appl. Phys.* **42**, 3741.
Simhony, M., and Shaulov, A. (1972). *Appl. Phys.* **21**, 375.
Simhony, M., Shaulov, A., and Lavi, S. (1973). *Appl. Phys. Lett.* **22**, 99.
Singer, B. (1979). Invited paper of the *Int. Symp. Appl. Ferroelectrics, Minneapolis, Minnesota*, June 13-15 1979.
Singer, B., and Lalak, J. (1976). *Ferroelectrics* **10**, 103-107.
Singer, B., Crowell, M., and Conklin, T. (1974). *IEEE Trans. Electron. Devices* **21**, 744.
Smith, W. A. (1979). Invited paper at the *Int. Symp. Appl. Ferroelectrics, Minneapolis, Minnesota*, June 13-15 1979).
Southgat, P. D. (1974). *Proc. IEEE* **62**, 540-541.
Steinhage, P. W., Schmidt, V., Mester, V., and Kunze, C. (1971). *Eur. Infrared Symp. 3rd, Malvern*, May 17-21.
Steinhage, P. W., and Zeyfang, R. R. (1976). *Ferroelectrics* **11**, 301-304.
Szigeti, B. (1975). *Phys. Rev. Lett.* **35**, 1532.

Szigeti, B. (1976). *Phys. Rev. Lett.* **37**, 792.

Talmi, Y. (1978). *Appl. Opt.* **17**, 2489–2501.

Taylor, R. G. F., and Boot, H. A. T. (1973). *Contemp. Phys.* **14**, 55.

Thiboumery, A. (1972). *Int. Electron.* **27**, 42.

Tompsett, H. F. (1971). *IEEE Trans. Electron. Devices* **ED-18**, No. 11, 1070.

Van der Ziel, A. (1973). *J. Appl. Phys.* **44**, 546.

Van der Ziel, A., and Liu, S. T. (1972). *Physica G* **1**, 589.

von Boguslawski, S. (1914). *Phys. Z.* **15**, 283–288.

Watton, R. (1974). *Phys. Med. Biol.* **10**, 21–121.

Watton, R. (1976). *Ferroelectrics* **10**, 91–98.

Watton, R. (1978). *Infrared Phys.* **18**, 73.

Watton, R., Smith, G., and Harper, B. (1973). *Electron. Lett.* **9**, 534.

Watton, R., Smith, G., Harper, B., and Wreathal, W. M. (1974a). *IEEE Trans. Electron. Devices* **21**, 462–469.

Watton, R., Jones, G. R., and Smith, C. (1974b). *Electron. Lett.* **10**, 469–470.

Watton, R., Jones, G. R., and Smith, C. (1976). *Adv. Electron.* **A40**, 301–312.

Watton, R., Burgess, D., and Harper, B. (1977). *J. Appl. Sci. Eng.* **2**, 47.

Wreathall, W. M., and Chaplin, I. (1972). *Electron. Lett.* **8**, No. 2.

Wreathall, W. M., and Savage, S. D., (1971). *Eur. Infrared Symp. 3rd Malvern*, May 17–21.

Wurfel, P., and Batra, I. P. (1973), *Phys. Rev. B* **8**, 5126.

Wyncke, B., Serrier, J., Brehat, F., and Hadni, A. (1978). *J. Phys. C.: Solid State Phys.* **11**, 2639.

Yacobi, B. G., and Brada, Y. (1976). *J. Appl. Phys.* **47**, 1243–1247.

Yamaka, E., and Teranish, A. (1978). *Ferroelectrics* **19**, 171.

Yamaka, E., Hayashi, T., and Matsumoto, M. (1971). *Infrared Phys.* **11**, 247.

Yamaka, E., Teranish, A., Nakamura, K., and Nagashim, T., (1976). *Ferroelectrics* **11**, 305–308.

Zeyfang, R. R., Sehr, W. H., and Kiehl, K. V. (1976). *Ferroelectrics* **11**, 355–358.

Zheludev, I. S. (1971). "Physics of Crystalline Dielectric." Plenum Press, New York.

Ziel, A., and Liu, S. T. (1972). *J. Appl. Phys.* **43**, 4260.

Zook, J. D., and Liu, S. T. (1976). *Ferroelectrics* **11**, 371.

Zook, J. D., and Liu, S. T. (1978). *J. Appl. Phys.* **49**, 4604–4606.

Zwicker, W. K., Dought, J. P., Delfino, M., and Ladell, J. (1976). *Ferroelectrics* **11**, 347–350.

CHAPTER 4

Photon Drag Detection

A. F. Gibson

Rutherford Laboratory
Chilton, Didcot, Oxfordshire,
England

and

M. F. Kimmitt

University of Essex
Colchester, Essex
England

I. Introduction

The photon drag effect leads to the generation of electric fields in semi-conductors, the magnitude of the field being a linear function of intensity. It was first observed in germanium at microwave frequencies by Barlow (1958) who gave a classical description based on the Hall effect. Further early contributions were made by Gurevich and Rumyantsev (1967) and by Gulyaev (1968), but the first observations using lasers were by Danishevsky et al. (1970) and Gibson et al. (1970a). These two groups, working independently, both used germanium rods and Q-switched CO_2 lasers and, again independently, chose to call the effect photon drag by analogy with phonon drag, which was familiar from the study of thermoelectric effects in semi-conductors (Smith, 1979a). The latter work fortuitously coincided with the development by Beaulieu in Canada of TEA-type CO_2 lasers. Immediately there was a need for fast response, simple, however, robust detectors that would work at room temperature; sensitivity was no longer at a premium. Photon drag detectors filled this role and commercial devices based on p-type germanium appeared in the autumn of 1970. Since then photon drag has been studied, particularly at the CO_2 laser wavelength, in a number of other semiconductors including silicon (Serafetinides et al., 1978), GaAs (Doviak and Kothari, 1974; Schneider and Hubner, 1975), GaP (Gibson et al., 1977), InAs (Patel, 1971) and Te (Hammond et al., 1972; Panyakeow et al., 1972; Ribakovs and Gundjian, 1977). Tellurium, particularly, is more sensitive than p-type germanium at 10.6 μm, but crystals are more difficult to prepare in a controlled fashion and this appears to have inhibited commercial development. Because of its commercial importance and ease of fabrication, germanium has received the most detailed attention and will form the primary subject of this chapter.

II. Tensor Properties

In addition to the photon drag effect, an electric field proportional to the radiation intensity can be generated in crystals lacking a center of inversion symmetry by optical rectification (Zernike and Midwinter, 1973). In some materials and at some wavelengths, e.g., GaP at 3 μm and Te at 10 μm, optical rectification generates fields equaling or exceeding the photon drag field and can be exploited in detectors. Naturally, no optical rectification occurs in the elemental semiconductors Si and Ge. In other materials the two contributions can be distinguished by their tensor properties, since optical rectification is described by a third-rank tensor and photon drag by a fourth-rank tensor (Valov et al., 1972b; Hattori et al., 1973a)

$$E_i = I[\mathbf{R}_{ijk}\mathbf{p}_j\mathbf{p}_k + \mathbf{T}_{ijkl}\mathbf{q}_j\mathbf{p}_k\mathbf{p}_\ell] \tag{1}$$

where $\mathbf{E}_i$ is the generated field, I the intensity, $\mathbf{p}_j\mathbf{p}_k\mathbf{p}_\ell$ are components of the unit polarization vector of the radiation, and $\mathbf{q}_j$ a unit vector in the propagation direction. The first term then describes optical rectification and the second photon drag. For crystals with $\bar{4}3$ m symmetry, $\mathbf{R}_{ijk}$ is given by a single coefficient D, and $\mathbf{T}_{ijkl}$ contains only three nonzero, unequal, components conventionally designated by S, P and Q. E_i is then given by

$$E_i = I[2Dp_{i+1}p_{i+2} + Sq_i + (P - S)q_ip_i^2 + 2Qp_i(p_{j+1}q_{j+1} - p_{j+2}q_{j+2})]$$
$$(2)$$

where D and Q are both zero in Si and Ge. In addition the term in Q vanishes even in III–V compound semiconductors if the radiation propagates along one of the principal crystallographic directions.

Experimental measurements of D, S, P, etc., are usually made using rodlike samples similar to those used in detectors and illustrated in Fig. 1. The measured quantity is usually the open circuit voltage developed between the longitudinally (V_L) or transversely (V_T) positioned contacts. In practice absorption in the finite length of rod between the contacts and the end

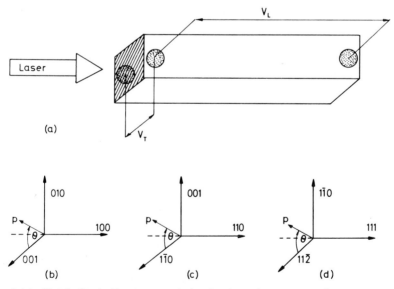

(a)

(b) (c) (d)

FIG. 1 (a) Sketch of typical bar-type sample showing electrode arrangement for measurement of longitudinal and transverse voltages, V_L and V_T. (b), (c), and (d) the three main orientations for the study of photon drag and optical rectification. The line p represents the polarization direction of the incoming light.

surfaces has to be taken into account, but if this is neglected, the generated voltages, including multiple reflections, can be shown to be

$$V_L = \frac{(1 - R)I[1 - \exp(-KL)]}{K} \left\{ \frac{[D]}{1 - R\exp(-KL)} + \frac{[P, S]}{1 + R\exp(-KL)} \right\},$$

(3)

and

$$V_T = \frac{(1 - R)Iy}{1 - R^2 \exp(-KL)}$$

$$\times \{[D][1 + R\exp(-2KL)] + [P, S][1 - R\exp(-2KL)]\}, \quad (4)$$

where R is the surface reflection coefficient, K the absorption coefficient, and L and y the length and width of the rod. The factors $[D]$ and $[P, S]$ are the appropriate $E(\mathbf{q}, \mathbf{r})$ factors from Eq. (2) where $\mathbf{q}$ and $\mathbf{r}$ specify the crystal directions along which the radiation propagates and the voltage is measured. A number of these factors have been tabulated by Doviak and Kothari (1974) and are given in Table I. The differing denominators under $[P, S]$ and $[D]$ in Eq. (3) and equivalent multipliers in Eq. (4) arise from the different effect of multiple reflections for photon drag and optical rectification: Light beams traveling in opposite directions produce drag signals of opposite sign due to the q dependence and tend to cancel, while optical rectification signals are additive.

TABLE I

Photon Drag and Optical Rectification Factors

Light direction	Contact direction	Combined photon drag and optical rectification field	Figure reference
100	100	$I(S + D \sin 2\theta)$	1(b)
100	001	0	1(b)
110	110	$I\left(S + \dfrac{P - S}{2} \cos^2\theta\right)$	1(c)
110	$1\bar{1}0$	$-I(Q\cos^2\theta + D\sin 2\theta)$	1(c)
111	111	$I\left(\dfrac{P + 2S}{3} - \dfrac{D}{3^{1/2}}\right)$	1(d)
111	$11\bar{2}$	$I\left(\dfrac{S - P}{3(2^{1/2})}\cos 2\theta + \dfrac{2Q}{6^{1/2}}\sin 2\theta + \dfrac{2D}{6^{1/2}}\cos 2\theta\right)$	1(d)

III. Photon Drag in Germanium and Silicon

A. GENERAL FEATURES

As noted above, $D = 0$ for these materials. The expressions in Table I then simplify considerably and, in particular, the longitudinal voltage in 100 oriented rods is proportional only to S,

$$E_{100}^{100} = SI, \qquad (5)$$

and is independent of radiation polarization. Rods oriented along a 111 direction should also produce a polarization-independent longitudinal voltage, even though a transverse field is generated. Unless the electrodes have transverse symmetry, however, a polarization-dependent longitudinal signal can be observed due to second-order cross-terms proportional to the gradient of the longitudinal field, as demonstrated by Gatenby and Kar (1978).

Photon drag signals may be generated in any type of optical transition: free carrier absorption, impurity photoionization, or interband transitions; although if the latter are due to single photon processes, the absorption coefficient may be so high that practical restrictions on contact size preclude measurement of longitudinal voltages. Drag due to two photon interband transitions has been considered by Brynskikh and Sagdullaeva (1978). The Leningrad group have made extensive studies of photon drag due to impurity photoionization (see, for example, Valov et al., 1972a). For linear detectors operating at room temperature, however, only free carrier effects may be used. Such transitions are subject to the usual requirements of energy and wave vector conservation, namely

$$E_f - E_i = \hbar\omega, \qquad (6)$$

and

$$\mathbf{k}_f - \mathbf{k}_i = \mathbf{q}, \qquad (7)$$

where E_f, $\mathbf{k}_f$, E_i, and $\mathbf{k}_i$ are the final and initial energies and wave vectors respectively and $\hbar\omega$ and $\mathbf{q}$ are the photon energy and wave vector. The conservation rules do not apply, however, (and indeed have no meaning) when the quantities involved are small compared with the uncertainties in their values. The uncertainty in the energy of an electron is of order h/τ, where τ is the scattering time. Hence, energy conservation is not required if $\omega\tau < 1$, and a similar conclusion applies to wave vector conservation. For $\omega\tau < 1$ photon drag effectively reduces to classical radiation pressure, except that the interacting particles are imbedded in a medium of refractive index $n \neq 1$. The electronic band structure becomes irrelevant and the tensor components S and P are expected to become equal. We will show later that in the limit of $\omega\tau \ll 1$,

$$S = n\sigma/ec \qquad (\omega\tau \ll 1), \qquad (8)$$

where σ is the absorption cross section of the charge carriers involved, e is the electronic charge, and c the velocity of light in vacuum. At sufficiently long wavelengths the only source of absorption is the free carriers, so $\sigma = K/N$ where N is the electron or hole density. Since the band structure is irrelevant, Eq. (8) applies equally to n- and p-type material. The situation for $\omega\tau \gg 1$ is more complicated and is described in the succeeding sections.

B. n-Type Germanium and Silicon

Optically induced transitions in single, near-parabolic, bands require (for $\omega\tau \gg 1$) the cooperation of a third particle such as a phonon to conserve energy and wave vector. The multivalley nature of the conduction bands of Si and Ge does not alter this statement, and hence electron absorption cross sections in these materials are relatively small in the near infrared (2–20 μm wavelength) although they rise at longer wavelengths.

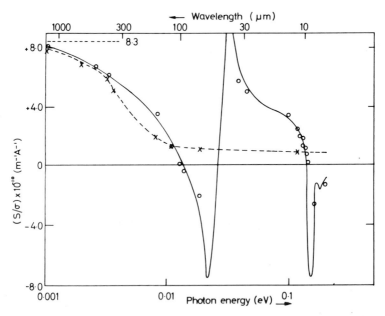

FIG. 2   The photon drag coefficient (S/σ) of n-type (dashed line) and p-type (full line) germanium at long wavelengths. Note that the coefficients become equal at very long wave-lengths, when $\omega\tau$ becomes less than unity (after Gibson et al., 1979).

In addition, the photon momentum is shared between the electron and the cooperating particle so that S and P are small on both counts; in germanium at 10 μm they are two orders of magnitude less than the corresponding quantities in p-type material of comparable doping concentration (Gibson *et al.*, 1970b). Near-intrinsic and slightly n-type germanium is dominated by the minority holes at the CO_2 laser wavelength (Gibson *et al.*, 1970a). In consequence, n-type Ge and Si are of little interest for detectors in the near infrared and have received less attention than p-type devices.

The above conclusions are not valid for $\omega\tau < 1$ when phonon assistance in a transition is not required and the absorption cross section of electrons exceeds that of holes by the ratio of the mobilities (Smith, 1979b). We now expect Eq. (8) to apply and S to be large. The condition $\omega\tau = 1$ is reached in n-type germanium at a wavelength of about 700 μm and in n-type silicon at about 400 μm. Figures 2 and 3 show the experimentally observed variation of S/σ for these two materials from 10 μm to 1.2 mm; the rise in S/σ at long wavelengths to reach the limiting value given by Eq. (8) is evident.

C. p-Type Germanium and Silicon

The general form of the valence band structure of Ge, Si, and III–V compounds is well known: two bands degenerate at or near zero wave

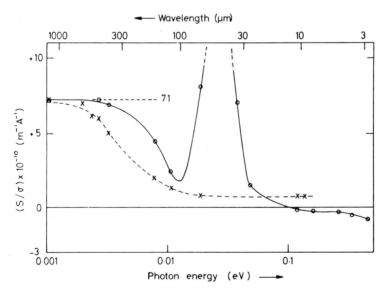

FIG. 3   The photon drag coefficient (S/σ) of n-type (dashed line) and p-type (full line) silicon at long wavelengths. Note that the coefficients become equal at very long wavelengths, when $\omega\tau$ becomes less than unity (after Gibson *et al.*, 1979).

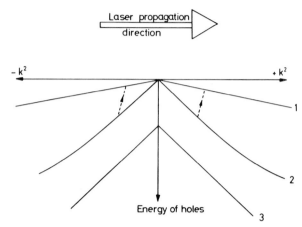

FIG. 4   Sketch showing the valence band structure of a semiconductor like germanium. Possible transitions between bands 2 and 1 at a given photon energy are indicated. The magnitude of the radiation wave vector has been exaggerated for clarity.

vector and a spin-orbit split-off band, as illustrated in Fig. 4. In p-type material, states in bands 1 and 2 with energies $\sim kT$ are occupied by holes, and hence electron transitions can take place between bands $2 \to 1$, $3 \to 2$, and $3 \to 1$. It is well known that these transitions lead to relatively high hole absorption cross sections (10^{-16} cm$^2 < \sigma < 5 \times 10^{-16}$ cm^2) in the wavelength range 2–20 μm and, for semiconductors with suitable spin-orbit splitting energies (e.g., Ge and GaAs), an absorption spectrum with observable structure (Briggs and Fletcher, 1953; Braunstein and Kane, 1962). The high-absorption cross section of holes and the occurrence of direct valence band transitions in p-type germanium can lead to large photon drag signals and accounts for the commercial popularity of this material for CO_2 laser detectors. However, the complex band structure of this material and the strong coupling between holes and optic phonons (Paige, 1966) leads to quite dramatic variations in sensitivity with temperature and wavelength. The former is well exemplified by the observation by Danishevskii et al. (1970) that the sign of the photon drag effect at 10.6 μm in p-type germanium reversed on cooling below about 180 K. This effect was analyzed theoretically by Grinberg (1970), and the data extended by Gibson and Walker (1971) who showed that the sign reversal did not occur in heavily doped samples. Thus it was demonstrated that the reaction of the lattice could be sufficient to allow holes to flow against the photon stream, but this reaction was reduced by impurity scattering. These effects are illustrated in Fig. 5. Perhaps even more striking is the variation in photon drag sensitivity, expressed in terms of S/σ, with wavelength as shown in Fig. 6. This figure has been con-

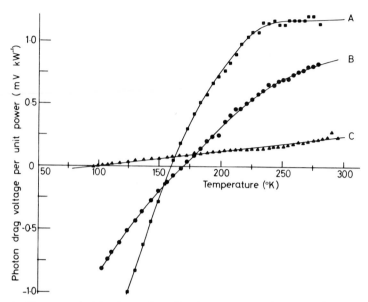

FIG. 5 Sign reversal of the photon drag effect in p-type germanium on cooling. Laser wavelength 10.6 μm. Hole densities 2.9 × 10^{14} cm^{-3}, 1.6 × 10^{15} cm^{-3}, and 7.4 × 10^{15} cm^{-3} in samples A, B, and C, respectively (after Gibson and Walker, 1971).

structed from various published sources (Cameron *et al.*, 1975; Gibson and Serafetinides, 1977; Al-Watban and Harrison, 1977; Gibson *et al.*, 1979) and includes a theoretical curve due to Gibson and Montasser (1975). It will be seen that the coefficient S/σ reverses in sign between 2 and 100 μm no less than seven times. Such rapid variations in sensitivity are clearly undesirable in devices, but do imply that a wealth of information is in principle available from photon drag studies. We discuss the interpretation of these data in the following sections.

D. THE PHOTON DRAG SPECTRUM OF p-TYPE GERMANIUM

The photon drag spectrum of p-type germanium between 2 and about 40 μm (Fig. 6) is dominated by intervalence band transitions. An electron excited into a state f contributes an increment of current proportional to its charge e, group velocity V_F, and momentum relaxation time τ_f, with a similar but negative term for the removal of the electron from its initial state i. Equations (6) and (7), together with the band structure that determines $E(k)$ in each band, define the energy and wave vector of the initial and final states. The total photon drag current is then determined by summing over all states between which transitions are possible at a photon energy $\hbar\omega$, each transition being weighted by the transition rate, p_{if}. The measured quantity, namely

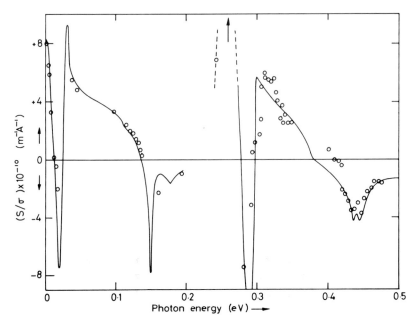

FIG. 6   The photon drag coefficient (S/σ) of p-type germanium as a function of photon energy. The curve drawn is theoretical and based on the work of Gibson and Montasser (1975). The theory is unreliable in the 0.2–0.28 eV region and, in particular, predicts an infinite positive coefficient at 0.26 eV (see text). The points are experimental, and were obtained using a variety of laser sources by the authors referred to in the text.

the photon drag field, can then be found by multiplying the current by the resistivity ρ. The transition rate p_{if} contains matrix elements that are not accurately known for many semiconductors; therefore, it is convenient, as indicated before, to calculate S/σ and P/σ rather than S and P as the absolute values of the matrix elements, and then cancel. Since σ can be measured experimentally, no information is lost by this procedure and comparison with experiment is facilitated.

The above argument leads to the following theoretical expression for S/σ

$$\frac{S}{\sigma} = \frac{Be\rho \sum_k p_{if}[V_f(\mathbf{k})\tau_f(\mathbf{k}) - V_i(\mathbf{k})\tau_i(\mathbf{k})]}{\hbar\omega \sum_k p_{if}/N}, \tag{9}$$

where N is the hole density, B is a numerical constant of proportionality to be discussed later, and the term $\hbar\omega$ appears in the denominator because the measured photon drag field is defined in terms of incident energy rather than photon flux. The resistivity ρ is given by

$$\rho = [Ne^2\langle\tau/m\rangle]^{-1}, \tag{10}$$

where $\langle \, \rangle$ indicates an average over the hole distribution function, so we may anticipate that B will contain a term of the form $\langle \tau/m \rangle/(\tau/m)$. The wave vector dependence of the group velocities is determined by the band structure, but that of the scattering times depends on the relative strengths of the possible scattering mechanisms, optic and acoustic phonons, impurities, etc., and their respective energy dependencies. Finally, the transition rate p_{if} is proportional to the density of occupied states in band i, the density of unoccupied states in band f, and the relevant matrix element W_{if} [compare Kane, 1956, (74)]. If the holes in band i can be neglected, $p_{if}(k)$ is given by

$$p_{if}(k) = W_{if} k_i^2 k_f^2 \exp(-E_f/kT). \tag{11}$$

For a complex valence band structure such as that of Ge and Si (Kane, 1956; Fawcett, 1965) Eq. (9) can only be solved numerically. This was done by Gibson and Montasser (1975) and is the origin of the theoretical curve shown in Fig. 6. It is, however, possible to account qualitatively for all the main spectral features shown in Fig. 6 by a simplified model that assumes one dimensional, parabolic bands. Only two transitions are then possible, one each side of $k = 0$ as indicated in Fig. 4, and the summation in Eq. (9) contains only four terms. An analytic solution may then be obtained, which gives a physical picture of the processes involved and which we now describe.

As a further simplification, suppose initially that τ and V are invariant with k. The photon drag field is then zero if

$$p_{if}(L) = p_{if}(R), \tag{12}$$

where L and R refer to the left and right of $k = 0$ in Fig. 4. Equation (12) is equivalent, for parabolic bands characterized by effective masses m_i and m_f, to the condition

$$\hbar\omega = kT[(m_f^2 - m_i^2)/m_f m_i], \tag{13}$$

where kT is Boltzmann's constant times the absolute temperature and determines the distribution of holes in the two bands. For $\hbar\omega$ less than given by Eq. (13) and $m_f > m_i$, the photon drag field is positive, i.e., the holes are displaced in the same direction as the radiation wave vector but the field is reversed for larger values of $\hbar\omega$. The sign reversals of the photon drag field at 0.136 and 0.42 eV in Fig. 6 can be identified with this effect where the final and initial states are in bands 1 and 2, and 1 and 3, respectively. The two sign reversals are separated by the spin-orbit splitting energy, namely 0.28 eV, and inserting the photon energies into Eq. (13) gives the mass ratios 6:1:1 for bands 1, 2, and 3, respectively, in reasonable agreement with known values (Kane, 1956). For low k values the mass in band 2 is less than that in band 3, so $3 \rightarrow 2$ transitions give rise to a sign reversal in which the photon drag field is positive for $\hbar\omega$ greater than the reversal energy. This is seen in Fig. 6 at 0.3 eV.

The electron group velocity varies only slowly with energy or wave vector so the approximation of invariance is fairly good at all k values. However, τ varies rapidly with energy if optic phonon scattering is strong and the energy of the initial or final state is near the energy of an optic phonon of a small wave vector. It is possible, for example, for the energy of the final state in band 1 to be less than the optic phonon on the left of $k = 0$ and greater on the right (Fig. 4). When this occurs, electrons on the right can emit optic phonons while those on the left cannot, and τ(right) $\ll$ τ(left). For one-dimensional bands there is only a narrow range of photon energies over which this state of affairs can exist and a sharp feature can be expected in the spectrum. In three dimensions, with warped bands, "τ features" are broadened or even split (Gibson and Montasser, 1975). The negative-going peak at about 0.15 eV is a τ feature in the band 2 to band 1 transition region as a result of the final state in band 1 being near the optic phonon energy in germanium, namely 0.037 eV. The split, negative-going peak at about 0.44 eV has the same basic origin, except that band 3 to band 1 transitions are now involved. The positive peak and sign reversal at about 0.025 eV is also a "τ feature," this time due to coincidence of the optic phonon energy with the initial state in band 2 of 2 → 1 transitions.

A final qualitative aspect of the photon drag spectrum which requires interpretation is the large positive peak shown at 0.26 eV in Fig. 6. A similar peak, with the same basic origin, is shown by p-type silicon at 0.03 eV in Fig. 3. Neither region has received much experimental study. Strong positive peaks are expected theoretically, however, as these regions correspond to transitions between bands that are parallel in k space when separated by the appropriate photon energy. Bands 2 and 3 run parallel to one another in both germanium and silicon when they are separated in energy by about 0.26 and 0.033 eV, respectively. Unfortunately, at these points even the full, three-dimensional numerical analysis of Gibson and Montasser (1975) fails as the Taylor expansions used diverge. Kane (1956) found the same problem when trying to account for the absorption spectrum in this region. Thus theory predicts that both S/σ and σ approach infinity at these singularities and a more sophisticated treatment is required. No theoretical prediction is shown in this region in Fig. 6, except for an arrow to show the point of singularity. Additional experimental data would also be valuable. Cameron *et al.* (1975) obtained a measurement at about 0.24 eV in p-type germanium and observed a large, positive signal, and Hattori and Umeno (1975) observed a large, negative signal at 0.228 eV. The latter is not included in Fig. 6. Compared with the study of S/σ, the coefficient P/σ has received considerably less attention. The values of S and P as a function of wavelength in the tech-

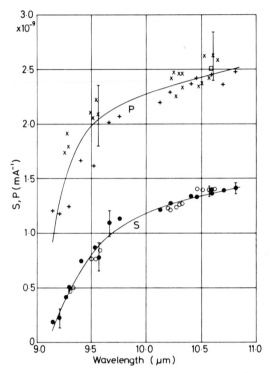

FIG. 7   The photon drag coefficients S and P for p-type germanium in the CO_2 laser wavelength band (after Cameron *et al.*, 1975), with points indicated ⊡ taken from Hattori *et al.* (1973a).

nically important CO_2 laser range are shown in Fig. 7. To calculate P/σ theoretically requires that the k dependence and the polarization dependence of the transition matrix elements be taken into account. This has been done by Montasser (1976a) who demonstrated good agreement between theory and experiment in the 10-μm region.

E.   THE PHOTON DRAG SPECTRUM OF p-TYPE SILICON

The spectra dependence of S/σ for p-type silicon is shown in Fig. 3. The qualitatively interesting features are the sign reversal at about 0.1 eV and the broad positive peak between 0.02 and 0.035 eV. We consider the sign reversal first.

Except for very small values of k, bands 1 and 2 run nearly parallel for band 1 energies up to several kT with a separation of about 0.025 eV. Hence, the only direct interband transitions which are possible for photon energies greater than 0.025 eV are $3 \rightarrow 1$ and $3 \rightarrow 2$ transitions, with the former predominating (Kane, 1956). The sign reversal is due primarily to the changeover in predominant transition rates from one side of $k = 0$ to the other (Fig. 4) and analogous to the 0.42 eV sign reversal of p-type germanium. The energy at which it occurs can be estimated from Eq. (13) by adding the spin-orbit splitting energy of 0.036 eV. If we take the average mass ratio (neglecting warping) of bands 1 and 3 as 3.0 (Elliott and Gibson, 1974), we predict a sign reversal at 0.101 eV which is in fortuitously good agreement with experiment.

Part of the origin of the broad positive peak between 0.02 and 0.035 eV is that bands 2 and 3 run parallel at their minimum separation of 0.033 eV (Kane, 1956), and as discussed in the previous section theory predicts an infinite positive peak with respect to the 0.26 eV peak in the drag spectrum of p-type germanium. However, the peak in silicon is broader and extends to lower photon energies than can be accounted for by this factor alone. The high positive signal is sustained by the near parallelism of bands 1 and 2 alluded to above. These bands remain nearly parallel until their separation falls below about 0.025 eV and the photon drag coefficient falls for photon energies less than this value. The subsequent rise in S/σ at even lower photon energies is due to $\omega\tau$ becoming less than unity and is discussed more fully later.

F. The Temperature Dependence of Photon Drag in Germanium and Silicon

Experimental measurements of the temperature dependence of the drag coefficients have been made primarily with CO_2 laser irradiation at 10.6 μm wavelength. The results of Valov et al. (1972a) show that the photon drag current in lightly doped n-type Ge and Si increases on cooling, initially slowly due to an increase in scattering time, and more rapidly at lower temperatures due to carrier freeze out and the onset of drag by photoionization of the impurities. Due to the variation of resistivity with temperature the photon drag field is almost independent of temperature for modest cooling and rises dramatically when freeze-out occurs.

We have already noted that at 10.6 μm the photon drag signal in p-type germanium reverses in sign on cooling. First observed by Danishevsky et al. (1970) in lightly doped samples, Gibson and Walker (1971) showed that the transition temperature was reduced, and the sign reversal finally removed, in heavily doped samples. The temperature variation of S and P separately was measured by Hattori et al. (1973a).

The theory of photon drag in p-type germanium given above shows immediately that the sign reversals due to varying transition rates will be temperature sensitive, but the "τ features" will only increase in magnitude and not change in spectral position. Valov et al. (1978) have measured the temperature at which the photon drag signal due to band 2 to band 1 transitions reverses at four wavelengths between 12.8 and 9.5 μm. The results can be compared with the approximate Eq. (13); but good agreement cannot be expected since band 2 is far from parabolic over the relevant range of energies, and band 1 is, of course, warped (Kane, 1956). From Valov *et al.*'s data for the slope $d(\hbar\omega)/d(kT)$, we find a mass ratio for band 1 to band 2 of 6.0, in good agreement with theory, but the absolute value of the reversal temperature at one wavelength, namely 10.6 μm, indicates a mass ratio of 7.5. Using the full three-dimensional theory, Montasser (1976b) has calculated the temperature dependence of S and P separately in the same wavelength region and obtained good agreement with the results of Hattori *et al.* (1973a). Equation (13) cannot, because of the approximations used, explain the variation with impurity concentration. This effect was ascribed by Gibson and Walker (1971) to the fact that impurity scattering at a given concentration reduces the scattering time τ in band 1 much more than in band 2.

A sign reversal, from a small negative to a large positive signal, has been observed by Bassioni (1975) on cooling p-type silicon. This cannot be due to varying transition rates and, as the reversal temperature depends on the activation energy of the doping impurity, is believed to be due to carrier freeze-out.

G. THE ABSOLUTE VALUE OF THE PHOTON DRAG COEFFICIENTS OF GERMANIUM AND SILICON

Equation (9) is an incomplete description inasmuch as there remains an undetermined constant of proportionality B. Determination of the absolute magnitude of S or P at any one wavelength determines both at all wavelengths, and the obvious choice is S at wavelengths such that $\omega\tau < 1$. We have already argued (Section A) that for $\omega\tau < 1$ the photon drag coefficients will become independent of the electronic band structure and reach a classical limit. Achievement of this limit experimentally will be characterized by the coefficient S/σ becoming the same for both n- and p-type material and independent of wavelength. Inspection of Figs. 2 and 3 shows, as expected, that this condition is met at wavelengths $\gtrsim 1$ mm. Saturation occurs earliest in p-type silicon, since holes in Si have the shortest mean free time, and is only marginally reached in n-type germanium at the longest wavelength used (1.2 mm). To find S or S/σ in the long wavelength limit, we use a simple classical argument as follows.

The energy absorbed from the radiation per unit time per unit volume is IK, and hence per charge carrier $I\sigma$. The number of photons absorbed per carrier per second is then $I\sigma/\hbar\omega$, and if each photon imparts momentum p, the rate of momentum absorption per charge carrier is $Ip\sigma/\hbar\omega$. Under open-circuit, steady-state conditions, an electric field E is created to oppose the motion, so equating forces

$$S = E/I = p\sigma/eh\omega. \tag{14}$$

To continue, we require a value for p. However, the value of the momentum to be associated with a photon in a dielectric of refractive index n has been a matter for dispute for many years (Burt and Peierls, 1973; Gordon, 1973; Jones, 1978). Three theoretical expressions for photon momentum in a non-dispersive dielectric, derived respectively by Abraham (1914), Minkowski (1910), and Peierls (1976), have been discussed in the literature. These can be written

$$\text{Abraham} \quad h/n\lambda_o \tag{15}$$

$$\text{Minkowski} \quad nh/\lambda_o \tag{16}$$

$$\text{Peierls} \quad (h/2n\lambda_o)[1 + n^2 - \alpha(n^2 - 1)^2] \tag{17}$$

where n is the refractive index, λ_o the wavelength in free space, and α a constant, estimated by Peierls to be 0.2, which is a measure of the field gradient at an atom compared with the macroscopic field. Abraham's expression derives from $E \times H$, is consistent with the relation $E = mc^2$, and is believed to describe the momentum to be associated with the electromagnetic field. Minkowski's is based on $D \times B$, is consistent with de Broglie's relationship, and contains a "mechanical" component due to the reaction of the lattice (Jones, 1978). In Peierls' view neither of the earlier expressions give a complete description (although they are applicable in some circumstances) and the total momentum is given by Eq. (17).

The problem is not so much which expression is "correct" but which is applicable in given experimental conditions, particularly in photon drag. The Abraham force has been observed under suitable conditions (Walker and Lahoz, 1975), but the results of Ashkin and Dziedic (1973) are in qualitative, and those of Jones and Leslies (1978) quantitative, agreement with Minkowski's expression. Gordon (1973) has derived a theorem giving the conditions under which Minkowski's momentum will be appropriate and has shown that the latter results are covered by his theorem.

As regards photon drag, Eq. (13) is based on the conservation of wave vector, so that the predicted spectrum (apart from its absolute magnitude) depends only on the assumption that p is proportional to $1/\lambda_o$, which is satisfied by all three expressions. However, the magnitude and even the sign of Eqs. (15), (16), and (17) differ considerably when n is large, as it is in Ge and Si. Näively, we might expect Abraham's expression to apply to photon drag, since the interaction is with charged particles. We would not expect Peierls' expression to apply, since the bulk reaction of the lattice is communicated to the crystal mounting and is not included in the measurement. Minkowski's expression might be appropriate, but it is not obvious to us that Gordon's theorem applies to the photon drag situation.

In the circumstances we must resort to experiment and, as n is large, an unambiguous answer can be obtained. Substituting the three expressions for p in Eqs. (15), (16), and (17) into Eq. (14) and solving numerically, we predict the following values for S/σ at very long wavelengths in Ge and Si.

	$S/\sigma \times 10^{-10} \mathrm{\ m^{-1}\ A^{-1}}$	
	Germanium	Silicon
Abraham	+0.52	+0.61
Minkowski	+8.33	+7.1
Peierls	−7.3	−3.0

Inspection of Figs. 2 and 3 and comparison with the above then shows clearly that the absolute value of the photon drag coefficient is determined by the Minkowski expression. Substitution of Eq. (16) into Eq. (14) then gives Eq. (8), and gives the numerical constant B in Eq. (13) as

$$B = 2[(\tau/m)_o/\langle \tau/m \rangle], \tag{18}$$

where $(\tau/m)_o$ is the limiting value of τ/m in band 1 at very low energies and $\langle \tau/m \rangle$ is the average over the distribution function (Section D). This value of B was adopted by Gibson and Montasser (1975) for the derivation of the theoretical curve given in Fig. 6, but on less complete evidence than presented here.

IV. Photon Drag and Optical Rectification in Gallium Phosphide

We have shown that direct, intervalence band transitions lead to strong photon drag signals in p-type Ge and Si. Interband transitions by free-charge carriers are not, of course, restricted to valence bands; it has been known for many years that the same process gives rise to a strong absorption band in n-type gallium phosphide at around 3 μm. The conduction band of GaP is like that of silicon, and the lowest minima occur near the Brillouin zone edge in the 100 crystal directions. These minima are believed to have a "camel's back" type structure (Lawaetz, 1975) and further minima lie about 0.3 eV above them. A number of determinations of the energy separation of the minima have been made by absorption studies, and illustrative experimental values are those of Wiley and DiDomenico (1970), namely 0.276 (± 0.007) eV and the 0.355 (± 0.003) eV obtained by Onton (1971).

By analogy with p-type Ge and Si, we may anticipate large photon drag signals in n-type GaP around 3 μm. Furthermore, because the lower minimum is nonparabolic, the electric polarization will be a nonlinear function of the applied field and optical rectification will also occur. Apart from its intrinsic interest, n-type GaP is also potentially valuable for use as detectors with HF lasers in a wavelength region in which Ge shows strong two-photon absorption leading to nonlinear behavior in drag detectors at high intensities (Gibson et al., 1976).

Figure 8 shows the optical rectification coefficient D and the photon drag coefficient P of n-type GaP in the 1–10 μm region (Gibson et al., 1977). The

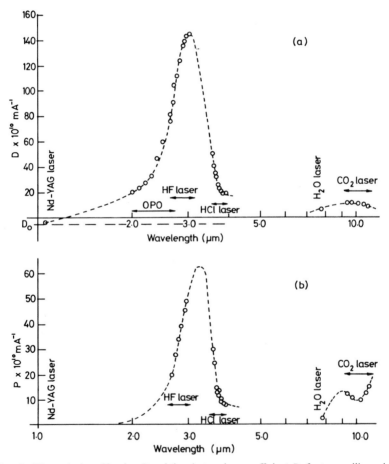

FIG. 8   The optical rectification D and the photon drag coefficient P of n-type gallium phosphide with an electron concentration of 2.4×10^{16} cm^{-3}. The dashed line labelled D_o indicates the optical rectification coefficient of semi-insulating GaP deduced from the measurements of Nelson and Turner (1968). Diagram after Gibson et al. (1977).

S coefficient was found to be about an order of magnitude less than P and could not be determined very accurately, but it appeared to show a peak at 3 μm similar to that shown by D and P. The large value of D at 3 μm is particularly striking, and to the best of our knowledge is the largest optical rectification coefficient observed in any solid at any wavelength. It is also of interest as an unusually unequivocal example of optical rectification due to free carriers, since the 3 μm peak does not occur in semi-insulating GaP (Fig. 8) and the area under the peak in doped material is proportional to the electron concentration (Gatenby et al., 1979).

No complete theoretical treatment of the photon drag coefficients of n-type GaP has been published. Gibson et al. (1977) showed, on the basis of Lawaetz's (1975) band structure and assuming that D was proportional to the third derivative of the electron energy with respect to wave vector, that

$$D/\sigma = C[(\hbar\omega)^2 - \varepsilon^2]^{1/2}/(\hbar\omega)^5, \qquad (19)$$

where C is an arbitrary constant of proportionality and ε is the energy separation of the minima. Putting $\varepsilon = 0.335$ eV, Eq. (19) describes the behavior of a heavily doped sample (electron concentration 2.4×10^{16} cm^{-3}) remarkably well, as illustrated in Fig. 9, but completely fails to describe the

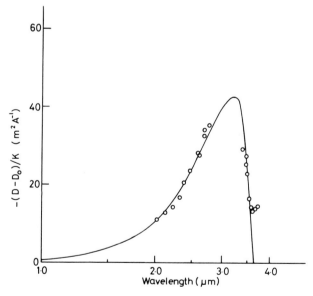

FIG. 9   Free carrier contribution $D-D_o$ to the optical rectification coefficient of n-type GaP divided by the absorption coefficient K of the sample. Curve is theoretical [Eq. (19)], assuming an energy separation between band minima of 335 meV and a scaling factor to fit the experimental point at 2.4 μm.

resonance in more lightly doped material when the 3 μm peak splits into two partially resolved peaks (Gatenby et al., 1979).

p-type gallium phosphide has received rather less attention than n-type. Serafetinides et al. (1978) showed that D, P, and S for p-type GaP were all relatively small at wavelengths around 2.8 μm, but were much larger in the CO_2 wavelength region. Using the absorption data of Wiley and DiDominico (1971) we can estimate, for example, D/σ to be about 0.3×10^{10} m^{-1} A^{-1} at 2.8 μm, 10.2×10^{10} m^{-1} A^{-1} at 9.3 μm, and 19×10^{10} m^{-1} A^{-1} at 10.7 μm. Qualitatively similar behavior is seen in the photon drag coefficients. The data, therefore, suggest the existence of a peak response beyond 11 μm, possibly comparable with the 3 μm peak in n-type material.

Interpretation of photon drag and optical rectification data for p-type GaP in the 2–15 μm region is complicated by the fact that transitions between all three valence bands contribute (Wiley and DiDominico, 1971). However, by analogy with Ge and Si we expect a large positive peak in the drag coefficients when two bands run parallel in K space. This occurs for bands 2 and 3 when their separation is about 0.08 eV. If, additionally, one or both bands is strongly nonparabolic, a large optical rectification effect will occur about the same wavelength. We therefore anticipate a large peak in D and P for p-type GaP at a wavelength of about 15 μm. Though consistent with existing data, this prediction awaits experimental investigation.

V. Photon Drag and Optical Rectification in Tellurium

Tellurium is a particularly interesting semiconductor in which to observe photon drag and optical rectification effects. Early studies were made by Hammond et al. (1972) and Panyakeow et al. (1972), but in a series of papers Ribakovs and Gundjian (1974, 1977, 1978) report very detailed experimental measurements and have explained these theoretically. Tellurium, unlike germanium, has light and heavy hole bands which are split but, as this split corresponds to a wavelength of about 11 μm, CO_2 laser radiation has sufficient energy to produce transitions between the two bands.

Interpreting results for tellurium is very difficult. As a result of its point group 32 trigonal symmetry, there exist five, finite, third-rank tensor coefficients, two of which are independent, and 25, finite, fourth-rank tensor coefficients, of which three are independent. Apart from this, the conduction and optical absorption of tellurium are highly anistropic, and it is difficult to prepare stress-free samples. However, Ribakovs and Gundjian have overcome these daunting problems and their work on tellurium is probably the most rigorous of all studies of photon drag and optical rectification. In summary, they show a peak photon drag coefficient of approximately 10^{-9} m A^{-1}, and an optical rectification coefficient that varies with the sample doping and reaches a maximum roughly equal to the highest drag coefficient. These values compare favorably with the drag coefficients of p-Ge at 10.6 μm shown in Fig. 7. Unlike germanium, on cooling to nitrogen temperature there is no sign change in the photon drag coefficients.

Optical rectification in tellurium is of interest. As with n-GaP, it requires free carriers in the material. However, it is suggested by Ribakovs and Gundjian (1977) that the large effect is related to nonlinear behavior in the conductivity. They also report an emf that could not be characterized by any third- or fourth-rank tensor component. It occurred only in undoped samples and was essentially zero at room temperature, but rose to be as high as 3 V MW^{-1} cm^{-2} on cooling to 110 K. This signal must be compared with a responsivity of $\simeq 0.1$ V MW^{-1} cm^{-2} for a typical photon drag detector. It is suggested that this voltage arises from dislocations and stresses caused by a damage mechanism in the relatively weak, single crystal of tellurium, and this appears to be confirmed by the observation that the effect could appear in a crystal after several cooling cycles. The response time seems to be as fast as for photon drag and optical rectification, and it does not change sign when the direction of the light is reversed.

Ribakovs and Gundjian (1978) give details of a longitudinal optical rectification detector with a rather higher response than commercially available p-Ge detectors at 10.6 μm. Although the maximum photon drag and optical rectification coefficients of tellurium are approximately equal, the absorption of tellurium is much higher in the direction suitable for large optical rectification voltages. The response of this material at wavelengths other than 10.6 μm is certainly worthy of study.

VI. Photon Drag and Optical Rectification Devices

A. RANGE OF INSTRUMENTS

Detectors are clearly the most important devices employing photon drag and optical rectification but on-line monitors, using either longitudinal or transverse voltages, are also of interest. In this section we will initially discuss design considerations for detectors and monitors, and then deal with actual detectors to cover the entire infrared and submillimeter region. Unlike most detection processes, photon drag is direction sensitive, and this property provides the basis for optical bridges that can measure pulse length and study the characteristics of nonlinear absorbers. We conclude with a discussion of these bridges.

B. LONGITUDINAL DETECTORS

1. *General Considerations*

The optimum design of a photon drag detector for a specific application can be quite a complex exercise. Commercially available devices for CO_2

wavelengths are manufactured by a number of firms, but only Rofin Limited in England and the A. F. Ioffe Institute in the U.S.S.R. offer a wide range of detectors. Monitors, and a gallium phosphide detector primarily for 2.5–3 μm are also produced by Rofin. The criteria for choice of a photon drag detector are much the same as for any other type. The most important are: (i) the voltage responsivity; (ii) noise equivalent power; (iii) wavelength range; (iv) response time; and (v) detector area. The linearity of the detector and the intensity at which it damages may also be significant, particularly when used with very high power lasers.

Photon drag has been used with cw lasers; however, except in special circumstances which will be discussed later, there is little to commend it when compared with conventional thermal and pyroelectric detectors. We shall confine our initial discussion to considering its use for high-speed detection of pulsed, Q-switched, and mode-locked lasers.

2. Choice of Material

The first, and still the most widely used, material at 10.6 μm is p-type germanium. The reason for its initial choice was high absorption per carrier due to intravalence band transitions. However, as can be seen in Fig. 6, it was rather fortuitous that it gave a signal of the expected sign and approximately the expected magnitude at 10.6 μm.

It is convenient in the understanding of the design of detectors to separate the effect of the photon drag and optical rectification coefficients from the other semiconductor parameters. This is easily done by initially restricting consideration to the longitudinal voltage from a 100 rod of an elemental semiconductor such as germanium with the light also traveling in the 100 direction. In this case there is only an S component and we will also assume that Eq. (8) is valid. Substituting Eq. (8) into Eq. (3), and neglecting reflection terms we obtain

$$V_{\text{L}} = (n\sigma/ec)\{I[1 - \exp(-KL)]/K\}. \tag{20}$$

But since

$$\sigma = K/N, \tag{21}$$

where N is the number of absorbing carriers, and $I = W/A$ where W is the power falling on the sample and A is the cross-sectional area, then

$$V_{\text{L}}/W = n[1 - (\exp - KL)]/NeAC. \tag{22}$$

For a correct value of the voltage responsivity V_{L}/W the right-hand side of Eq. (22) must be multiplied by a "quality factor" which relates the true value of S to the classical value given by Eq. (8). However, for a specific value of S, it is clear from Eq. (22) that a good photon drag detector will have a high absorption per carrier. Moss (1971) showed the importance of a high absorption coefficient and listed values for a number of semiconductors at 10.6 μm.

In Eq. (22) only the effect of one type of charge carrier has been considered. In near-intrinsic semiconductors there will be an opposite sign voltage due to the effect of the minority carriers, which will reduce the responsivity. For this reason, most photon drag detectors are made from material in which one carrier is predominant. Equation (22) must be treated with some caution, although it is very convenient for assessing the potential voltage responsivity. The mobility of the semiconductor is also of importance. If we again assume only one type of charge carrier then

$$N = 1/\rho e \mu, \qquad (23)$$

where ρ is the resistivity of the material. Substituting Eq. (23) into Eq. (22), we find that

$$V_{\rm L}/W = n\mu\rho[1 - (\exp - KL)]/AC. \qquad (24)$$

Increasing the mobility and decreasing the resistivity by the same ratio will yield an identical voltage responsivity but from a lower resistance detector. As the noise voltage from photon drag detectors is predominantly Johnson noise, which is proportional to $R^{1/2}$ where R is the detector resistance, and R has a fixed relationship to ρ for a detector of given dimensions, it is clear from Eq. (24) that the noise voltage varies as $\mu^{-1/2}$. The noise equivalent power (NEP) of a detector is given by dividing the noise voltage by the voltage responsivity, so we find that the NEP is proportional to $\mu^{-1/2}$.

Equation (23) shows that $\rho \propto N^{-1}$, and thus from Eq. (21), $\rho \propto \sigma \propto R$ again for a detector of given dimensions. We find therefore that NEP $\propto$ $\mu^{-1/2}\sigma^{-1/2}$. Moss (1971) has shown that in semiconductors such as n-type germanium where absorption is phonon assisted, μ and σ are not independent and that increasing the mobility produces a proportionate decrease in the absorption cross section. However, as was shown earlier (Section III.B), this type of material is not useful for photon drag detectors at short wavelengths, as much of the momentum is taken up by the phonons. Where direct intraband transitions dominate, as in p-type germanium at 10.6 μm, there is no predictable relationship between μ and σ. It must be emphasized that the above considerations of semiconductor properties have to be considered in conjunction with the magnitude of the photon drag and optical rectification coefficients. There is little point in finding a semiconductor with high absorption coefficient and mobility unless it also has reasonable values of S, P, etc., at the required wavelength.

Returning to Eq. (3) we must next consider the optimum absorption length for a detector. As can be seen from this equation, it is not the same for photon drag as for optical rectification, because reflections from the back of the sample cause a decrease in photon drag but an increase from the optical rectification signal. Since the reflectivity depends on the refractive index, the optimum length will vary from one semiconductor to another, but in practice the differences are rather small and the minimum NEP is obtained with detectors having KL values of 1.5–2. A further parameter of importance is the actual resistance of the detector. This is significant for two reasons. The speed of response is usually limited by the detector resistance and lead capacitance (CR time constant), so that if the fastest response times are required, it is convenient to make the detector resistance 50 Ω. However, because it is difficult to make wide bandwidth amplifiers with an equivalent noise input as low as 50 Ω, it is often better where the ultimate response time is not important to have higher resistance detectors. From Eq. (16) it is clear for a given semiconductor that if, for example, we halve the number of carriers, this will lead to a doubling of the length (for an unchanged value of KL) and an increase of four in resistance and of two in the voltage responsivity. The NEP is therefore unchanged with the higher source resistance. One of the problems that arises is that the intrinsic carrier concentration of several potentially useful semiconductors, for example, indium antimonide and tellurium, is high because of the small energy gap. This leads to a high absorption per unit length and a low resistance device unless the area is made small. Moss (1971) discussed the advantages of cooling InSb and Panyakeow et al. (1972) made a p-type Te detector with a very high response at 77 K, but this does negate the great convenience of room temperature operation.

As mentioned at the beginning of this section, p-type germanium is still the most widely used material for CO_2 lasers. The reasons for this are clear from the above arguments and can be summarized by (i) its absorption per carrier at 10 μm is high at 6×10^{-16} cm^2, (ii) the mobility at 1800 cm^2 V^{-1} s^{-1} is good, (iii) the resistance of a convenient length and area detector with an absorption length of $\simeq 1.5$ is about 50 Ω. For example, a 5×5 mm cross-section detector, 30-mm long, of 5 Ω cm material has $KL = 1.5$ and $R = 48\ \Omega$. (iv) The photon drag coefficients are high because direct valence band transitions dominate the absorption. However, the coefficients fall rapidly below 10 μm and at the shortest CO_2 wavelength of 9.2 μm are rather low. Further advantages are that it is easily and reasonably cheaply available in the required resistivities, it is convenient to cut and polish, and contacts can be made without difficulty. Discussion of alternative detectors for 10 μm and suitable materials for other wavelengths is considered in detail in Section E.

3. *Speed of Response*

As mentioned above, the response time of photon drag detectors is usually determined by the CR time constant. Agavonov et al. (1974), in a detailed paper on the design of photon drag detectors, suggest that the transit time of the light down the detector may be the limiting factor. However, even with the high refractive index, light travels at 75 mm ns^{-1} in germanium, and as a typical detector is $\simeq 20$-mm long, the transit time is <0.3 ns. In the authors' experience it is difficult to reduce the CR time constant to much less than 0.5 ns for a reasonable responsivity detector. Except possibly in cooled detectors, the momentum relaxation time is very much shorter, being of the order of 10^{-12} s. Agavonov and his colleagues (1974) have also shown that the time for the establishment of an equilibrium in the system with a drag current and an electric field is effectively the dielectric relaxation time, and for this they calculate a value of 10^{-10}–10^{-11} s for p-type Ge. The response of a photon drag detector to a signal from a passively mode-locked laser is shown in Fig. 10.

4. Damage and Linearity

Two problems that arise with very high intensity radiation falling on semiconductors are damage to the surface and nonlinear response. The level at which damage occurs varies from one semiconductor to another and is dependent on the length of the laser pulse and particularly on the surface quality of the material. For well-polished germanium, intensities of

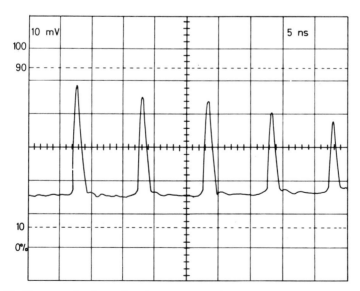

FIG. 10   Response of a photon drag detector to part of a mode-locked pulse from a CO_2 laser recorded on a Tektronix 7904 oscilloscope. Time base 5 ns per division (after Gibson *et al.*, 1974).

> 100 MW cm^{-2} are acceptable and even in less easily polished semiconductors, such as gallium phosphide, > 10 MW cm^{-2} is achievable. At these intensities linearity of the detectors is a more significant problem.

Nonlinear dependence of photon drag voltage on incident power density was first reported by Kamibayashi et al. (1973). Bishop et al. (1973) showed that the absorption of p-type germanium decreased at high intensities due to saturation of the absorption process, while the resistivity increased because of reduced mobility of the holes. As the first of these effects reduces the signal voltage and the second increases it, careful detector design leads to a linear response at intensities greater than 40 MW cm^{-2}. Bishop and her colleagues assumed homogeneously broadened absorption, but it was later shown (Keilmann, 1976; Phipps and Thomas, 1977) that the broadening is inhomogeneous. At the intensities available to Bishop et al. the degree of nonlinearity due to the absorption change was insufficient to distinguish between homogeneous and inhomogenous broadening, and so, conversely, only at their highest power levels are their results and conclusions significantly in error. Grave et al. (1978) have studied nonlinearity, and with the inclusion of inhomogenous broadening have obtained good agreement between theoretical predictions and experimental results.

It is a matter for discussion as to how important nonlinearity is in photon drag detectors, as it only becomes significant at intensities greater than $\simeq 5$ MW cm^{-2}. As the voltage responsivity of detectors is typically 0.1 V MW^{-1} cm^{-2}, actual outputs of ≈ 0.5 V are obtainable before nonlinearities occur, and if in a practical situation a detector is producing more than 0.5 V output, one should either attenuate the beam or use a larger area detector. The initial interest is attempting linearization to a level of a few volts was to be able to drive a direct-access high-speed oscilloscope with a bandwidth greater than 1 GHz. However, the development in recent years of very wide bandwidth amplifiers has reduced the need for high responsivity detectors. An alternative method of high-speed detection is to use the change in mobility of the carriers in p-type germanium to provide a photoconductive detector. Such a photoconductive effect was first observed (Moss and Hawkins, 1960) using conventional sources and was subsequently discussed as a detection process for lasers by a number of authors. A practical detector using a pulsed high-voltage bias to increase the photoconductive voltage responsivity has been described by Gibson et al. (1975) and has been marketed by Rofin under the name of a "hot hole detector." Output voltages of 10 V with a rise time of < 500 ps are obtainable.

C. LONGITUDINAL MONITORS

One distinct advantage of photon drag detection is the ability to make on-line monitors. These will typically absorb 20–25% of the power with

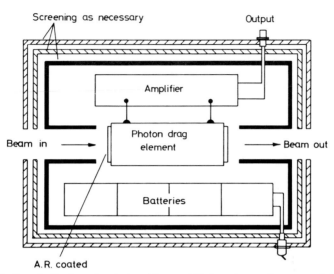

FIG. 11 Longitudinal monitor designed for use at 10.6 μm close to the output of a low-power pulsed laser (after Kimmitt *et al.*, 1972).

longitudinal monitors and 1–15% with transverse devices. Kimmitt *et al.* (1972) have described a simple monitor for use close to a pulsed CO_2 laser; their design is shown in Fig. 11. As photon drag detectors are normally low impedance and have no biasing circuits, pickup problems are not severe. However, near the hostile environment of a high-power pulsed TEA laser, sensible screening precautions must be taken, particularly if the monitor requires subsequent amplification.

Even with the highest resistivity *p*-type germanium that can be used, without significant opposite sign signals due to the minority carriers, the absorption is significant. For example, 30 Ω cm *p*-type Ge has an absorption coefficient of 0.07 cm^{-1} at 10 μm, and a 5 cm length will absorb 30% of the signal. If this monitor is to observe all the signal from a 5 × 5 cm cross-section laser (without focusing the output), it needs to have a 5-cm-square cross section itself, and the monitor will be a cube with contacts at opposite faces. Because of its shape, the field set up by radiation near the center will be significantly different from that produced by the same power at the edge. For this reason, it is convenient with large area monitors to utilize the transverse photon drag effect.

D. TRANSVERSE DETECTORS AND MONITORS

Detectors and monitors exploiting transverse photon drag and optical rectification coefficients suffer from the major disadvantage that the responsivity is polarization dependent. However, if the laser is polarized, they

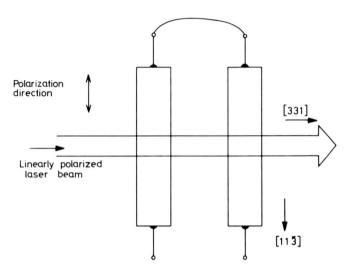

FIG. 12   Schematic diagram of a design for a high-sensitivity photon drag detector proposed by Hattori *et al.* (1973b). Note that maximum response is obtained in *p*-type germanium with the electrodes and light polarization in the 11$\bar{3}$ direction and the laser beam propagating in a 332 direction.

offer significant advantages. As we pointed out (Section B.2) a number of potentially useful semiconductors have such a high intrinsic carrier concentration that they cannot be used conveniently in longitudinal detectors. By using very thin plates, it can be seen from Eq. (4) that a high-impedance detector with good voltage responsivity can be made from such material. Alternatively, it is possible to make monitors that absorb only a very small percentage of the laser power.

Hattori *et al.* (1973b) suggested making a very high responsivity detector by the method shown in Fig. 12. The light from the laser passes through a large number of the antireflection coated plates, which are electrically connected in series. At 10.6 μm, using 1000 plates each made from 20 Ω cm *p*-type germanium 0.1-mm thick, they calculated a voltage responsivity of 400 μV W^{-1}, which is several hundred times greater than from a longitudinal detector. However, it would be a very expensive and complex device to construct. Ribakovs and Gundjian (1978) have constructed a very useful transverse monitor from tellurium. They have divided their detector into strips in the form shown in Fig. 13. The advantage of this can be seen if we rewrite Eq. (4) in terms of the voltage responsivity when, neglecting the reflection terms,

$$V_T/W = (y/A)[D + (P, S)];\qquad(25)$$

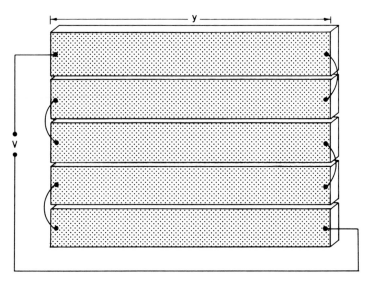

FIG. 13   Design for a high-responsitivity tellurium monitor (after Ribakovs and Gundjian, 1978). In the example shown here, the voltage responsivity would be increased by a factor of five.

y increases as the number of sections and the responsivity increases proportionately. It should be noted that the resistance increases as y^2 so that the NEP is unchanged. However, this method of increasing the source resistance is particularly useful with lower resistivity material.

Gatenby *et al.* (1979) have proposed two methods (Fig. 14) by which an n-type GaP monitor can be used within the cavity of a HF laser. Such monitors would not be possible with photon drag where the reverse signal would cancel the forward one, but as Fig. 8 indicates, optical rectification is the dominant component in n-type GaP. The arrangement shown in Fig. 14a was successfully used by Gatenby and his colleagues, but that shown in Fig. 14b was unsuccessful due to imperfections in the surface quality and parallelism of the GaP plate. Intercavity devices of this type provide a particularly elegant way of monitoring the output of lasers. However, pickup problems can be considerable when using them with high power pulsed lasers.

E.   SPECIFIC DETECTORS FOR 1–1000 μm

Photon drag detectors have been used over the entire wavelength range from 1 to 1200 μm. At a specific wavelength it is not difficult to choose an optimum detector, but even then the choice is dependent on availability. For example, tellurium is marginally better than p-type germanium at 10 μm, but it is not easy to obtain in the correct doping, correct orientation is complex, and it is easily damaged. Therefore, germanium remains the favored

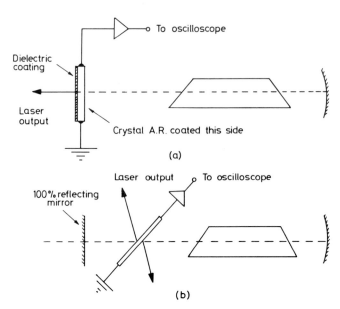

FIG. 14   Two methods of using optical rectification for intercavity monitoring of HF lasers
proposed by Gatenby *et al.* (1979).

detector. It is more difficult to decide on a detector for a range of wavelengths
where uniform response is usually the major consideration. Added to this
are the questions of whether high responsivity is more important than high
speed, and there are problems of damage, linearity, etc. However, we have
drawn together the available data for useful detectors to cover the entire
infrared and this information is summarized in Figs. 15 and 16. In these two
diagrams the solid sections of the curves cover actual data points. The dotted
sections are in regions where experimental results are not yet available, but
the shape can be predicted with reasonable confidence from theory.

For all the detectors the cross-sectional area is standardized at $4 \times 4\,\text{mm}$;
there are no antireflection coatings, and the contact region, where light is
lost with no voltage produced, is taken as 1 mm at each end of the detector.
The resistance of the short wavelength detectors is designed to be 50 Ω, but
this criterion was not adhered to at long wavelengths, where laser powers
tend to be lower and speeds of response of a few, rather than one nanosecond,
are usually acceptable. At the shortest wavelengths, p-GaAs is the best
detector and as can be seen from Fig. 15 it should cover the whole range of
1–11 μm with a reasonable responsivity. For 2–11 μm, n-GaP is a good
choice. The peak response around 3 μm, due to optical rectification, makes
it particularly useful for the important HF laser wavelengths. Severe lattice

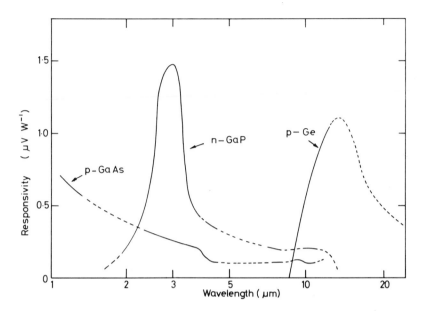

FIG. 15 Responsivity of 3.2 Ω cm p-Ge, 5 Ω cm n-Gap, and 2.5 Ω cm p-GaAs longitudinal detectors. All the detectors have an active area of 4 × 4 mm, a resistance of 50 Ω, and are oriented in a 111 direction). Responsivity of 1 μV W^{-1} is equivalent to a NEP of $\simeq 10^{-3}$ W Hz$^{-1/2}$.

absorption sets in at 12 μm and this extends to around 25 μm. It is possible to design detectors to operate in the lattice absorption region by using a high enough doping for the free carrier absorption to dominate. However, this usually leads to an inconveniently low resistance and low-voltage responsivity detector. p-type germanium cannot be used with high-power lasers at wavelengths below about 3 μm, because of the onset of two-photon absorption. In the 3–9 μm range the rapid change in responsivity and the five sign reversals (shown in Fig. 6) make it an inconvenient detector, but at specific wavelengths it can be usefully employed. Above 10 μm the responsivity rises, but lattice absorption becomes significant at $\simeq 16$ μm and remains quite high· to above 100 μm (Loewenstein et al., 1973). Even so, n-type Ge is a good detector to beyond 20 μm as shown in Fig. 15.

At longer wavelengths (Fig. 16) p-type silicon appears to be the best available detector. Its mobility and absorption cross section are reasonable, and as shown earlier (Section III.E) the nature of the direct valence band transitions leads to high photon drag coefficients. Lattice absorption in silicon extends from about 15–35 μm, but is not high enough between 24 and 35 μm to prevent useful detectors being made. At very long wavelengths

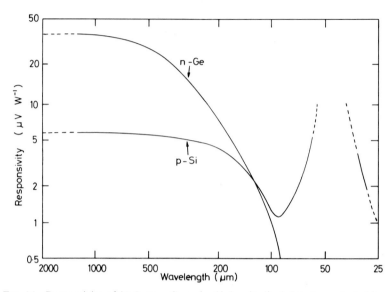

FIG. 16   Responsivity of 30 Ω cm n-Ge and p-Si, longitudinal detectors oriented in a 100 direction. The active area is 4 × 4 mm and the detector resistances are n-Ge, 250 Ω, and p-Si, 350 Ω. Responsivity of 10 μV W^{-1} is equivalent to a NEP of $\simeq 2 \times 10^{-4}$ W Hz$^{-1/2}$ for n-Ge and $\simeq 2.5 \times 10^{-4}$ W Hz$^{-1/2}$ for p-Si.

the higher mobility of n-type Ge coupled with a high absorption per carrier (Birch *et al.*, 1974) makes it the best detector.

F.   CALIBRATION STANDARDS FOR 100–2000 μm

A specific problem associated with submillimeter and short millimeter waves is lack of convenient devices for measuring power. Kimmitt *et al.* (1978) suggested that photon drag could provide standard detectors for pulsed and low-power cw lasers at wavelengths greater than 100 μm. The advantages are: (i) detectors with areas large compared with the wavelength, (ii) uniform responsivity between detectors, (iii) fast response time, (iv) room temperature operations, and (v) the detectors are robust and easily shielded from pickup. This suggested use for photon drag can now be given further consideration using the experimental results reported by Gibson *et al.* (1980).

At very long wavelengths where $\omega\tau < 1$, we have shown (Section III.A) that photon drag reduces to classical radiation pressure and, when this is the case, the voltage responsivity is correctly given by Eq. (24). As the absorption of free carriers is high at these wavelengths, it is easy to arrange for the exponential term to be near to zero, and for this condition

$$V_{\mathrm{L}}/W = n\mu\rho/AC. \tag{26}$$

It is therefore possible to design detectors for this region with responsivities that depend only on well-known parameters of the semiconductor. Moreover, the response is independent of wavelength and Fig. 16 shows that the experimental results provide confirmation of this situation.

The important question is whether the detectivity of photon drag devices is sufficient to provide useful calibration standards in this region. If we consider a 4×4 mm area detector made from n-type germanium with the responsivity shown in Fig. 16, the calculated NEP at 1 mm is 6×10^{-5} W $Hz^{-1/2}$. With such a detector it should be possible to detect pulsed laser powers of a few watts with a bandwidth of 10^8 Hz and cw powers of $< 100 \, \mu W$. Results achieved by the authors and their colleagues using a pulsed laser at 1.2 mm, and by Röser (1979) with cw lasers in the submillimeter region, show that in practice powers of < 10 W pulsed and < 1 mW cw can be measured with accuracy better than 10%. Although n-type germanium gives the best NEP at 1 mm, it can be seen from Fig. 16 that for a uniform response in the submillimeter region, p-type silicon is a better choice. The reason is that $\omega \tau < 1$ at a shorter wavelength, but this uniform response detector has a poorer NEP.

A further advantage of silicon is that resistivities of $> 1000 \, \Omega$ cm are readily available, and with these it is possible to make large on-line monitors for long wavelengths. In this case the exponential term in Eq. (24) must be considered, and the responsivity will change with wavelength as the absorption changes. A monitor with a diameter of 3 cm and 10 cm long made from 2000 Ω cm n-type silicon has a responsivity of 2.5 $\mu V \, W^{-1}$, NEP of 2.7 $\times 10^{-3}$ W $Hz^{-1/2}$, and absorbs about 20% of the laser power at 1 mm wavelength. One difficulty of such a monitor is to provide antireflection coatings at these long wavelengths. With a monitor of similar specification, but without AR coatings, the authors have obtained an excellent signal to noise at 490 μm with a pulsed power of 1 KW and a bandwidth of 10^7 Hz.

G. OPTICAL BRIDGES

One interesting feature of photon drag (but not optical rectification) detectors is that the sign of the signal depends on the direction of the light. Patel (1971) described a method of pulse length measurement using an optical bridge in which the laser beam was split by a semireflecting mirror and the two halves passed in opposite directions through an indium arsenide photon drag detector. When the detector crystal was positioned to receive the two pulses simultaneously, there was zero signal. However, the speed of such a linear optical bridge is limited by the bandwidth of the display system. Patel suggested ways of overcoming this speed limitation using high-speed diodes to rectify the photon drag voltages, and a similar bridge was proposed by Agavonov et al. (1974). However, such bridges require not only a very fast

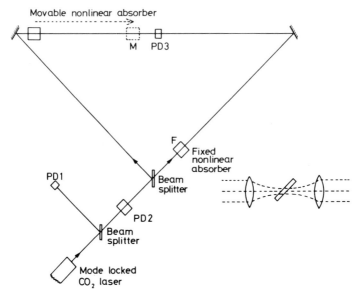

FIG. 17 Schematic diagram of an optical bridge for measuring laser pulse lengths and studying the characteristics of nonlinear absorbers (after Gibson *et al.*, 1972). The inset shows the arrangement for focusing the light onto the nonlinear absorbers.

rectifying system but voltages sufficient to use the diodes in their linear range.

An alternative approach to optical bridges was employed by Gibson *et al.* (1972) to determine the length of mode-locked CO_2 laser pulses. This bridge relies on the saturation of the absorption of *p*-type germanium at high power, and a schematic diagram is shown in Fig. 17. To understand the principal of the bridge, it is convenient to consider it initially without the nonlinear absorbers, that is, when it is similar to that discussed by Patel (1971). A small part of the laser output is extracted by the beam splitter and focused onto a photon drag detector PD1, which triggers an oscilloscope. The main part of the beam is monitored by PD2, then split into two equal parts and recombined at the monitor PD3. As the pulses are traveling in opposite directions, and assuming that their arrival at PD3 is coincident in time, there will be zero output from the detector. As with Patel's arrangement the resolution is limited by the response time of the display equipment, and with post-detector integration, the detector PD3 can be displaced from the center of the bridge and still give zero output.

For pulse duration measurements the bridge is completed by the inclusion of two (ideally identical) nonlinear absorbers. These components must have a response time considerably less than the duration of the pulses to be measured. p-type germanium, with a saturation power level of about 5 MW cm^{-2} and a scattering time well below 10^{-11} s at room temperature, was used in this bridge. Each absorber was mounted, as shown in the inset of Fig. 17, with two lenses to bring the beam to a focus and increase the intensity on the germanium. Now assume that PD3 is not at the center of the bridge and has a slow-speed integrator. The nonlinear absorber M can be moved to cover a range that includes the bridge center, and initially it is in a position well away from the center. In this situation there is a null output from PD3. As the absorber M is moved to the center of the bridge, the pulses, traveling in opposite directions, overlap in time to an increasing extent and the pulse from the right reduces the absorption of the absorber as seen by the pulse from the left. The bridge goes out of balance, the maximum out of balance signal being given when the moveable nonlinear absorber is at the center of the bridge. It is of interest to note that the discrimination factor of the bridge is ideally infinite but, in fact, is limited by noise. The out-of-balance signal is, to a good approximation, directly proportional to the second-order correlation function, provided certain relatively simple precautions are taken. With 3 ns pulses, the shortest that could be accurately measured on the oscilloscope, no difference was found between the shape measured using the bridge and that measured directly.

The bridge can also be used to study the characteristics of a nonlinear absorber by making it the moveable absorber far from the center of the bridge. The fixed, nonlinear absorber F is replaced by a linear, variable absorber and the bridge balanced at low-power levels. Any out-of-balance at high powers is a direct measure of the change in the absorption coefficient of the nonlinear material. With a typical CO_2 mode-locked pulse chain there are 20 to 30 peaks, varying in power by a factor of ten or more, and it is possible to measure the nonlinearity as a function of power with a single laser pulse.

REFERENCES

Abraham, M. (1914). "Theorie de Elektrizitat," 3rd ed., Sect. 38 and 39. Teubner, Leipzig.
Agavonov, V. G., Valov, P. M., Ryvkin, B. S., and Yaroshetskii, I. D. (1974). *Sov. Phys.-Semicond.* 7, 1540.
Al-Watban, F. A., and Harrison, R. G. (1977). *J. Phys. D: Appl. Phys.* 10, L249.
Ashkin, A., and Dziedzic, J. M. (1973). *Phys. Rev. Lett.* 30, 139.
Barlow, H. E. (1958). *Proc. IRE* 46, 1411.
Bassioni, M. (1975). Private communication.
Birch, J. R., Bradley, C. C., and Kimmitt, M. F. (1974). *Infrared Phys.* 14, 189.
Bishop, P. J., Gibson, A. F. and Kimmitt, M. F. (1973). *IEEE J. Quant. Electron.* QE-9, 1007.
Braunstein, R., and Kane, E. O. (1962). *J. Phys. Chem. Solids* 23, 1423.
Briggs, H. B., and Fletcher, R. C. (1953). *Phys. Rev.* 91, 1342.
Brynskikh, N. A., and Sagdullaeva, S. A. (1978). *Sov. Phys. Semicond.* 12, 467.

Burt, M. G. and Peierls, R. (1973). *Proc. R. Soc. London Ser. A* **333**, 149.

Cameron, K., Gibson, A. F., Giles, J., Hatch, C. B., Kimmitt, M. F., and Shafik, S. (1975). *J. Phys. C: Solid State Phys.* **8**, 3137.

Danishevsky, A. M., Kastal'skii, A. A., Tyvkin, S. M., and Yaroshetskii, I. D. (1970). *Sov. Phys. JEPT* **31**, 292.

Doviak, J. M., and Kothari, S. (1975). *Proc. Int. Conf. Phys. Semicond., 12th, Stuttgart, 1974* p. 1257.

Elliott, R. J., and Gibson, A. F. (1974). "Solid State Physics and Its Applications," Chapters 2 and 6. Macmillan, New York.

Fawcett, W. (1965). *Proc. Phys. Soc.* **85**, 931.

Gatenby, P. V., and Kar, A. K. (1978). *Opt. Quant. Electron.* **10**, 153.

Gatenby, P. V., Kar, A. K., and Kimmitt, M. F. (1979). *IEEE J. Quant. Electron.* **QE-15**, 69.

Gibson, A. F., and Montasser, S. (1975). *J. Phys. C: Solid State Phys.* **8**, 3147.

Gibson, A. F., and Serafetinides, A. P. (1977). *J. Phys. C: Solid State Phys.* **10**, L107.

Gibson, A. F., and Walker, A. C. (1971). *J. Phys. C: Solid State Phys.* **4**, 2209.

Gibson, A. F., Kimmitt, M. F., and Walker, A. C. (1970a). *Appl. Phys. Lett.* **17**, 75.

Gibson, A. F., Kimmitt, M. F., and Walker, A. C. (1970b). *Proc. Int. Conf. Phys. Semicond., 10th* p. 690. U.S. Atomic Energy Commission.

Gibson, A. F., Rosito, C. A., Raffo, C. A., and Kimmitt, M. F., (1972). *J. Phys. D: Appl. Phys.* **5**, 1800.

Gibson, A. F., Kimmitt, M. F., and Norris, B. (1974). *Appl. Phys. Lett.* **24**, 306.

Gibson, A. F., Kimmitt, M. F., Maggs, P. N. D., and Norris, B. (1975). *J. Appl. Phys.* **46**, 1413.

Gibson, A. F., Hatch, C. B., Maggs, P. N. D., Tilley, D. R., and Walker, A. C. (1976). *J. Phys. C: Solid State Phys.* **9**, 3259.

Gibson, A. F., Hatch, C. B., Kimmitt, M. F., Kothari, S., and Serafetinides, A. P. (1977). *J. Phys. C: Solid State Phys.* **10**, 905.

Gibson, A. F., Kimmitt, M. F., Koohian, A. D., Evans, D. E., and Levy, G. F. D. (1980). *Proc. R. Soc. London Ser. A* **370**, 303.

Gordon, J. P. (1973). *Phys. Rev. A* **8**, 14.

Grave, T., Wurz, H., Schneider, W., and Hubner, K. (1978). *Appl. Phys.* **15**, 89.

Grinberg, A. A. (1970). *Sov. Phys. JEPT* **31**, 531.

Gulyaev, Yu. V. (1968). *Radiat. Eng. Electron. Phys.* **13**, 599.

Gurevich, L. E., and Rumyantzev, A. A. (1967). *Sov. Phys.-Solid State* **9**, 55.

Hammond, C. R., Jenkins, J. R., and Stanley, C. R. (1972). *Opto-Electron.* **4**, 189.

Hattori, H., and Umeno, M. (1975). *Jpn. J. Appl. Phys.* **14**, 35.

Hattori, H., Umeno, M., Jimbo, T., Fujitani, D., and Miki, S. (1973a). *J. Phys. Soc. Jpn.* **35**, 826.

Hattori, H., Umeno, M., Jimbo, T., Fujitani, D., and Miki, S. (1973b). *IEEE J. Quant. Electron.* **QE-9**, 663.

Jones, R. V. (1978). *Proc. R. Soc. London Ser. A*, **360**, 365.

Jones, R. V., and Leslie, B. (1978). *Proc. R. Soc. London Ser. A* **360**, 347.

Kamibayashi, T., Yonemochi, S., and Miyahawa, T. (1973). *Appl. Phys. Lett.* **22**, 119.

Kane, E. O. (1956). *J. Phys. Chem. Solids* **1**, 83.

Keilmann, F. (1976). *IEEE J. Quant. Electron.* **QE-12**, 592.

Kimmitt, M. F., Tyte, D. C., and Wright, M. J. (1972). *J. Phys. E: Sci. Instrum.* **5**, 239.

Kimmitt, M. F., Serafetinides, A. A., Roser, H. P., and Huckridge, D. A. (1978). *Infrared Phys.* **18**, 675.

Lawaetz, P. (1975). *Solid State Commun.* **16**, 65.

Loewenstein, E. V., Smith, D. R., and Morgan, R. I. (1973). *Appl. Opt.* **12**, 398.

Minkowski, H. (1910). *Math. Ann.* **68**, 472.

Montasser, S. S. (1976a). Ph.D. Thesis, Univ. of Essex.

Montasser, S. S. (1976b). *J. Phys. C: Solid State Phys.* **9**, L93.
Moss, T. S. (1971). *Phys. Status Solidi (a)* **8**, 223.
Moss, T. S. and Hawkins, T. D. F. (1960). *Proc. Phys. Soc.* **76**, 565.
Nelson, D. F., and Turner, E. H. (1968). *J. Appl. Phys.* **39**, 3337.
Onton, A. (1971). *Phys. Rev. B* **4**, 4449.
Paige, E. G. S. (1966). *J. Phys. Soc. Jpn. Supple.* **21**, 397.
Panyakeow, S., Shirafugi, J., and Inuishi, Y. (1972). *Appl. Phys. Lett.* **21**, 314.
Patel, C. K. N. (1971). *Appl. Phys. Lett.* **18**, 25.
Peierls, R. (1976). *Proc. R. Soc. London Ser. A.* **345**, 343.
Phipps, C. R., and Thomas, S. J. (1977). *Opt. Lett.* **1**, 93.
Ribakovs, G., and Gundjian, A. A. (1974). *Appl. Phys. Lett.* **24**, 337.
Ribakovs, G., and Gundjian, A. A. (1977). *J. Appl. Phys.* **48**, 4601, 4609.
Ribakovs, G., and Gundjian, A. A. (1978). *IEEE J. Quant. Electron.* **QE-14**, 42.
Röser, H. P. (1979). Private communication.
Schneider, W., and Hubner, K. (1975). *Phys. Lett.* **53A**, 87.
Serafetinides, A. A., Kar, A. K., and Kimmitt, M. F. (1978). *Opt. Laser Technol* **8**, 243.
Smith, R. A. (1979a). "Semiconductors," 2nd ed., p. 157. Cambridge Univ. Press, London and New York.
Smith, R. A. (1979b). "Semiconductors," 2nd ed., p. 299. Cambridge Univ. Press, London and New York.
Valov, P. M., Ryvkin, B. S., Ryvkin, S. M., Titova, E. V., and Yaroshetskii, I. D. (1972a). *Sov. Phys: Semicond.* **6**, 99.
Valov, P. M., Ryvkin, B. S., Ryvkin, S. M., and Yaroshetskii, I. D. (1972b). *Phys. Status Solidi* **53**, 65.
Valov, P. M., Vasil'ev, B. I., Dyad'kin, A. P., and Yaroshetskii, I. D. (1978). *Sov. Phys. Semicond.* **12**, 762.
Walker, G. B., and Lahoz, D. G. (1975). *Nature (London)* **253**, 339.
Wiley, J. D., and DiDomenico, M. Jr. (1970). *Phys. Rev. B* **1**, 1655.
Wiley, J. D., and DiDomenico, M. Jr. (1971). *Phys. Rev. B* **3**, 375.
Zernike, F., and Midwinter, J. E., (1973). "Applied Nonlinear Optics." Wiley, New York.

CHAPTER 5

Electrically Excited Submillimeter-Wave Lasers

*F. K. Kneubühl and Ch. Sturzenegger**

Physics Department
Eidgenössische Technische Hochschule
Zurich, Switzerland

* Present address: Sprecher and Schuh, AG, CH-5036, Oberentfelden, Switzerland.

I. History

A. THE GAP BETWEEN THE INFRARED AND MICROWAVES

Studies in the far infrared and in the submillimeter-wave region started with the work of Rubens and co-workers (Rubens 1922a, b; Palik, 1960, 1971, 1977; Barr, 1965) who reached this spectral area with thermal sources from the infrared end at the beginning of this century. The most remarkable achievement was the experimental determination of the temperature dependence of the relative radiance of a blackbody at numerous near- and far-infrared wavelengths (Rubens and Kurlbaum, 1900, 1901; Rubens and Michel, 1921), and its theoretical interpretation by Planck (1900, 1901) on the basis of cavity oscillators with quantized energies. In the 1920s the initial attempts to reach the submillimeter-wave region from the microwave end with Hertz oscillators were reported by Glagolewa–Arkadiewa (1924) as well as by Nichols and Tear (1923, 1925). They accomplished a first, although unsatisfactory, bridging of the microwaves and the far infrared (Palik, 1977).

Since then, a large number of gifted high-frequency and microwave engineers examined almost every phenomenon in physics for possible application to the problem of generating coherent submillimeter waves. The deplorable state of the art of generating submillimeter waves around 1960 was competently reviewed by Coleman (1963) and later by Moser et al. (1968, 1969). In spite of numerous and serious efforts, the submillimeter-wave power gap between the infrared and the microwave remained unchallenged until the early 1960s. The history of coherent submillimeter waves began around 1963–1964 with extension of the "Carcinotron type 0" millimeter-wave backward-wave oscillators of the Compagnie Générale de Télégraphie sans Fil (CSF) down to submillimeter waves, and with the discovery of the noble-gas and molecular far-infrared gas lasers.

B. THE FIRST SUBMILLIMETER-WAVE LASERS

Within a year of its discovery (Javan et al., 1961) the He–Ne laser was extended toward the far infrared, out to a wavelength of 28 μm in 1962 (McFarlane et al., 1964a). Research on submillimeter-wave lasers started two years later. In 1964 continuous noble-gas lasers with Ne attained 85 μm in January (McFarlane et al., 1964b) and 133 μm in March (Patel et al., 1964). Simultaneously, the first polyatomic molecular pulsed submillimeter-wave lasers came into operation. Stimulated emissions of H_2O out to 78 μm were reported in January (Crocker et al., 1964) and out to 118 μm in April (Gebbie et al., 1964a), as well as those of D_2O out to 72 μm. Continuous operation of the H_2O laser was achieved a few months later (Witteman and Bleekrode, 1964). Probably the most promising discovery in this art was the first observation of the relatively strong laser emission of HCN and related molecules in May (Gebbie et al., 1964b). This definitely proved that population inversion and lasing can be achieved by a glow discharge in a gas of polyatomic molecules. The range of stimulated emissions from electrically excited polyatomic molecules could be expanded out to 538 μm (Steffen et al., 1966a) and out to 774 μm (Steffen et al., 1966b) with a gas containing ICN in 1966. In the same year continuous operation of the 337-μm HCN laser was achieved (Gebbie et al., 1966; Flesher and Mueller, 1966). Later, only a few further polyatomic molecules which give stimulated submillimeter-wave emissions in a glow discharge were discovered, SO_2 in 1968 (Dyubko et al., 1968; Hard, 1969; Hassler and Coleman, 1969), H_2S in 1969 (Hassler and Coleman, 1969), and H_2CO in 1972 (Okajima and Murai, 1972).

The list of molecules that lase in the far infrared when excited electrically is still rather small. While definitive studies would be required to answer the questions of why so few molecules are present in this list, one could mention as limiting factors the problems of relative excitation cross sections, unfavorable relaxation rates, excessive dissociation, etc. These problems will be discussed in Sections E and IV.

C. EARLY APPLICATIONS

The 337-μm HCN laser was relatively intense compared to other sources available at the time of its discovery. Hence, it was used for making submillimeter-wave measurements of absorption coefficients and indexes of refraction shortly after its first operation. As examples, we may mention the determination of the 337-μm complex dielectric constant of monohalogen-substituted benzenes in the liquid phase (Gebbie et al., 1965) and of the 337-μm atmospheric transmission which gave a clear-atmosphere attenuation of 50 dB km^{-1} at 0°C and saturated vapor pressure (Burroughs et al., 1966). In solid state physics, early applications of the HCN laser were the cyclotron resonance in semiconductors (Button et al., 1966, 1968; Murotani and Nisida, 1970) and high-resolution electron paramagnetic resonance spectroscopy (epr) in holmium ethyl sulfate (Boettcher et al., 1968).

Submillimeter-wave laser epr spectroscopy was also performed with gaseous O_2 (Evenson et al., 1968). Another interesting early application of the HCN laser was the spatially resolved measurement of the electron density in a helium steady-state plasma with a 337-μm interferometer (Kon et al., 1968).

The HCN and H_2O lasers represented not only the first emitters of intense submillimeter waves but also the first coherent sources in this part of the electromagnetic spectrum. Hence, they allowed the first precise frequency and wavelength measurements in the far infrared. Hocker et al. (1967)

succeeded in harmonically mixing in a silicon diode the 12th harmonic of a klystron working at around 74,230 MHz and the 337-μm laser line of HCN. By precise counting of the beat note, they found a frequency of 890,759.5 MHz for the 337-μm HCN emission. This work was extended to the H_2O laser by Frenkel et al. (1967), who determined a frequency of 2,527,952.8 MHz for the 118-μm emission.

Simultaneously with the frequency measurements performed by U.S. scientists, Gebbie et al. (1967) demonstrated that the HCN laser can be used as a coherent source for submillimeter-wave heterodyne detection. Laser harmonic frequency mixing of two different far-infrared laser lines was achieved by Hocker and Javan (1968a,b) the following year.

D.   LASER RESONATORS

An interesting aspect of the first submillimeter-wave lasers is the resonator paradox. Both the noble-gas lasers and the molecular lasers were equipped with resonators of very low Fresnel numbers $N = a^2/\lambda b$, where λ indicates the wavelength of the laser radiation, a the radius, and b the length of the resonator. Examples to be mentioned are $N \simeq 0.9$ for the 85-μm Ne laser (McFarlane et al., 1964b) and $N \simeq 0.3$ for the 337-μm HCN laser (Gebbie et al., 1964b). Hence the diffraction losses of the resonators used in these lasers were tremendous. In spite of this fact and of the low gain of the laser gases, these submillimeter-wave lasers worked well. Nobody seemed to be aware of this paradox. As late as 1967, Schwaller et al. (1967) first noticed and demonstrated by laser resonator interferometry the discrepancy between diffraction theory (Bergstein and Schachter, 1964) and the actually observed resonator modes and losses in submillimeter-wave lasers. Subsequently, Steffen and Kneubühl (1968) proved experimentally and theoretically that this discrepancy was due to waveguiding effects of the dielectric glass walls of the plasma tubes used in these lasers. Their theory was analogous to the calculations and suggestions of Marcatili and Schmeltzer (1964) who had proposed the construction of waveguide lasers for the visible and the near infrared.

By accident, these early submillimeter-wave lasers evloved as the first waveguide lasers in operation (Kneubühl, 1977; Kneubühl and Affolter, 1979). The first optical waveguide laser devised along the guidelines of Marcatili and Schmeltzer was built and operated some years later by Smith (1971).

E.   LASER TRANSITIONS AND MECHANISMS

The assignment of the submillimeter-wave emissions of the noble-gas lasers to the corresponding atomic transitions was performed at the moment they were discovered (McFarlane et al., 1964b; Patel et al., 1964). On the other hand, the assignment of the submillimeter-wave lines obtained from the new molecular-gas lasers was puzzling and controversial (Garrett, 1965, 1966). The spectroscopy of H_2O and HCN had been studied for many years, and the energy levels of these molecules was well known for the low-lying states. Furthermore, one was aware of the fact that the bending modes of a triatomic molecule relax faster than the stretching modes; hence a population inversion might exist between these two modes even for nonselective excitation. Also, for transition probabilities between normal modes to exist, some interaction was required to couple them together, e.g., Coriolis coupling. Consequently, all the pieces were available to make laser line assignments in H_2O and HCN as early as 1964. Yet, three years passed before Lide and Maki (1967) succeeded in making the first reliable line assignment for a submillimeter-wave molecular-gas laser (HCN).

The first model for the laser emissions of HCN and related molecules was proposed by Chantry et al. (1965) in which the 337-μm line was attributed to a rotational transition of the CN radical. This theory was discussed passionately by a number of authors (Mathias et al., 1965; Broida et al., 1965; Steffen et al., 1966a–c; Hocker et al., 1967). This affair was settled when Lide and Maki (1967) attributed the lines at 310, 311, and 337 μm to vibrational and rotational transitions in the electronic ground state of the HCN molecule, e.g., 11^10, $J = 10 \rightarrow 04^00$, $J = 9$ for the 337-μm emission. Absolute frequency measurements by Hocker and Javan (1967) confirmed this identification.

When Lide and Maki (1967) made the line assignments in HCN, they were guided by the idea of irregular perturbations. Two levels of a bending and a stretching mode of a molecule having the same J, identical symmetry, and about equal energy can have Coriolis interaction. The corresponding wave functions mix, which results in a drastic increase of the transition probability between the levels associated with the perturbations. Laser lines are then looked for where irregular perturbations occur. This principle allowed laser line assignments to be made for H_2O (Hartmann and Kleman, 1968; Benedict et al., 1969) and SO_2 (Hubner et al., 1971; Hassler et al., 1973). It will probably also apply to the H_2S laser (Hassler and Coleman, 1969) where assignments have yet to be made. Nobody yet knows whether the scheme of irregular perturbations applies also to the laser emissions out to 774 μm obtained from a mixture of ICN and H_2 (Steffen et al., 1966a, b; Steffen and Kneubühl, 1968). According to Coleman (1977) this represents the first three-atom molecular chemical laser. However, the problem was not pursued as a chemical laser at the time.

Present data on far-infrared molecular lasers allow the following statements [see also Coleman (1973)]:

(i) Few three- and four-atom molecules can be pumped in glow discharge.

(ii)   For producing far infrared, optical pumping with the present fixed-frequency pumps is more probable with five or more atom molecules.

Once the laser lines are assigned, the energy levels that must be populated by the glow discharge are known, but, in general, the excitations and transition probabilities remain unknown. Unfortunately, the basic problems associated with the glow-discharge pumping of infrared and submillimeter-wave molecular lasers involve plasma chemistry of polyatomic molecular-gas discharges, which is a difficult subject (Bekefi, 1976).

One clue to the excitation problem is the measurement of the time delay of the laser signal versus the discharge current (Steffen and Kneubühl, 1968; Yamanaka et al., 1971; Coleman, 1973; Kunstreich and Lesieur, 1974). The time delay is determined by the laser mechanism, as well as by the temporal variation of the plasma parameters, e.g., index of refraction (Steffen and Kneubühl, 1968; Withbourn et al., 1972; Tait et al., 1973) and the density decay and dynamics of electrons (Turner and Poehler, 1971; Turner, 1977). It was found that of the four molecules, HCN, H_2O, H_2S, and SO_2, only H_2O lases during the exciting current pulse (Coleman, 1973). This suggests that H_2O is directly excited, while the other molecules are excited indirectly. Calculations made on a simple model for the H_2O emission (Pichamuthu et al., 1971; Coleman 1973), assuming direct electron impact excitation, correlate fairly well with experiments. For SO_2 an excitation mechanism involving excitation to the first excited singlet electronic state, followed by crossing to the highly vibrationally excited levels of the ground electronic state, with subsequent selective vibrational relaxation, is consistent with the experimental data (Hassler et al., 1973).

F.   OPTICALLY PUMPED SUBMILLIMETER-WAVE LASERS

The first optically pumped submillimeter-wave laser was operated by Chang and Bridges (1970) by exciting the CH_3F molecule with 9.55-μm CO_2 laser radiation. Within a few years a large number of new, optically pumped laser emissions in the far infrared were reported (Chang, 1974). Additional impetus to this development was given by Hodges and Hartwick (1973) who introduced for the first time a waveguide resonator (Marcatili and Schmeltzer, 1964; Kneubühl, 1977; Kneubühl and Affolter, 1979) in an optically pumped submillimeter-wave laser. Today, hundreds of stimulated emissions (Yamanaka, 1976; Rosenbluh et al., 1976; Chang, 1977a; Knight, 1978) between 0.5443 μm (Byer et al., 1972) and 1965.34 μm (Chang and McGee, 1976) can be obtained by optical pumping with lasers. Furthermore, Oka and Shimizu (1970) have achieved a cw microwave emission from CD_3CN at 19,075.7 μm by double-photon pumping with a K-band klystron.

In optically pumped far-infrared lasers, waveguide resonators have become a common feature (Yamanaka, 1977).

Although the generation of a rich spectrum of lines has not proven to be difficult, the production of intense far-infrared radiation by optically pumping polar molecules is a difficult task (Hodges 1978). This difficulty resides in the basics of the excitation process and the small energy difference between laser states. Energy storage is not possible because laser-state lifetimes are essentially equal. To approach the quantum limit, one submillimeter-wave photon per pumping photon, a detailed knowledge of the fundamental laser mechanism is required. Hence laser emission characteristics have been extensively analyzed using rate equation models (Temkin, 1977). Good agreement is reported in certain cases. However, quantum mechanical two-photon effects have recently been shown to play an important role in the laser performance (Seligson et al., 1977; Petuchowski et al., 1977; Wiggins et al., 1977). When the laser is off resonance, submillimeter-waves may be emitted in a two-photon process corresponding to stimulated Raman emission. In addition, the dynamic and ac Stark effect can significantly modify the emission characteristics of both cw and pulsed (Chang, 1977b) lasers.

The realization of high pulse power (MW) and simultaneous narrow emission line width (25 MHz) in an optically pumped submillimeter-wave laser requires a master oscillator–power amplifier combination and a Fox–Smith resonator geometry for the oscillator (DeTemple, 1977). On the other hand, the development of strong optically pumped cw lasers has concentrated on improved waveguide resonators and output coupling techniques (Yamanaka, 1977). For the present, the optically pumped laser process is efficient in only a small number of tabulated cases (Hodges, 1978).

G.   Transversely Excited Submillimeter-Wave Lasers

Optically pumped lasers can generate submillimeter-wave pulses of about 150 ns duration and a peak power up to 1 MW, corresponding to a pulse energy near 150 mJ (Woskoboinikow et al., 1978b). Pulses of slightly lower energies, with reproducible shapes and a duration of 8 μs, may be obtained by transverse excitation of the 337-μm HCN laser emission (Kneubühl, 1978; Sturzenegger et al., 1978, 1979). The transverse electric excitation of gas lasers has been known since the discovery of the TEA–CO$_2$ laser by Beaulieu (1970). It has proven to be an effective means of producing intense laser pulses from high-pressure gas mixtures (Wood, 1974). This excitation scheme was first applied to far-infrared lasers by Wood et al. (1971) who attained emissions out to 28 μm. Unfortunately, their transversely excited laser did not yield much power, and it was assumed that the TEA laser would not be anywhere as effective with far-infrared molecules as with

CO_2 (Coleman, 1973). On the other hand, Sharp and Wetherell (1972), as well as Jassby et al. (1973), suggested that a substantial improvement in HCN laser power might be obtained using transverse electric excitation. A first attempt was made by Lam et al. (1973) with an auxiliary dc discharge in a reaction chamber outside the resonator. The first successful proper transverse electric excitation of the 337-μm HCN laser emission without an auxiliary discharge was reported by Adam et al. (1973). They found by experiment that the usual electrode configurations of infrared TEA lasers do not propagate submillimeter waves. Therefore, they introduced particular waveguide resonators, which include the transverse electrode structure (Adam et al., 1973; Adam and Kneubühl, 1975). With this arrangement they achieved the transverse excitation of the 337-μm HCN emission from sealed-off gas mixtures at pressures up to 5 Torr.

In the usual longitudinally excited low-pressure submillimeter-wave lasers the separation of the longitudinal modes is wider than the gain profile. Hence their laser resonators have to be tuned to the resonance length, and the laser pulse shape strongly depends on the resonator length (Steffen and Kneubühl, 1968). The transverse excitation of the HCN laser allows an increase in the pressure of the laser gas, which widens the gain profile beyond the mode separation. At a total gas pressure of a few Torr the laser pulse obtained from transverse excitation becomes independent of the resonator length and, as a consequence, reproducible. In addition, the increase of the gas pressure shortens the pulse duration. The peak power of the laser pulse reaches a maximum at an optimum pressure which is higher than the usual working pressure of longitudinally excited HCN lasers (Adam et al., 1973; Adam and Kneubühl, 1975).

Sturzenegger et al. (1977) introduced UV preionization in the transversely excited HCN laser in order to increase the working gas pressure and the laser pulse energy. With a special electrode structure and UV preionization they achieved homogeneous glow discharges in HCN laser gas mixtures at 1 atm, but no stimulated emission was observed. On the other hand, UV preionization lead to an increase of the pulse energy of a transversely excited HCN laser with pin-type electrodes (Adam et al., 1973) by a factor of two (Sturzenegger et al., 1977). Fortunately, a drastic improvement of the HCN pulse energy was achieved subsequently by introducing a flow system in the UV preionized transversely excited HCN laser (Sturzenegger et al., 1978, 1979). At present, research concentrates on this type of laser.

H. Bridging the Gap

As mentioned before, the first bridging of infrared and microwaves was improvised in the 1920s with the aid of thermal and Hertzian broadband

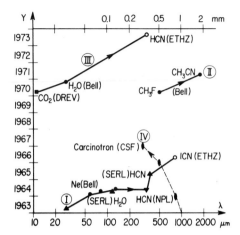

FIG. 1 Bridging of the microwave and infrared spectral region by coherent sources: (I) longitudinally excited lasers, (II) optically pumped lasers, (III) transversely excited lasers, (IV) backward-wave microwave oscillators.

incoherent sources (Section A). It took another 40 years until a permanent reliable bridge over the submillimeter-wave power gap could be established by coherent monochromatic sources. As illustrated in Fig. 1, longitudinally excited submillimeter-wave gas lasers (Ne, H_2O, HCN, ICN) and backward-wave oscillators (Carcinotrons, CSF) take credit for this achievement (Section B).

At the beginning of the 1970s, the bridge between infrared and microwaves was considerably strengthened by the development of optically pumped (Section F) and transversely excited (Section G) far-infrared lasers. Today, the old and fascinating problem of closing the submillimeter-wave gap is solved.

Longitudinally excited far-infrared lasers are reviewed in publications by Garrett (1965, 1966), Yoshinaga et al. (1967), Steffen and Kneubühl (1968), Coleman (1973), Kunstreich and Lesieur (1974), and Belland (1975). Transversely excited submillimeter-wave lasers were discussed by Adam and Kneubühl (1975) and Sturzenegger et al. (1977, 1979). The submillimeter-wave emissions obtained from electrically excited gas lasers are listed in Table I. By virtue of their significance, we should also mention the reviews and tabulations on optically pumped submillimeter-wave lasers by Chang (1974, 1977a), Yamanaka (1976, 1977), Rosenbluh et al. (1976), Gallagher et al. (1977), and Röser and Schultz (1977).

TABLE I

ELECTRICALLY EXCITED GAS LASERS WORKING AT WAVELENGTHS BEYOND 100 μm[a]

Emitter	Vacuum wavelength	Operation	Transition
He	216.3	cw	$4p^1P$–$4d^1D$
Ne	106.07	cw	$10p[\frac{1}{2}]_0$–$9d[\frac{3}{2}]_1^0$
	124.52		$9p[\frac{3}{2}]_1$–$8d[\frac{5}{2}]_2^0$
	or	cw	or
	124.76		$9p[\frac{3}{2}]_2$–$8d[\frac{5}{2}]_3^0$
	126.1	cw	?
	132.8	cw	?
H_2O	115.32	p and cw	$(020)8_{35}$–$(020)8_{26}$
	118.591	p and cw	$(001)6_{42}$–$(020)6_{61}$
	120.08	p	$(001)6_{42}$–$(001)6_{33}$
	220.230	p and cw	$(100)5_{23}$–$(020)5_{50}$
D_2O	103.33	p	?
	107.731	p and cw	$(100)11_{66}$–$(020)11_{75}$
	107.91	p	$(100)13_{68}$–$(100)13_{59}$
	108.88	p	$(100)11_{65}$–$(020)11_{74}$
	110.49	p	$(100)12_{66}$–$(020)12_{75}$
	111.74	p	$(100)13_{68}$–$(020)13_{77}$
	170.08	p	$(020)11_{47}$–$(020)11_{38}$
	171.67	p and cw	$(100)11_{0,11}$–$(020)11_{38}$
	218.5	p	?
H_2S	103.3	p	?
	108.8	p	?
	116.8	p	?
	126.2	p	?
	129.1	p	?
	130.8	p	?
	135.5	p	?
	140.6	p	?
	162.4	p	?
	192.9	p	?
	225.4	p	?
SO_2	139.80	p	$(020)27_{16}$–$(020)26_{15}$
	140.78	p and cw	$(100)27_{14}$–$(020)26_{15}$
	140.88	p and cw	$(100)26_{14}$–$(020)25_{15}$
	141.98	p	$(020)26_{16}$–$(020)25_{15}$
	149.99	p	$(100)28_{15}$–$(100)27_{14}$
	151.19	p and cw	$(100)28_{15}$–$(020)27_{16}$
	151.31	p and cw	$(100)27_{15}$–$(020)26_{16}$
	192.71	p and cw	$(100)28_{16}$–$(100)28_{15}$
	206.44	p	$(100)26_{15}$–$(100)26_{14}$
	215.33	p and cw	$(001)27_{10}$–$(020)27_{16}$

(cont.)

TABLE I (*continued*)

Emitter	Vacuum wavelength	Operation	Transition	
OCS	123	p	?	
	132	p	?	
HCN	126.164	p	$(12^20)-(05^10)$	$R(26)$
	128.629	p and cw	$(12^20)-(05^10)$	$R(25)$
	130.838	p	$(12^00)-(05^10)$	$R(25)$
	134.932	p	$(12^00)-(05^10)$	$R(24)$
	284	p and cw	$(11^10)-(11^10)$	$R(11)$
	309.7140	p and cw	$(11^10)-(11^10)$	$R(10)$
	310.8870	p and cw	$(11^10)-(04^00)$	$R(10)$
	335.1831	p and cw	$(04^00)-(04^00)$	$R(9)$
	336.5578	p and cw	$(11^10)-(04^00)$	$R(9)$
	372.5283	p and cw	$(04^00)-(04^00)$	$R(8)$
HCN or CN	101.257	p	?	
	112.066	p	?	
	116.132	p	?	
or ?	201.059	p	?	
	211.001	p and cw	?	
	222.949	p	?	
DCN	181.789	p	$(22^00)-(22^00)$	$R(22)$
	189.9490	p and cw	$(22^00)-(09^10)$	$R(21)$
	190.0080	p and cw	$(22^00)-(22^00)$	$R(21)$
	194.7027	p and cw	$(22^00)-(09^10)$	$R(20)$
	194.7644	p and cw	$(09^10)-(09^10)$	$R(20)$
	204.3872	p and cw	$(09^10)-(09^10)$	$R(19)$
HCN^{15}	110.240	p	?	
or ?	113.311	p	?	
	138.768	p	?	
	165.150	p	?	
ICN	538.2	p	?	
or HCN	545.4	p	?	
or CN	676	p	?	
or ?	773.5	p	?	
H_2CO	101.9	p	?	
	119.6	p	?	
	122.8	p	?	
	125.9	p	?	
	155.1	p	?	
	157.6	p	?	
	159.5	p	?	
	163.8	p	?	
	170.2	p	?	
	184.4	p	?	

[a] From Steffen and Kneubühl (1968); Pressley (1971); Hassler *et al.* (1973); Horiuchi and Murai (1976).

II. Design

The experimental arrangements used for the first observation of the H_2O and D_2O stimulated far-infrared emissions (Crocker et al., 1964; Gebbie et al., 1964a) demonstrated for the first time that monochromatic and relatively intense submillimeter-waves can be generated by simple laser systems. This gave a drastic impetus to research on submillimeter waves where strong monochromatic sources had been aimed for decades (Coleman, 1963). Therefore, a number of infrared-physics laboratories started the development and design of longitudinally excited submillimeter-wave lasers. Consequently, construction of such lasers in a large variety of designs was reported in the following years, e.g., by Kneubühl et al. (1965), Akitt et al. (1966), Flesher and Muller (1966), Hadni et al. (1967, 1968), Kon et al. (1967), Minoh et al. (1967), Yoshinaga et al. (1967), Steffen and Kneubühl (1968), Yamanaka et al. (1968), McCaul (1970), Ulrich et al. (1970), Robinson and Whitbourn (1971), Sharp and Wetherell (1972), Belland and Veron (1973), Bicanic and Dymanus (1974, 1975), Belland (1975), Kunstreich and Lesieur (1975), Belland et al. (1976), Nizishawa et al. (1976), Vanderkooy and Kang (1976), and Bruneau et al. (1978). Characteristic designs will be reported in Section A.

Proper transverse excitation of a submillimeter-wave stimulated emission was first achieved by Adam et al. (1973) by introducing transverse electrodes in a waveguide structure. Later, Sturzenegger et al. (1977) improved the transversely excited, 337-μm HCN laser with the aid of UV ionization and a flow system. (Sturzenegger et al., 1978; Sturzenegger, 1979). This will be discussed in Section B.

A. Longitudinally Excited Submillimeter-Wave Lasers

The first successful designs (Gebbie et al., 1964a, Kneubühl et al., 1965) of longitudinally excited pulsed submillimeter-wave molecular lasers were simple, as demonstrated in Fig. 2. They consisted of glass tubes with a diameter of 2–10 cm, a length of 3–10 m, and two pin-type tungsten electrodes sealed through side arms of the tube. The glass tube was evacuated by an oil diffusion pump before operation of the laser, and then filled with the feasible gas mixture to a pressure of 0.01–3 Torr. This gas was excited by a longitudinal discharge in the glass tube by application of electric pulses with a peak voltage of 5–50 kV, a peak current of the order of magnitude of 100 A, and a duration of a few microseconds. Since these lasers are characterized by a gain width smaller than the separation of the fundamental resonator modes, the resonators were usually devised as interferometers with plane-parallel mirrors that allowed tuning by moving one mirror with parallelism. The submillimeter-wave stimulated emission was coupled out of the resonator

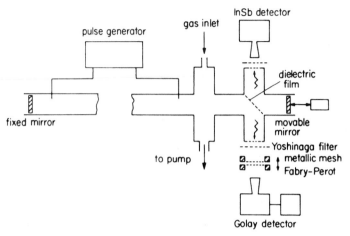

FIG. 2 Experimental arrangement of a pulsed submillimeter-wave molecular laser, according to Steffen and Kneubühl (1968).

with the aid of a diagonal dielectric film or a hole in one mirror. Before reaching the detector, the radiation emitted by the laser was filtered and analyzed by Yoshinaga filters (Yamada *et al.*, 1962; Yoshinaga, 1965), a Michelson interferometer, a metallic-mesh Fabry–Perot (Ulrich *et al.*, 1963), or a grating spectrometer. Golay cells were used as detectors for the search for new stimulated submillimeter-wave emissions, for wavelength determinations, and for the study of resonator modes. For the measurement of pulse shapes, the slow Golay cells were replaced by He-cooled InSb detectors and other fast solid–state devices.

Figure 2 shows a longitudinally excited submillimeter-wave laser equipped with a movable plane resonator mirror and an external metallic-mesh Fabry–Perot. This experimental arrangement was devised by Steffen and Kneubühl (1968) with the .aim of performing diagnostics of the emission spectrum and the resonator modes by laser resonator interferometry. As illustration, we demonstrate in Fig. 3 the diagnostics of the pulsed emission of the long-wavelength ICN laser. On top we find the laser resonator interferogram, which is dominated by the fundamental modes of three emissions. Below we notice two interferograms produced by the external metallic-mesh Fabry–Perot for fixed resonator lengths corresponding to fundamental modes. The experimental data are as follows: peak discharge voltage 15 kV; electric peak power 1.2 MW; pulse repetition rate 6 s^{-1}; vapor pressure 0.04–0.05 Torr; resonator length 6.47 m; plane parallel mirrors of 75 mm diameter; Yoshinaga filters with BeO, ZnO, NaF, and KCl; Unicam Golay detector; signal-amplifier time constant 1 s.

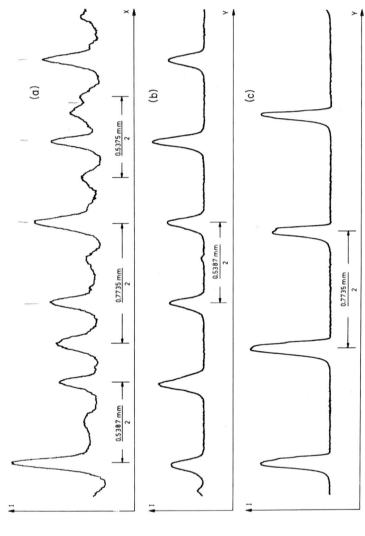

FIG. 3 Laser resonator interferogram and related metallic-mesh Fabry-Perot interferograms of pulsed submillimeter-wave ICN laser emissions, according to Steffen et al. (1966b). (a) Laser resonator interferogram (Steffen and Kneubühl, 1968). (b) Resonator length tunes to the 538.7-μm emission, interferogram measured with external metallic-mesh Fabry-Perot, resonances observed up to 70th order. (c) Resonator length tunes to the 773.5-μm emission, interferogram measured with external metallic-mesh Fabry-Perot, resonances observed up to 40th order.

Another example, which illustrates the laser resonator interferogram of the fundamental and higher modes of almost pure 337-μm emission from a HCN waveguide laser, is presented in Fig. 18 of the review on infrared and submillimeter-wave waveguides by Kneubühl and Affolter (1979).

Typical problems involved in the continuous (cw) operation of longitudinally excited submillimeter-wave lasers are the cathode (Schötzau et al., 1971; Belland and Veron, 1973; Woskoboinikow, 1974; Bicanic and Dymanus, 1975; Belland, 1975; Belland et al., 1975, 1976; Vanderkooy and Kang, 1976; Bruneau et al., 1978), the stabilization of the resonator, and the cooling of the plasma tube (Belland and Veron, 1973; Bicanic and Dymanus, 1975; Belland et al., 1975, 1976; Vanderkooy and Kang, 1976; Bruneau et al., 1978).

The conditions on the electrodes of a longitudinally excited submillimeter-wave lasers is that they provide a stable discharge within the capabilities of the power supply. The most common shape used for electrodes is cylindrical. Though there may be general agreement as to what is the best shape for electrodes, the materials used to make them have varied widely: tungsten, stainless steel, copper, brass, aluminum, and even carbon. There are a couple of reasons why the cylindrical shape is very popular. First, the shape provides a large surface area because both the outside and the inside surfaces are exposed to the current flow. Second, the cylinder allows the negative electrode to run as a hollow cathode. This makes more efficient use of the electrons within the cylinder as they are forced to be reflected back and forth between the inside surface. Keeping the electrons active within the cylinder allows them to make many more ionizing collisions than they normally would. Therefore, they provide an efficient supply of charged carriers for the laser discharge.

In Fig. 4 we show the design by Schötzau et al. (1971) of a water-cooled hollow cathode for continuous HCN lasers. The cathode is made of Pyrex glass, quartz glass, and stainless steel. This special arrangement focuses the discharge on the inner side of the stainless steel tube, thus avoiding sparking

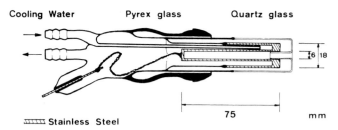

FIG. 4　Design of a water-cooled hollow cathode used in a continuous HCN laser (Schötzau et al., 1971).

at the edge of the tube and providing a very stable discharge. With this cathode the discharge current can be varied between 0.1 and 2.5 A at HCN-laser gas pressures of 0.1–0.6 Torr. For a distance of 1.2 m from cathode to anode, the corresponding voltage ranges from 1.8–1.2 kV.

A compact cw HCN laser with high stability was constructed by Belland and Veron (1973). This laser was equipped with a cavity 135 cm in length, which was kept constant within ± 0.1 μm by four temperature-regulated silica tubes. The glow discharge was produced in a 5-cm-diameter Pyrex tube over a length of 1 m. The discharge current was between 1 and 2 A for a voltage drop of about 2 kV and a gas pressure near 2 Torr. The Pyrex tube had an oil jacket for temperature control. The wall of the tube could be maintained at any temperature between 40°C and 250°C by thermostatically controlled oil flow in its jacket. Belland and Veron (1973) demonstrated that the cw 337-μm HCN laser output strongly depends on this temperature. For a $N_2:CH_4$ mixture without He they found a maximum laser power at 100°C. Addition of He to the mixture shifted this maximum to higher temperatures.

Figure 5 illustrates a recent design of a high-performance longitudinally excited cw DCN laser by the same research team (Bruneau et al., 1978). The laser cavity consists of a Pyrex tube with a 3.2-cm inner diameter, a length of 70 cm, and plane-parallel reflectors at the ends. One of the reflectors is a copper mesh, which acts as output coupler. For the 195-μm emission, the maximum laser power is found for a gas mixture $N_2:CD_4:He$ of volume ratio 1:3:12 at a pressure of 1.4 Torr, a discharge current of 1.4 A, and an

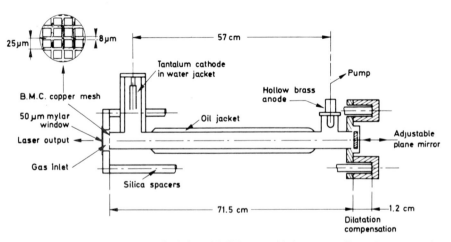

FIG. 5   Design of continuous far-infrared DCN waveguide laser, according to Bruneau et al. (1978).

oil-jacket temperature of 140°C. Under these conditions, no brown polymer deposit on the wall is observed.

B. TRANSVERSELY EXCITED SUBMILLIMETER-WAVE LASERS

The experimental arrangement of an early transversely excited HCN laser (Adam *et al.*, 1973; Adam and Kneubühl, 1975) without auxiliary discharge is shown in Fig. 6. The discharge tube consisted of the particular waveguide structure illustrated in Fig. 7. The resonator was formed by two plane mirrors at the end of the tube. The length of the resonator could be varied by moving one mirror in parallel with a motor-driven translation stage. The laser-gas mixture was excited by pulsed discharges from pin-type electrodes, as used in early TEA CO_2 lasers. The cathode was formed by about 2200 current-limiting resistors of 1 kΩ and 2 W. The anode was a commercially available flat aluminum profile. The anode and the cathode were glued to the Plexiglas with epoxy. A spark-gap switched 0.5 or 1 μF capacitor charged to 20–40 kV generated the high-voltage pulses. The gap was free running and had a repetition frequency of about 1 s^{-1}. The total current reaches its maximum of 5–10 kA after 2 μs. The voltage across the discharge is fairly low. This can be explained by the large voltage drop over the circuit inductance of 5 μH. The current density in the discharge of the present laser is of the same order of magnitude as observed for longitudinally excited lasers, namely, 4–8 A cm^{-2}.

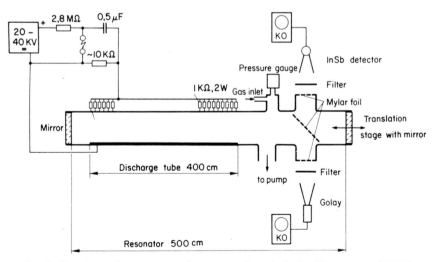

FIG. 6 Experimental arrangement of a transversely excited submillimeter-wave HCN laser without UV preionization, according to Adam and Kneubühl (1975).

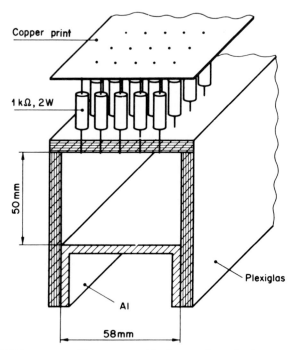

FIG. 7   HCN laser waveguide structure with pin-type electrodes for transverse excitation without UV preionization, according to Adam and Kneubühl (1975).

The design of the discharge tube shown in Fig. 7 represents a compromise between a transverse electrode structure and a waveguide resonator, since the usual pin-type electrode structures of TEA CO_2 lasers severely disturb the propagation of submillimeter waves. The cathode pins consist of the wire ends of the current-limiting resistors. In order to avoid disturbance of the submillimeter waves, these pins enter the waveguide by less than 0.5 mm.

The described transversely excited HCN laser was allowed to increase the working pressure of the laser gas up to 5 Torr in a closed system, with corresponding reduction of pulse duration, increase of power, and pressure broadening of the gain width. The latter became larger than the separation of the fundamental resonator modes, which resulted in reproducible pulses independent of the resonator length. The pulse duration could be varied between 60 and 10 μs by changing the working-gas pressure.

Performance and power of transversely excited HCN lasers have been improved by introducing UV preionization (Sturzenegger et al., 1977, 1978) and, even more, by replacing the sealed-off system by a flow system (Sturzenegger et al., 1979). Figure 8 shows the waveguide and discharge structure used for the operation with UV preionization.

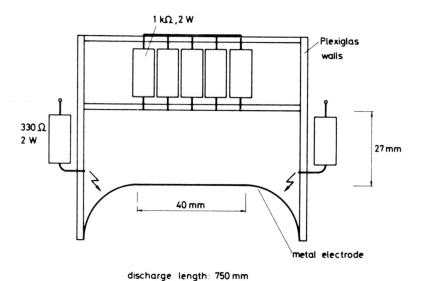

FIG. 8 HCN laser waveguide structure with pin-type electrodes for transverse excitation with UV preionization, according to Sturzenegger *et al.* (1977).

The introduction of UV preionization and flow systems has resulted in a power increase of a factor 15–20 over the output of the first transversely excited HCN laser (Sturzenegger *et al.*, 1978). A detailed description of the design and the performance of the latest version of the transversely excited HCN laser is presented by Sturzenegger *et al.* (1979). Hitherto, only pin-type transverse electrode structures have been applied in transversely excited HCN lasers. It is possible that the use of other electrode structures, such as those incorporated in TEA CO_2 lasers, will further improve the performance of transversely excited submillimeter-wave lasers.

C. Output Coupling

The output power spectrum of a submillimeter-wave laser depend, among many other parameters, on the properties of the output coupler. The methods chosen for coupling the radiation out of the laser depend on the particular purpose:

(i) search for new laser emissions,
(ii) investigation and selection of resonator modes,
(iii) investigation and selection of the field polarization,
(iv) maximum output power for a particular wavelength,
(v) suppression of unwanted wavelengths and modes.

A good solution would be a partially transparent solid-state mirror as a coupler. Unfortunately, this does not yet exist in the submillimeter-wave range. Hence a variety of output coupling systems has been devised and applied successfully:

(i)   coupling with a thin dielectric film,
(ii) interferometric Michelson coupling,
(iii) hole coupling,
(iv) annular coupling,
(v)   coupling with a metallic mesh,
(vi) interferometric Fabry–Perot coupling,
(vii) coupling with a tilted mirror.

A simple and reliable output coupling can be realized by mounting a dielectric film with a tilting angle of 45° to the axis into the resonator (Gebbie *et al.*, 1964a; Kneubühl *et al.*, 1965; Hadni *et al.*, 1967; Kon *et al.*, 1967, Steffen and Kneubühl, 1968; Belland and Veron, 1973; Kunstreich and Lesieur, 1974) as shown in Fig. 2. The film material used is polyethylene, Mylar, Hostaphan, etc. of a thickness 5 and 20 μm. As an absorption coefficient of this material we mention 1.55×10^{-3} μm^{-1} at a wavelength of 190–195 μm for Mylar (Löwenstein and Smith, 1971; Bruneau *et al.*, 1978). The reflection coefficient of these foils varies considerably with polarization. For a Hostaphan foil of 12-μm thickness, the 45° reflection coefficients at a wavelength of 337 μm have been measured as $R_{\parallel} = 0.064$ and $R_{\perp} = 0.16$ (Steffen and Kneubühl, 1968). Consequently, the output coupling by dielectric film provides two polarized exit beams. The output characteristics do not vary noticeably over a wide range of wavelengths, and the coupling is effective and homogeneous over the entire cross section of the resonator. Hence, output coupling by a dielectric film has considerable advantage in the search for new submillimeter-wave emissions and in mode studies (Steffen *et al.*, 1966a, b; Schwaller *et al.*, 1967; Steffen and Kneubühl, 1968).

An interferometric Michelson output coupler with a dielectric film as beam splitter is very effective (Gebbie *et al.*, 1964a; Smith, 1965; Wells *et al.*, 1971; Adam and Kneubühl, 1975). By varying the length of one arm of the interferometer, the output can be varied between 0 and $4r^2$, where r is the single amplitude reflection factor of the dielectric film at a tilting angle of 45°. However, the full variation can be observed only if the extracted amount of power is negligible in comparison to the total power in the cavity. This disadvantage has its origin in the interdependence of the output coupling and gain of the laser-active medium.

Many experimenters have used a small coupling hole located at the center of one resonator mirror, (Patel *et al.*, 1964; Witteman and Bleekrode, 1964; Akitt *et al.*, 1965; Brannen *et al.*, 1967; Pollack *et al.*, 1967a,b; Kon *et al.*, 1967;

Sharp and Wetherell, 1972; Sturzenegger et al., 1977). In this arrangement the modes, with the field concentrated along the resonator axis, are preferably coupled out. This results in a well-defined laser output. Yet this method fails for nonaxial modes and is not suited for mode studies. In addition, the coupling strongly depends on the wavelength. The influence of small holes in the resonator mirrors of confocal geometry on the modes was studied by McCumber (1965). Later, this problem was studied by Li and Zucker (1967) for plane parallel mirrors.

A suitable method for observing weak laser emissions was developed by Mathias et al. (1965). A small annular concentric slit is left open between the circular mirror and the plasma glass tube. By diffraction, some radiation is extracted from the resonator and collected by a concave reflector behind the resonator mirror. Metallic-mesh reflector grids (Ulrich et al., 1963; Vogel and Genzel, 1964; Ulrich, 1967; Sakai et al., 1969) have been introduced as output couplers in submillimeter-wave lasers by Prettl and Genzel (1966). In this coupling system a plane resonator mirror is replaced by a plane metallic-mesh grid which is selected according to the laser wavelength and the degree of coupling desired (Fig. 5). Since their introduction metallic-mesh couplers have been applied successfully in many lasers (Yoshinaga et al., 1967; Yamanaka et al., 1968; Belland et al., 1975; Bruneau et al., 1978).

An improvement of metallic-mesh couplers was achieved by Ulrich et al. (1970) who added a second grid parallel behind the first metallic-mesh grid. In this arrangement the two metallic-mesh grids form a Fabry–Perot interference filter (Ulrich et al., 1963). The reflectance of this filter depends critically on the spacing of the two grids. As an output coupler it can be tuned conveniently by adjusting the spacing by means of a micrometer screw.

Finally, we should mention output coupling by tilted mirrors. Wells (1966) suggested controlling the coupling from the laser resonator by tilting one mirror. The influence of the tilt of one mirror on the modes of resonators with small N was calculated for rectangular apertures. Instead of tilting a resonator mirror, a very small tilted mirror may be introduced in the resonator (Kon et al., 1967). The advantage of this is the ease of positioning, angle adjustment, and replacement.

III. Plasmas

As shown in Table I, low-pressure discharges of a number of molecular-gas mixtures produce stimulated emissions of submillimeter-waves. HCN laser emissions are generated by electric excitations of a large group of mixtures with molecules containing C, N, and H. The 337-μm emission from the transition (11^10) $J = 10$ to (04^00) $J = 9$ is among the most intense submillimeter-wave radiations obtained from electrically excited lasers. In most cases the HCN laser discharge is operated with nonpoisonous gas mixtures, e.g., $CH_4 + N_2$. Therefore, the HCN molecules responsible for the laser emission have to be formed in the low-pressure discharge by chemical reactions. In order to understand these reactions and the laser relaxation processes, knowledge of the parameters of the HCN laser plasma is required. For this reason, they have been investigated experimentally and theoretically (Turner and Poehler, 1971; Schötzau and Kneubühl, 1975a, b; Schötzau and Veprek, 1975). In the following we shall focus our attention on the plasma of the HCN laser, since no or scarce data are available for the plasmas of other electrically excited submillimeter-wave lasers.

A. Continuous Discharges

Schötzau and Kneubühl (1975a, b) performed a detailed study on the plasma of a cw HCN laser. For its characterization a standard dc discharge in a $CH_4:N_2 = 1:1$ gas mixture was used. The laser produced the maximum output on the 337-μm emission at a pressure of 0.5 Torr and a dc discharge current of 0.8 A. A variety of plasma parameters were measured or calculated, e.g., the electron temperature T_e, the electron density n_e, the neutral-gas

TABLE II

PARAMETER OF 337-μm HCN LASER PLASMA PRODUCED BY A CONTINUOUS ELECTRIC DISCHARGE OF 0.8 A IN A GAS MIXTURE $CH_4:N_2 = 1:1$ AT A PRESSURE OF 0.5 TORR[a]

Temperature	$T_e = 16000$ K		$T_g = 450$ K	
Particle density	$n_i \simeq n_e = 2.7 \times 10^{10}$ cm^{-3}		$n_g = 10^{16}$ cm^{-3}	
Mobility	$b_e = 6.6 \times 10^5$ cm^2 (V s)$^{-1}$		$b_g \simeq b_i = 10^3$ cm^2 (V s)$^{-1}$	
Free-path length	$\lambda_e = 3 \times 10^{-2}$ cm	$\lambda_g = 8.5 \times 10^{-3}$ cm	$\lambda_i = 1.2 \times 10^{-2}$ cm	
Coefficient of ambipolar diffusion	$D_{ambi} = 14{,}750$ cm^2 s^{-1}			
Degree of ionization	$\alpha_i = 10^{-6}$			
Degree of dissociation	$\alpha_d = 10^{-1}$			
Energy of ionization	$eV_i = 10$ eV			
Energy of dissociation	$eV_d = 5$ eV			
Electric field strength	$E_{ax} = 7.8$ V cm^{-1}			
Ratio of energies	$E_{ion}:E_{gas}:E_{diss} = \alpha_i\,eV_i:kT_g:\alpha_d\,eV = 10^{-5}:10^{-1}:1$			

[a] Schötzau and Kneubühl (1975a).

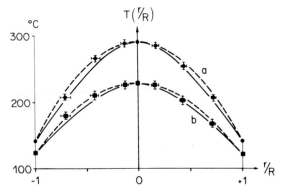

FIG. 9 Radial distribution of the neutral-gas temperature of the gas mixture ($CH_4 : N_2 \simeq$ 7:7 Torr liter min^{-1}) at a total gas pressure of 0.5 Torr and a dc discharge current of 0.8 A according to Schötzau and Kneubühl (1975a). (R: tube radius; *solid lines*: calculated for a constant electron density; *dashed lines*: calculated for an electron density $n_e \propto J_0(r/R)$; *curve a*: measured with a Pt/Pt–Rh thermocouple; *curve b*: measured with a Pt/Pt–Rh thermocouple covered by quartz glass.)

temperature T_g, the neutral-gas density n_g, the degree of ionization α_i and of dissociation α_d, the chemical composition, etc. These parameters are listed in Table II. Figure 9 shows the radial distribution of the neutral-gas temperature $T_g(r/R)$. Laser intensity and chemical composition of the plasma are illustrated in Fig. 10 as functions of the dc discharge current.

The above experimental conditions lead to the assumption that the laser processes and the chemical reactions take place in the positive column of the

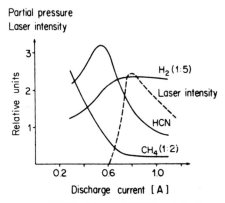

FIG. 10 Partial pressures of CH_4, H_2, and HCN and the laser intensity of the 337-μm emission of the gas mixture ($CH_4 : N_2 \simeq$ 7:7 Torr liter min^{-1}) as a function of the discharge current at a total pressure of 0.5 Torr, according to Schötzau and Kneubühl (1975b).

low-pressure discharge. The theory of the positive column was developed by Schottky (1924). This theory is valid if the diameter of the laser tube $2R$ is large compared to the free path length λ_i of the ions, λ_e of the electrons, and λ_g of the neutral-gas particles. This condition is valid for all standard HCN laser tubes. It is assumed that the axial electric field strength E_{ax} remains constant over the tube cross section. The theory is based on the quasi-neutrality ($n_i = n_e$) of the positive column. The electrons and ions diffuse together to the wall with an ambipolar diffusion coefficient D_{ambi}. Recombination takes place on the wall, and the resulting loss of charge carriers is balanced by the production of ions and electrons in the discharge. Under these assumptions the radial distribution of the electron density $n_e(r)$ can be expressed by the zero-order Bessel function J_0.

$$n_e(r) = n_e^0 J_0(r[\alpha b_e E_{ax} D_{ambi}^{-1}]^{1/2}) = n_e^0 J_0(\rho), \tag{1}$$

where α is the electron ionization coefficient, b_e the mobility of the electrons, and n_e^0 the electron density on the tube axis. The first zero point of $J_0(\rho)$ determines the boundary condition.

$$\rho_0 = 2.4 = R(\alpha b_e E_{ax} D_{ambi}^{-1})^{1/2}. \tag{2}$$

This condition allows us to test the above assumption. For a tube radius $R = 4$ cm, $\alpha = 10^{-3}$ (von Engel and Steenbeck, 1934), and the plasma parameters listed in Table II one finds $\rho_0^{exp} = 2.3$. Therefore, we can assume that Schottky's theory of the positive column of the low-pressure discharge is valid for the HCN laser. This is in agreement with studies on plasmas in other lasers (Kleen and Müller, 1969).

The discharge current I of the glow discharge is related to the electron density n_e. Since the mobility of the electrons is much higher than the mobility of the ions (Table II), we can assume the following relation

$$I = eb_e E_{ax} n_e^0 \int_0^R J_0(\rho) 2\pi r \, dr = 0.43\pi R^2 eb_e E_{ax} n_e^0. \tag{3}$$

The average value of $\bar{n}_e$ is determined by

$$\bar{n}_e = 0.42 n_e^0 \int_0^{2.4} J_0(\rho) \, d\rho = 0.61 n_e^0 = 0.45 I (R^2 eb_e E_{ax})^{-1}. \tag{4}$$

Insertion of numerical data results in

$$\bar{n}_e = \bar{n}_i = 2.7 \times 10^{10} \text{ cm}^{-3}.$$

Microwave diagnostics of pulsed HCN laser discharges give values of the electron densities in the order of magnitude 10^{12} cm^{-3}. Taking into account that the discharge current of pulsed lasers is two or three orders higher than the current of the cw laser, one agrees with the above result.

Elastic collisions of the electrons with the heavy particles determine the neutral-gas temperature. The heat transfer to the heavy particles is proportional to the electron density (Kenty et al., 1951; Francis, 1956). The following differential equation defines the radial temperature distribution

$$\Delta T_g(r) + cn_e(r) = 0, \tag{5}$$

where c is a constant. With the assumption

$$n_e(r) = n_e^0 J_0(r) \cong n_e^0(1 - r^2/R^2), \tag{6}$$

where R represents the radius of the discharge tube, the solution is

$$T_g(r) = T_R + (T_0 - T_R)\left(1 - \frac{4r^2}{3R^2} - \frac{r^4}{3R^4}\right), \tag{7}$$

where T_R indicates the wall temperature, and T_o the temperature in the middle of the discharge tube.

The assumption that the electron density is constant over the tube cross section leads to another solution:

$$T_g(r) = T_0 - (T_0 - T_R)r^2/R^2. \tag{8}$$

In Fig. 9 the measured gas temperature and the theoretical curves defined by Eqs. (7) and (8) are compared. The experimental errors do not allow any discrimination between the two assumed electron density distributions. The measured temperatures correspond to a mean gas temperature of $\overline{T}_g = 450$ K. Similar temperature profiles have been published by Belland et al., (1974) for a HCN laser plasma tube with an oil jacket and higher dc discharge currents.

The neutral-gas density n_g at a gas temperature $\overline{T}_g = 450$ K and a pressure $p = 0.5$ Torr can be calculated by the relation of Boyle, Mariotte, and Gay-Lussac:

$$\overline{n}_g = p/k\overline{T}_g = 10^{16} \text{ cm}^{-3}. \tag{9}$$

The degree of ionization is defined by the following equation:

$$\alpha_i = \overline{n}_i/\overline{n}_g = \overline{n}_e/\overline{n}_g = 2.5 \times 10^{-6}. \tag{10}$$

Therefore, the positive column is weakly ionized.

Although real electron energy distributions of laser discharges (Bletzinger and Garscadden, 1971; Kagan and Mitranov, 1971; Novgorodov et al., 1971) often deviate from Maxwell's distributions, a Maxwellian distribution may be still assumed in the HCN laser plasma as a first approximation. By extension of the diffusion theory, Schottky (1924) and von Engel and Steen-beck (1934) found a relation between the mean electron energy kT_e and the ionization of the gas,

$$(eV_i/kT_e)^{-1/2} \exp(eV_i/kT_e) = c. \tag{11}$$

The constant c depends on the pressure, the tube radius, and the gas. For the value $eV_i \simeq 15$ eV, the electron temperature T_e of the HCN plasma is $T_e = 15,000$ K.

The electron temperature is determined by the equation (Weizel, 1958)

$$T_e = (5.5 \times 10^3)(\lambda_e E_{ax} \kappa_e^{1/2}) \quad \text{K}, \tag{12}$$

where λ_e is the mean-free path length of the electrons, and κ_e is the accommodation coefficient. The measured value of the electric field strength $E_{ax} = 7.8$ V cm^{-1} agrees well with values of pure N_2 or H_2 discharges, $E_{ax}(N_2) = 9$ V cm^{-1} and $E_{ax}(H_2) = 5$ V cm^{-1}, published by von Engel and Steenbeck (1934). With their parameters

$$\lambda_e(15,000 \text{ K}) = 3 \times 10^{-2} \text{ cm}, \qquad \kappa_e(15,000 \text{ K}) = 6 \times 10^{-3},$$

one obtains the electron temperature $T_e = 16,000$ K. This and the above electron temperature are in good agreement.

The degree of dissociation α_d is illustrated by the measurement of the neutral gas temperature. This temperature depends on the translational gas temperature, and on heat production on the surface of the thermocouple, which is dominated in a low-ionized molecular plasma by the recombination of the dissociated particles. The Pt/Pt–Rh-thermocouple has a good catalytic efficiency, with a recombination coefficient of about 1; the quartz-glass surfaces covered with HPO_3 show the smallest coefficient. The measured temperatures are shown in Fig. 9. The difference between the two profiles demonstrates the dissociation over the total tube cross section. Measurements of the degree of dissociation of similar gas discharges (Crew and Hulbart, 1927; Veprek et al., 1967; Veprek, 1972) indicate that this $\alpha_d \leq 0.1$.

The energy in the dc discharge is transformed into different energy components, e.g., ionization, dissociation, excitation of the molecules and radicals, heating effects of the gas and the wall, as well as radiation. It is possible to calculate only the ratios of ionization, dissociation, and the translational energies.

$$E_{\text{ion}} : E_{\text{trans}} : E_{\text{diss}} = eV_i \alpha_i : kT_g : eV_d \alpha_d = 10^{-5} : 10^{-1} : 1. \tag{13}$$

Therefore, the main part of the energy is stored in the dissociation of the molecular species.

Mass spectroscopy of the HCN laser plasma formed by the dc discharge in the gas mixture $CH_4 : N_2 = 1:1$ shows the formation of new molecules (Schötzau and Kneubühl, 1975b): H, H_2, H_3; NH_x, N_2H_y, N; CH_x, C^2H_y;

CN, HCN. The concentration of HCN molecules is small compared to the concentration of hydrogen, which is the main product of these reactions. The observation of the formation of higher hydrocarbons agrees well with other experiments (Eckert and Eckert-Reese, 1968).

Figure 10 shows a remarkable phenomenon concerning the relation between the 337-μm HCN laser intensity and the molecular components of the plasma of the gas mixture $CH_4 : N_2 = 1 : 1$. Laser intensity and concentration of CH_4, HCN, and H_2 as determined by mass spectroscopy are plotted versus the dc glow discharge current. The CH_4 concentration decreases with increasing current, whereas the hydrogen concentration increases. The HCN concentration achieves a maximum at the threshold current of the laser emission. The highest value of the laser intensity coincides with the maximum concentration of hydrogen molecules. It does not coincide with the maximum concentration of HCN.

B. PULSED DISCHARGES

The output of an electrically excited gas laser depends on the plasma produced in the lasing medium by the excitation current. Number and distribution of electrons determine the laser gain, the gain profile, and the refractive properties of the lasing medium. The index of refraction $\mathbf{n}$ away from any transition of the partially ionized gas in a longitudinally excited submillimeter-wave laser is given by (Turner, 1977)

$$\mathbf{n}(\lambda) = 1 + \sum_i k_i n_i - 4.5 \times 10^{-14} \lambda^2 n_e, \tag{14}$$

where n_i (cm^{-3}) is the density of a specific species, k_i (cm^3) is the specific refractivity multiplied by its weight, n_e (cm^{-3}) is the free electron density, and λ (cm) is the wavelength. The k_i at $\lambda = 337$ μm for CH_4 is 1.56×10^{-23} cm^3, 1.06×10^{-23} cm^3 for N_2, and 0.13×10^{-23} for He. According to Eq. (14) electron refractive effects are proportional to the square of the wavelength. Thus they can play an important role in the 337-μm HCN laser by changing the optical length of the cavity or the focal length of the mirrors. Because in a pulsed electrically excited 337-μm HCN laser the electron density decays in a time ($\simeq 5$ μs) that is short compared to the thermal diffusion time ($\simeq 5$ ms), the refractive effects due to the neutral gas, which are generally dominant at shorter wavelengths, can also be important in submillimeter-wave lasers. For a HCN laser gas mixture $CH_4 : N_2$ at a pressure of 1 Torr and an electron density of $n_e = 10^{10}$ cm^{-3}, the neutral gas and electron contributions to the refractive index $\mathbf{n}$ at $\lambda = 337$ μm are nearly equal.

The importance of electron refractive effects in long wavelength lasers was recognized early. Steffen and Kneubühl (1968) and many others (Whitbourn et al., 1972; Tait et al., 1973, 1974; Turner, 1977) demonstrated that the delay in pulsed laser emission can be attributed to the decaying electron density. A spacially uniform decaying electron density results in an increase in the optical length of the cavity. A change of the electron density n_e of 1×10^{10} cm^{-3} corresponds at 337 μm to an axial mirror displacement ΔL of about 1 μm. As a consequence of the electron refractive effect, the delay time of the pulsed laser emission is a function of the laser mirror separation L, at least for low excitation current. Figure 11 shows the measured delay time and peak intensity of the pulsed 337-μm HCN emission as function of the resonator length L. The emission was generated by a weak electric excitation of CH_3SCN in a resonator of approximate length $L = 647$ cm, Fresnel number $N = 0.647$, and equipped with plane-parallel mirrors.

For a 1-μs discharge of 15 kA in a $CH_4:N_2$ mixture, the initial electron density is greater than 10^{12} cm^{-3}, but within 5 μs it has decayed to less than 10^{-11} cm^{-3}, and by 40 μs it is less than 10^{-10} cm^{-3} (Turner and Poehler, 1971).

Gradients in the electron density n_e or neutral-gas concentration n_i will produce gradients in the refractive index that will bend a ray propagating through the medium. If the electron density gradient dominates and increases quadratically with radius, the laser medium will act like a converging lens to focus the rays (Yariv, 1971). A similar distribution of neutral gas, with a minimum density on the axis, will act like a divergent lens if it produces the dominant effect. Near a molecular transition, as in the case of the HCN

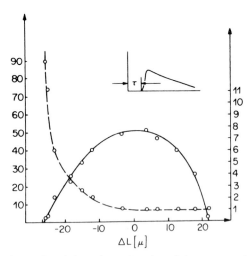

Fig. 11   Delay time and peak intensity as function of the resonator length L for a pulsed 337-μm HCN laser with weak longitudinal excitation, according to Steffen and Kneubühl (1968) (– –, τ = delay (μs); ———, relative intensity).

laser gas mixture $CH_4 : N_2$, the gain or absorption and the dispersion associated with the laser transition can also effect the propagation of the beam (McCaul, 1970; Yariv, 1971). The dispersion produces a change in the index of refraction that depends on the difference between the oscillator frequency and the center frequency of the molecular transition (Kramers–Kronig relation). If this difference is zero, the effect vanishes. If the oscillator frequency is higher than the HCN transition frequency and the density of the HCN molecules is a maximum on the axis, the medium will act like a focusing lens. On the other hand, if the oscillator frequency is lower than the HCN transition frequency, the medium acts like a defocusing lens. A radial gain profile with a maximum on axis that decreases quadratically with distance from the axis will also act like a focusing lens.

Turner (1977) performed detailed measurements of the transmission of a 337-μm HCN laser beam through a lasing gas mixture $CH_4 : N_2$ and the nonlasing gas CH_4 in order to understand the various influences on the pulsed laser output. He concluded that the refractive-index effects due to the electron density gradients are most likely to predominate over the effects of gain, dispersion, and neutral-gas molecules for the time interval of interest. Furthermore, he confirmed the previous statements by Turner and Poehler (1971) that the electron density is initially a maximum at the wall in the considered laser configuration.

A pulsed longitudinal electric discharge excites mechanical oscillations of the plasma tube, as well as standing acoustic waves of the laser gas (Born, 1968). Turner and Poehler (1971) demonstrated that the shock-excited motion of the tube is much greater than the motion imparted to it by the oscillating gas within the tube. The oscillation of a tube vibrating as a circular ring has the period

$$\tau_{tube} = 2\pi R(\rho/E)^{1/2}, \tag{15}$$

where R is the radius of the tube, E the modulus of elasticity, and ρ the density of the wall material. For a glass tube of diameter $2R = 15$ cm, density $\rho = 2.23$ g cm^{-3}, and modulus of elasticity $E = 6.55$ dyn cm^{-2}, Eq. (15) yields a period of $\tau_{tube} = 87$ μs. The periods of the standing acoustic waves in the gas are given by the equations (Born, 1968; Koslov and Shukhtin, 1969),

$$\tau_g = 2R/\alpha_{mn}c_s, \qquad c_s = (C_p/C_v)^{1/2}(kT_g/M)^{1/2}, \tag{16}$$

where c_s is the speed of sound in the gas, k is the Boltzmann constant, M the molecular weight, C_p/C_v the ratio of the specific heats, and α_{mn} the solution of $dJ_m(\pi\alpha)/d\alpha = 0$. For high-current longitudinal discharges, the induced standing acoustic waves cause a multiple-pulsed laser emission with the period τ_g (Turner et al., 1968; Turner and Poehler, 1971). A combination of

Eq. (16) allows us to relate the gas temperature $\overline{T}_g$ to the measured period of the multiple laser pulses.

$$k\overline{T}_g = (MC_v/C_p)(2R/\alpha_{mn}\tau_g)^2. \qquad (17)$$

For a laser gas mixture $CH_4:N_2 = 87:13$ in a laser plasma tube of radius $R = 7.5$ cm, Turner and Poehler (1971) observed a period of $\tau_g = 120\ \mu s$. On the basis of the above equation and with the assumption that the gas was oscillating in the lowest azimuthally symmetric mode, they estimated a gas temperature of $\overline{T}_g = 1700$ K, several hundred microseconds after the current pulse.

The performance of an electrically excited submillimeter-wave laser can be influenced by magnetic fields acting on the laser plasma. Weak fields can significantly increase the laser output (Minoh et al., 1967; Yamanaka et al., 1968), while stronger (Turner, 1974) or asymmetric fields (Brossier and Blanken, 1974) can reduce the output. The addition of an axial magnetic field reduces diffusion of electrons (Turner, 1977) so that they provide more efficient excitation of the molecules. This results in increased gain and higher output. As the magnetic field is increased, a resulting plasma instability destroys the symmetry of the gain profile and thus reduces the output. With a small length of axial magnetic field, electron refractive effects can be produced, which can be used to modulate the output.

IV. Transitions and Mechanisms

Electrically excited far-infrared laser transitions of noble gases were assigned immediately after discovery (McFarlane et al., 1964b; Patel et al., 1964). In comparison, the assignment of stimulated far-infrared molecular transitions is arduous, as mentioned in Section I.E. Thus, we restrict our considerations to molecular transitions (Coleman, 1973).

A. HCN Laser

Three years after the discovery of the so-called CN laser, Lide and Maki (1967) succeeded in explaining that the intense emissions at 337 μm and other nearby lines were transitions involving the 11^10 and 04^00 vibrational states of HCN, which are mixed by Coriolis perturbation. The same year Hocker and Javan (1967) made HCN lines at 373, 335, 310, and 284 μm oscillate cw as pure rotational cascade transitions pumped by the 337 and 311-μm emissions. Frequency measurements of three of these lines confirmed the identification corresponding to the scheme postulated by Lide and Maki (1967). Figure 12 shows the relevant energy level diagram of HCN. Splittings of the 11^10 levels are due to l-type doubling.

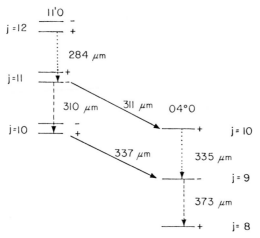

FIG. 12   Energy level diagram of the submillimeter-wave HCN laser, according to Hocker and Javan (1967).

After the successful assignment of the HCN laser emissions, Maki and Lide (1967) performed detailed infrared and microwave measurements on *l*-type resonance doublets in HCN and DCN. Further assignments of HCN laser emissions were achieved by Maki (1967) for lines near 130 μm and explained by a Coriolis resonance which affects the 12^00, 12^10, and 05^10 levels.

Research on the HCN laser mechanism has been rather complex from the beginning. The exact mechanism of the HCN laser is not yet known. One reason for this lack of knowledge may be the possibility that in such discharges there is not one, but several mechanisms of HCN formation and inversion of the molecular states involved in laser action. Another reason is the complexity of the superposition and interconnection of different phenomena occurring in the electrical discharge. First, there is an equilibrium between formation and destruction of HCN molecules by the discharge, which depends on discharge parameters like current, pressure, and wall temperature, and on the nature and constitution of the gases in the discharge. Second, the inversion of the molecular HCN states involved in laser emission may also be produced in the discharge independently of the mechanisms of the formation of the HCN molecules. Hence the process of HCN formation and the inversion mechanism cannot be observed directly, with one independent of the other. The upper and lower laser levels lie, respectively, in the predominantly CN stretching (e.g., 11^00) and bending (e.g., 04^00) vibrational modes of the ground electronic state. The bending modes relax more rapidly than the stretching modes. Hence they help to sustain the inversion. However, the process by which the modes containing the upper laser levels are excited is not well understood. With the hope of throwing new light on this matter, a variety of experiments was performed: pulse delay measurement (Yamanaka et al., 1971; Kunstreich and Lesieur, 1974), determination of the plasma parameters (Schötzau and Kneubühl, 1975a), variation of the chemical composition of the laser gas (Schötzau et al., 1972a), and mass spectroscopy of the laser plasma (Schötzau et al., 1972b; Schötzau and Kneubühl, 1975b). Experiments on the far-infrared HCN laser mechanism revealed a number of remarkable phenomena. The 337-μm HCN emission can be observed with many different gases and gas mixtures containing C, N, and H, e.g., CH_3CN, C_2H_5CN, $CH_4 + N_2$, $CH_4 + NH_3$, but the mixture $C_2H_2 + N_2$ is laser-inactive (Staffsudd et al., 1967; Schötzau et al., 1972a). However, the C_2H_2 + N_2 mixture can be made laser active by adding H_2 or D_2, but not by adding He with the same mass as D_2 (Schötzau and Kneubühl, 1975b). The mixtures $C_2H_2 + N_2$, $C_2H_4 + N_2$, $C_2H_6 + N_2$ have different hydrogen contents as well as different chemical bonds.

As mentioned before, the difference in hydrogen content can be corrected by adding H_2. Then the laser intensity is identical for all mixtures. Yet, it was found (Schötzau and Kneubühl, 1975b) that the electrical input power required for cw laser threshold is a linear function of the heat of formation of the hydrocarbons in the mixture. For pure HCN a pulsed, 337-μm emission is observed after the first electric discharge (Turner and Poehler, 1971; Schötzau and Kneubühl, 1975b), while a cw 337-μm emission occurs only in H_2 is added (Schötzau and Kneubühl, 1975b).

Mass spectroscopic studies demonstrate that the HCN molecules are not stable in a dc glow discharge since they are dissociated by electron impact (Stevenson, 1950; Moran and Hamill, 1963). Recently, gas mixtures containing pure HCN were excited in a waveguide structure by a transverse discharge and by photolysis (Sturzenegger et al., 1979). Strong far-infrared emission at 337 μm was found when excited by pulsed glow discharge. On the other hand, extremely weak 337-μm emission accompanied by strong near-infrared oscillation related to the CN radicals between 1 and 2.4 μm occurred when the laser gas was pumped by UV radiator pulses. Finally, it should be mentioned that the cw 337-μm emission can be influenced, or even caused, not only by the glow discharge but also by reactions of the plasma with the polymer deposits on the wall of the HCN laser plasma tube (Chantry, 1971; Belland and Veron, 1973; Kunstreich and Lesieur, 1975; Schötzau and Veprek, 1975).

Today, the almost general agreement on the far-infrared HCN laser mechanism states that the first step is the dissociation of molecules in the laser plasma by the impact with free electrons that have energies of about 10 eV. Nobody seems to believe in the dominant role of direct excitation of the higher vibrational or electronic levels of HCN by these collisions. Hence the excitation of the higher HCN levels originates in chemical reactions between constituents of the laser plasma, which leads to the formation of HCN. Hitherto, several reactions have been proposed. Pichamuthu (1974) suggested that the reaction of nitrogen atoms with organic radicals produces vibrationally excited HCN and leads to a population inversion. Consequently, he envisaged a HCN chemical laser where active nitrogen and hydrogen atoms react with an alkane + HCl mixture. Unfortunately, operation of this laser has not been reported thus far. Kunstreich and Lesieur (1974) proposed that the reaction $X - CN + H_2 \rightarrow HCN + X + H$ produces the HCN molecules involved in the far-infrared laser emissions. The same mechanism was found by Robinson (1978) for reaction pumping of a $(CN)_2/H_2$ mixture. These results are in reasonable agreement with Schötzau and Veprek (1975), who studied the wall processes in the cw HCN laser. They found that the reaction of the polymer deposit $(CN)_x$ with H_2 is forming HCN responsible for part of the 337-μm emission. Furthermore, spectroscopic experiments revealed that laser oscillation occurred only after formation on CN in the laser plasma (Kon et al., 1967).

B. DCN LASER

As in the interpretation of the far-infrared laser emissions of HCN in Section A, the stimulated far-infrared emissions of DCN were explained by transitions involving levels that are mixed by Coriolis resonance. Maki (1968) assigned the four DCN laser lines near 200 μm previously reported in the literature to transitions involving the $22^0 0$ and $09^{1c}0$ mixed at $J = 21$ about 4950 cm^{-1} above the ground state. Simultaneously, Hocker and Javan (1968b) discovered and assigned cw DCN lines at 190, 195, and 204 μm, oscillating as pure rotational cascade transitions pumped by previously known DCN lines also at 190 and 195 μm.

C. H$_2$O LASER

The water-vapor laser provides more than a hundred lines throughout the entire intermediate and far-infrared range from 7–220 μm, with typical pulse powers from a milliwatt to several hundred watts and cw power up to a few milliwatts (Benedict et al., 1969). Pollack et al. (1967b) discovered and identified competitive and cascade coupling between transitions for several cw laser lines in H$_2$O. Subsequently, Jeffers (1967) demonstrated that the use of a frequency-selective resonator results in enhanced stable pulse-power outputs and many new emissions, due to the removal of competitive interactions. He concluded that competition between lines was typical of water-vapor laser operation. In addition, he suggested that the competing lines were vibrational–rotational transitions between two close lying vibrational states in H_2O. The supposition that the H_2O molecule itself is the most likely to be responsible for laser action in H_2O discharges was also supported by the observation by Tomlinson et al. (1967) that the 118.6-μm transition has a very small effective magnetic moment, of the order of a nuclear magneton. These clues, and analogies with the previous analysis of the HCN laser system (Lide and Maki, 1967), led to three, almost simultaneous, successful efforts to identify the H_2O laser transitions by Hartman and Kleman (1968), Benedict (1968), and Pollack and Tomlinson (1968). These independent identifications covered a total of 26 lines of $H_2{}^{16}O$, and agreed almost completely. Later, Benedict et al. (1969) identified the majority of the known $H_2{}^{16}O$ and $H_2{}^{18}O$ laser transitions and observed new laser lines in $H_2{}^{18}O$ predicted by the perturbation model. Figure 13 shows the energy-level diagram of $H_2{}^{16}O$ evaluated by Benedict et al. (1969), with many of the observed laser transitions identified by the observed laser frequency per centimeter.

Although a large number of perturbations are known to exist in H_2O, the only perturbations that cause laser action are those between the stretching fundamental modes 001 and 100 and the fundamental 010 or the first overtone 020 of the bending mode. On the other hand, strong perturbations that do not result in laser action exist between the 100 and the 001 states. These results suggest that the discharge only produces significant inversions between the stretching and the bending modes, and only in their lower lying states. The precise mechanisms responsible for populating the laser states and producing inversion are not known (Benedict et al., 1969; Pollack, 1969).

As mentioned earlier, one clue to the excitation problem of far-infrared molecular gas lasers is the time delay of pulsed laser signals versus the exciting current pulse. Taking into account the four molecules HCN, H_2O, H_2S, and SO_2, only H_2O lases during the exciting current pulse (Coleman, 1973). This fact suggests that H_2O is directly excited, whereas HCN, H_2S, and SO_2 are indirectly excited. The excitation problem for H_2O has been partially studied by Pichamuthu et al. (1971). Calculations made on a simple model for H_2O, assuming direct electron impact excitation, correlate fairly well with experiments (Coleman, 1973). On the other hand, chemical pumping of the H_2O laser is also possible by flash photolysis of H_2 and O_3 mixtures (Downey and Robinson, 1976).

D. D_2O LASER

D_2O laser emissions from electric discharges were originally reported by Mathias and Crocker (1964), Jeffers and Coleman (1967), and Benedict et al. (1969), but cw operation has been obtained on only a few lines. About 30 lines of $D_2{}^{16}O$ were identified by Benedict et al. (1969). Since the knowledge

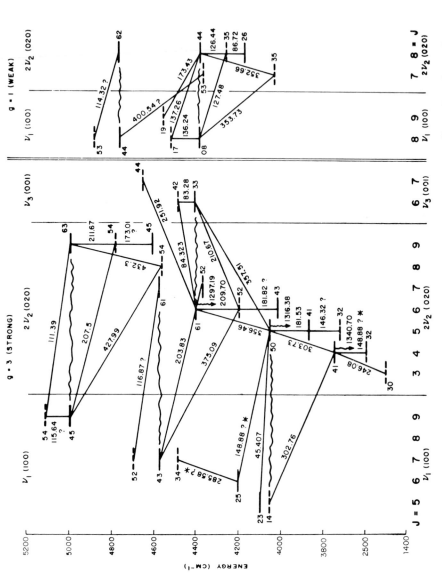

FIG. 13   Energy level diagram of $H_2^{16}O$ showing most of the observed laser transitions, according to Benedict $et\ al$ (1969).

of the energy levels of $D_2{}^{16}O$ from the absorption spectrum was not as complete as in the case of $H_2{}^{16}O$, all of the identification of the lines of D_2O involve levels of 020 first recognized in laser emission.

E. SO₂ LASER

Continuous wave laser action from discharges containing SO_2 and He was first reported by Dyubko et al. (1968). Subsequently, Hassler and Coleman (1968) found pulsed lasing from discharges containing SO_2 and N_2 or He, and Hard (1969) published wavelength measurements on the SO_2 emissions. Later, Hubner et al. (1970) discovered several new lasing lines from SO_2 and made studies of time behavior, relaxation phenomena, and line interactions. A definitive assignment of the laser transitions was performed by Hassler et al. (1973). The assignment involves a pair of irregular perturbations between 100, J, $K = 14$ and 020, J, $K = 16$ for $J = 26$ and 27. This assignment fits the selection rules that $\Delta J = 0$ and that ΔK is even and small for perturbed levels. The 215-μm laser emission represents an exception. Its assignment as a transition from the 001, $J = 27$, $K = 10$ to the 020, $J = 27$, $K = 16$ level is unusual (Hassler et al., 1973).

In the case of the H_2O laser, the output pulse occurs during the excitation pulse, and ends as the excitation stops. This indicates that lasing occurs through a population inversion produced by direct electron impact of the lasing levels (Pichamuthu et al., 1971). By contrast, the SO_2 laser output starts long after the excitation pulse has ceased (Hassler et al., 1973), as well as after the electron density has decayed (Hubner et al., 1970). This led to the conclusion that an indirect mechanism is responsible for lasing in SO_2. Thus, Hassler et al., (1973) suggested that the discharge excites somehow the high vibrational levels of the ground state with subsequent selective relaxation to produce population inversion.

F. REMARKS

Hitherto, no assignment of the laser emissions of ICN gas mixtures at extreme wavelengths up to 774 μm (Steffen et al., 1966a, b; Steffen and Kneubühl, 1968) was performed. The same is true for the stimulated emissions from H_2S on many lines between 33 and 225 μm (Hassler and Coleman, 1969). In both cases, assignments of the laser transitions could not be made due to lack of sufficiently precise spectroscopic data. According to Coleman (1973), the principle of irregular perturbations, which enabled laser line assignments in H_2O and SO_2, probably will also apply to H_2S.

V. Wall Processes

A common phenomenon in electrically excited HCN lasers is a brown-yellow polymer on the inner wall of the discharge tube or structure, which is formed by a discharge-activated polymerization of the gaseous hydrocarbons (Schötzau and Veprek, 1975). This deposit influences the HCN laser output, as demonstrated by Belland and Veron (1973) for the cw 337-μm emission. On many occasions (e.g., Chantry, 1971, Kunstreich and Lesieur, 1975) the 337-μm HCN laser emission was observed in a discharge tube with the polymer wall deposit when the discharge was maintained in pure H_2, H_2O, or NH_3. In a closed system, the 337-μm cw emission lasts for 30 min or longer. This phenomenon takes place because a sufficient number of excited HCN molecules is formed by interaction of the H_2, H_2O, or NH_3 plasma with the polymer deposit. This process influences the radial gain profile in a cw HCN laser with a flowing gas mixture (Schötzau and Veprek, 1975). The influence of the wall deposit on the HCN laser gain is also illustrated by the chemical selection of operating resonator modes (Schötzau and Kneubühl, 1974), which was demonstrated by laser resonator interferometry (Steffen and Kneubühl, 1968).

Chemical analysis shows that the polymer wall deposit in the discharge tube of a common cw HCN laser contains C, N, and H in a ratio of about 10:3:1 (Schötzau and Veprek, 1975). Infrared spectra of these wall deposits exhibit the strong, broad absorption band near 1500 cm^{-1} of cyanogen $(CN)_x$ according to Cotton and Wilkinson (1966), a strong H_2O band at 3300 cm^{-1}, and an extremely weak absorption at 2950 cm^{-1} by the CH stretching mode. Thus, the main constituent of the polymer deposit is cyanogen $(CN)_x$. Mass spectroscopy has proven that the dominant components of a cw HCN laser plasma are hydrogen and nitrogen, whereas the concentrations of hydrocarbons and HCN are rather low (Schötzau and Kneubühl, 1975b). The concentration of H atoms near the wall of the discharge tube is of the order of magnitude 10^{15} cm^{-3}, corresponding to a total flux towards the polymer deposit of about 10^{19} cm^{-2} s^{-1} (Schötzau and Veprek, 1975). The hydrogen impinging on the wall can react with the paracyanogen and form HCN:

$$(CN)_{x,\,solid} + H_{plasma} \longrightarrow (HCN)_{gas}.$$

According to the literature (Zinman, 1960), the collision efficiency of this reaction is about 10^{-3}. This results in a formation rate of HCN molecules on the wall of about 10^{16} cm^{-2} s^{-1} for the total surface in a laser tube. A laser output of 0.1 mW requires less than 10^{18} stimulated transitions. Thus, the polymer on the wall can provide a sufficient number of excited HCN molecules for the HCN laser action observed when pure gaseous H_2, H_2O, or NH_3 is filled in the discharge tube.

In well-known low-pressure gas lasers, such as the CO_2, the He–Ne, or the Ar laser, the amplification factor $\gamma(r)$ decays rapidly with increasing distance r from the tube axis (Herzinger et al., 1966; Franzen and Collins, 1972; Lüthi and Seelig, 1972). On the contrary, in the HCN laser $\gamma(r)$ is nearly

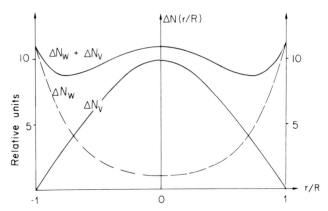

Fig. 14   Radial dependence of the population inversion $N(r)$ in the laser discharge tube of a cw HCN laser caused by wall deposits, according to Schötzau and Verpek (1975).

constant over the whole cross section of the discharge tube (Stafsudd and Yeh, 1969). This surprising phenomenon can be explained in terms of the formation of excited HCN molecules on the wall of the discharge tube (Schötzau and Veprek, 1975). Volume processes in the laser plasma induce a radial dependence of the population inversion ΔN_V and of the gain γ_V, which is described by the Bessel function J_0 (2.4 r/R), where R indicates the tube radius.

On the other hand, collision-induced deactivation of excited HCN molecules, which are formed on the tube wall and diffuse toward the axis, cause a radial dependence of the population inversion ΔN_W and the gain γ_W which is proportional to the Bessel function J_0 ($ir/R\sqrt{\delta R^2/D}$), where D is the diffusion coefficient of the excited HCN molecules, and δ the total probability of deexcitation. The resulting total population inversion $\Delta N = \Delta N_V + \Delta N_W$ and the corresponding amplification factor γ are nearly constant over the whole cross section, as demonstrated in Fig. 14. Thus we conclude that the combined effects of simultaneous volume and wall processes of the plasma make the HCN laser gain constant over the cross section of the discharge tube.

VI.   Pulses

The first stimulated submillimeter-wave emissions from molecular lasers were achieved by pulsed longitudinal excitation (Crocker et al., 1964; Mathias and Crocker, 1964; Gebbie et al., 1964a,b). Yet, three years passed before serious attention was paid to the properties of pulsed emissions

(Kon et al., 1967; Yoshinaga et al., 1967; Minoh et al., 1967; Steffen et al., 1967; Steffen and Kneubühl, 1968; Turner et al., 1968). A large variety of pulse shapes and a complex dependence on the parameters of the discharge current, the resonator, and the molecular plasma were found. For short current pulses with a duration of a few microseconds, a delay from 10 μs to over 100 μs of the stimulated pulsed emission was observed in HCN (Mathias et al., 1965; Steffen et al., 1967; Steffen and Kneubühl, 1968), DCN, H_2S, and SO_2 (Hassler et al., 1973; Coleman, 1973). Soon, remarkable differences between the emissions of weak and strong longitudinal excitations were recognized (Steffen and Kneubühl, 1968; Turner et al., 1968) and were subsequently explained by Steffen and Kneubühl (1968), McCaul (1970), Turner and Poehler (1971), and Turner (1977). In 1973, transverse excitation of the submillimeter-wave pulsed emission of the HCN laser was performed with (Jassby et al., 1973) and without an auxiliary dc discharge (Adam et al., 1973). Transverse electric excitation allows the generation of strong and reproducible submillimeter-wave pulses with HCN lasers (Adam and Kneubühl, 1975; Sturzenegger et al., 1978, 1979).

Magnetic fields influence the performance of submillimeter-wave lasers. Weak fields can significantly enhance the output (Minoh et al., 1967; Yamanaka et al., 1968), while stronger (Turner, 1974) or asymmetric fields (Brossier and Blanken, 1974) can reduce the output. According to Turner (1977), the addition of a magnetic field to an HCN laser acts to reduce diffusion of electrons, so that they provide more efficient excitation of the HCN molecules. This provides increased gain and output. Yet a high magnetic field causes plasma instabilities that destroy the symmetry of the gain profile and reduce the output. Magnetic fields of small extension can produce electron refractive effects that can be used for output modulation.

A. WEAK LONGITUDINAL EXCITATION

A detailed study of pulse shapes of stimulated HCN emissions generated by weak longitudinal electric excitations was performed by Steffen and Kneubühl (1968). The experimental arrangement is shown in Fig. 2. The resonator had planeparallel mirrors, a length of 647 cm, and a Fresnel number $N = 0.647$. The stimulated 337-μm HCN emission was excited in CH_3SCN by pulsed discharges with currents of 2 μs duration and peak values between 0.2 and 0.8 kA. The dependence of the pulse shapes on the resonator length was investigated. For this purpose one of the exit beams was directed onto a He-cooled InSb detector and the other was used for the observation of the laser-resonator interferogram with the aid of a Golay cell and a movable resonator mirror (Fig. 2). The result is shown in Fig. 15 (Steffen and Kneubühl, 1968). In this figure it should be noticed that the resonance lengths of the resonator correspond to waveguide modes (Marcatili and Schmelzer, 1964;

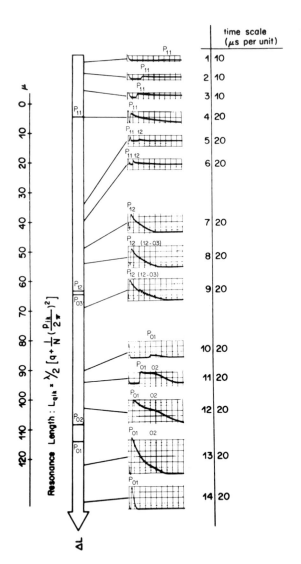

FIG. 15   Pulse shapes of the 337-μm HCN emission of CH_3SCN measured by Steffen and Kneubühl (1968). The resonance lengths were calculated according to waveguide theory.

Steffen and Kneubühl, 1968; Kneubühl and Affolter, 1979). The experiments revealed the following regularities:

(i) The laser pulse exhibits a simple maximum when a single mode is excited.

(ii) The delay of the radiation pulse versus the pulsed electric current, as well as the peak intensity, are asymmetric functions of the deviation of the resonator length from the resonance length (Fig. 11).

(iii) The laser pulse shows beats (Fig. 15, pulse No. 9) whenever the resonator length lies close to the resonance lengths of two nearly degenerate modes.

Figure 15 also demonstrates the remarkable phenomenon that the pulse from a longitudinally excited, molecular, submillimeter-wave laser changes shape when the resonator length of some meters is varied by a few microns. This unpleasant effect can be avoided by transverse electric excitation, which allows higher gas pressures (Adam et al., 1973; Adam and Kneubühl, 1975; Sturzenegger et al., 1978, 1979). The variation of the pulse delay and the peak intensity with the resonator length according to postulate (ii) and shown in Fig. 11 is caused by time-dependent variations of the index of refraction in the laser tube due to the electron and ion density variations in the discharge (Steffen and Kneubühl, 1968).

Q-switching of pulsed molecular submillimeter-wave lasers was attempted by Jones et al. (1969), Frayne (1969) and Yamanaka et al. (1971). Yamanaka et al. (1971) tried Q-switching with the aid of an optically polished stainless steel mirror rotating at frequencies up to 450 Hz. Practically no increase of the peak power was observed for discharge currents of 0.6 kA and 3 μs duration in CH_3CN and $CH_4 + N_2$. This result agreed with the observations of Jones et al. (1969) and contradicted Frayne (1969). Furthermore, the the experiments by Yamanaka et al. (1971) revealed the chemical nature of the HCN laser mechanism.

B. STRONG LONGITUDINAL EXCITATION

With strong pulsed excitation currents, the HCN laser emits 337- and 311-μm radiation in a series of discrete pulses of approximately 50-μs duration, in the order of 50-μs apart, and lasting for almost 0.5 ms (Turner et al., 1968; Turner and Poehler, 1971) as shown in Fig. 16. Jones et al. (1969) also observed multiple emission of pulses under operating conditions quite different from those of Turner et al. (1968). These pulses did not appear until 750 μs after the current pulse.

Initially, this phenomenon was attributed to the difference between the relaxation rate of the lower 04^00 level and the repopulation rate of the upper 11^10 level of the HCN molecule (Turner et al., 1968). Born (1968)

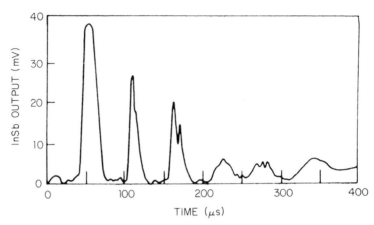

FIG. 16   Multiple pulsed 337-μm and 311-μm emission from an HCN laser according to Turner and Poehler (1971). Laser gas: 0.19 Torr CH_3CN plus He to 0.4 Torr; laser resonator: length 120 cm, diameter 10.3 cm; electric pulse: duration 1 μs, peak 5.5 kA.

suggested that radial acoustic oscillations, most likely produced in the laser gas by the large discharge current, modulate the rates of population and relaxation of the upper and lower laser levels as well as the inversion density sufficiently to produce the observed effect. McCaul (1969) suggested that the repeated pulses result from the decaying electron density in the afterglow plasma periodically tuning the cavity through the line. McCaul (1970) and Kasuya et al. (1969) have shown that in HCN and H_2O lasers the output depends on the radial variation of the index of refraction of the laser gas in the cavity. Steffen and Kneubühl (1968) demonstrated that repopulation of the upper level continues to be important long after the current pulse. Later, Turner and Poehler (1971) demonstrated that even for high-excitation currents, the change of the index of refraction due to the decay of the electron density after the current pulse is too small after 15 μs to affect the laser output or to produce multiple radiation pulses. On the other hand, they found that the current pulse excites standing radial acoustic waves in the laser tube, which at high currents above 9 kA appear to consist of more than a single mode. These acoustic waves have periods as short as 95 μs and produce density and temperature gradients in the gas that are responsible for the modulated laser output as stated by Born (1968).

Steffen and Kneubühl (1968), Withbourn et al. (1972), and Tait et al. (1973) have shown that the delay between electric excitation and HCN laser emission can be attributed to the decaying electron density. Sochor (1968) and McCaul (1970) have suggested that this delay is caused by a radial density gradient acting as a negative lens of sufficient strength to produce an unstable cavity during that interval. This interpretation was accepted by many authors (e.g., Sharp and Wetherell, 1972). Athough this is most probable for cw excitation, it is not necessarily the case for pulsed excitation (Turner and Poehler, 1971). Turner (1977) demonstrated that in a fast-pulsed HCN laser, the electrons responsible for the laser output move in from the wall to the axis. During this process a strong negative gradient in the electron density results and produces a strong focusing effect. Then, the delay between excitation and laser output is due to the time required for the high electron density region to reach the axis and to induce collisions required for the HCN inversion. At this time, the electron density distribution produces a defocusing effect. Since the laser gain is related to the densities of electrons and HCN molecules, gain and refractive effects of the medium are strongly tied to each other and to the radial dynamics of the plasma. As a result, gain profile and cavity refraction can be very complicated, particularly for high excitation currents.

C. TRANSVERSE EXCITATION

Pulsed transverse excitation of the stimulated emission of gases at high pressure was first reported by Beaulieu (1970) for the CO_2 laser. Transversely excited lasers represent significant progress compared to standard longitudinally excited high-power gas lasers. They provide stronger and shorter pulses. Consequently, Wood et al. (1971) applied transverse excitation to various atomic and molecular gas mixtures and thus obtained laser emissions at wavelengths between 0.8 and 28 μm.

Application of transverse excitation to the 337-μm HCN laser was suggested by Sharp and Wetherell (1972) in order to avoid column instabilities in pulsed lasers at high-current density. Later, Jassby et al. (1973) also proposed the use of a transverse discharge for a 337-μm HCN laser working at higher gas pressures. However, Coleman (1973) expressed serious doubts on the application of transverse excitation to molecular submillimeter-wave lasers. Nevertheless, Lam et al. (1973) achieved HCN laser oscillation by transverse excitation with the aid of an auxiliary discharge forming the laser plasma. The same year, the first standard transverse excitation of the HCN laser was reported by Adam et al. (1973). For this purpose they had developed a hollow waveguide with dielectric side walls and metal electrodes on top and bottom (Adam and Kneubühl, 1975). Their laser was operated successfully at pressures up to 5 Torr. Figures 6 and 7 show the laser and its waveguide resonator.

In low-pressure molecular submillimeter-wave lasers Doppler width and pressure broadening are smaller than mode separation (Steffen and Kneubühl, 1968). Consequently, the output varies strongly with the resonator length (Section A) and permits laser resonator interferometry. On the other hand, this phenomenon makes pulsed, longitudinally excited, molecular submillimeter-wave lasers unreliable. Transverse excitation of the HCN laser allows an increase of the working pressure, and the pressure broadening that results causes shorter pulsed emissions independent of the resonator length. This was demonstrated experimentally (Adam et al., 1973; Adam and Kneubühl, 1975).

In order to improve the performance of the transversely excited 337-μm HCN laser, Sturzenegger et al. (1977, 1978) applied UV preionization and introduced a flow system working with pure HCN as a constituent of the laser gas. Ultraviolet preionization has proven to be an effective means to obtain stable and uniform discharge in TEA CO_2 lasers (Wood, 1974). Sturzenegger et al. (1978, 1979) found that UV preionization allows an increase in the working pressure of the HCN laser gas up to 18 Torr. Unfortunately, the effect of an auxiliary discharge for UV preionization on the output was small. The pulse energy was enhanced only by a factor of 2 or 3, yet the pulse to pulse reproducibility was increased. On the other hand, the introduction of a flow system proved successful for gas mixtures $HCN:CH_4:He$. The flow system was required because the transverse discharge destroys molecules in the laser gas at a high rate. The introduction of the flow system resulted in an increase of the output power by a factor of the order of 10. Optimum output at a pulse repetition rate below 1 Hz was found for the flow rate

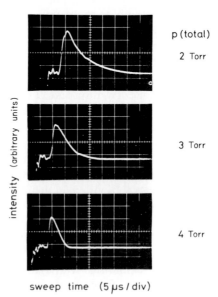

FIG. 17   Pulse shapes from a transversely excited 337-μm HCN laser with UV preionization, flow system, and operated with a gas mixture $HCN:CH_4:He$ (see text).

HCN:CH$_4$:He = 5:30:40 Torr liter min^{-1}, a pressure of 2 Torr, and an input energy of 56 J liter^{-1}. The pulse length of the laser output varied from 27 μs at 2 Torr to 8.5 μs at 4 Torr, as shown in Fig. 17. The pulse energy is between 1 and 15 mJ.

As a conclusion to the results obtained by Sturzenegger et al. (1978, 1979), we can postulate that introduction of transverse excitation, UV preionization, gas mixtures with HCN, and a flow system in the 337-μm HCN laser results in strong and stable output pulses as well as in a drastic increase of input and output pulse energy per volume.

VII.   Energy and Power

The absolute measurement of power and pulse energy of cw and pulsed submillimeter-wave lasers is still a matter of frequent disappointment and controversy. Hitherto, even well-equipped and experienced far-infrared laboratories have made mistakes in the determination of output power. Therefore, published data should be considered with care.

A variety of devices have been used for the measurement of the output power of cw submillimeter lasers: Golay cells, pyroelectric detectors, liquid crystal detectors, H$_2$O, and commercial calorimeters (e.g., Scientech). Typical 337-μm HCN lasers and their characteristics and measured output powers are listed in Table III. For comparison we include in this table the specific output power defined as output power per resonator volume. In parentheses we also present the output power per mode volume postulated by the authors. The data give some insight into the performance of these lasers. However, looking at the remarkable differences in specific power, one may question the ultimate reliability of the measurements. Recently similar results have been reported by Bruneau et al. (1978) on the output characteristics of a cw DCN waveguide laser with a cavity volume of 0.56 liter and an output power close to 8 mW. To our knowledge, today's strongest electrically excited molecular submillimeter-wave lasers reach a cw power of the order of 1 W at 337 μm.

Pulse energy and power measurements of the emission of pulsed submillimeter-wave lasers are based on the measurement of energy with Golay cells or pyroelectric detectors with special coatings, and on the determination of the laser pulse shape with a cooled detector, e.g., a He-cooled InSb detector (Turner and Poehler, 1971; Sharp and Wetherell, 1972). Recently, a serious attempt of absolute measurement has been reported by Woskoboinikow et al. (1979). Table IV shows the characteristics of three longitudinally and three transversely electrically excited HCN lasers. The performance of these lasers can be judged by the efficiency, defined as the ratio of the output pulse energy versus the electric input energy. For a longitudinally excited

TABLE III

Output Power and Other Data of cw 337-μm HCN Lasers

Authors	Volume (liters)	Resonator	Output coupling	Laser gas	Detector	Power (mW)	Specific power (mW liter^{-1})
Chamberlain et al. 1966	11.5	Concentric	Michelson	$CH_4:NH_3$	Liquid crystal	5	0.4
Kotthaus (1968)	51.0	Large radius mirrors	Hole	$N_2:(C_2H_5)_2O$	Calorimeter	600	11.8
Becklacke and Smith (1970)	3.0	Confocal	Hole	$CH_4:N_2$	Pyroelectric	7	2.4
Volk (1972)	30.6	Concentric	Hole	$CH_4:N_2:He$	Golay	50	1.6
Bicanic and Dymanus (1974)	31.4	Semiconfocal	Michelson	$CH_4:NH_3$	Pyroelectric	45	1.4
Belland and Veron (1973)	2.0	Concave	Beam splitter	$CH_4:N_2:He$	Calorimeter	20	10.2 (90)
Belland et al. (1975)	3.9	Waveguide	Beam splitter	$CH_4:N_2:He$	Calorimeter	100	25.4 (83)
Belland et al. (1976)	0.5	Waveguide	Metal mesh	$CH_4:N_2:He$	Calorimeter	3	6.2

TABLE IV

Pulse Energy, Average Pulse Powers, and Efficiency of Electrically Excited HCN Lasers

Authors	Excitation	Discharge volume (liter)	Input Energy (J)	Input Energy (J liter⁻¹)	Output Energy (mJ)	Output Energy (μJ liter⁻¹)	Power (W)	Efficiency
Turner and Poehler (1971)	Longitudinal	$\simeq 39$	160	$\simeq 4$			25	
Sharp and Wetherell (1972)	Longitudinal	$\simeq 60$	$\simeq 110$	$\simeq 1.8$	$\simeq 17$	280	1000	$\simeq 1.5 \times 10^{-4}$
Jassby et al. (1973)	Longitudinal	$\simeq 106$	$\simeq 156$	$\simeq 1.5$	3	30	300	$\simeq 2 \times 10^{-5}$
Lam et al. (1973)	Transverse	$\simeq 22$	$\simeq 180$	$\simeq 8$	$\simeq 0.045$	$\simeq 2$	9	$\simeq 2.5 \times 10^{-7}$
Adam et al. (1973)	Transverse	10	312	31	>0.1	>10	>10	$>3 \times 10^{-7}$
					<1	<100	<100	$<3 \times 10^{-6}$
Sturzenegger et al. (1979)	Transverse	1.6	90	56	>1	>620	>100	$>10^{-5}$
					<15	<9400	<1500	$<1.5 \times 10^{-4}$

HCN laser, an efficiency of the order of 10^{-4} is certainly unrealistic. Nevertheless, with the present techniques of electrically excited molecular lasers, reproducible submillimeter-wave pulses with a duration of about 10 μs, an average power up to 10 kW, and a pulse energy of 50 mJ can be generated.

VIII. Applications

Principal applications of electrically excited molecular submillimeter-wave lasers are in solid state spectroscopy and plasma diagnostics. In addition, potential applications lie in the fields of gaseous molecular spectroscopy (Shimuzu et al., 1968; Duxbury and Burroughs, 1970) and spectroscopy of molecules in interstellar space (Bahl, 1971; Robinson, 1976; Mikimoto, 1979).

A. SOLID STATE PHYSICS

As mentioned in Section I.E, electrically excited HCN lasers found early applications in solid state physics (Button, 1979). Cyclotron resonance with 337-μm HCN lasers was first performed in InSb (Button et al., 1966, 1968; Murotani and Nisida, 1970). Later, Därr et al. (1975) observed surface electron resonance in InSb inversion layers. Subsequently, von Ortenberg and Steigenberger (1977) investigated the surface cyclotron resonance in accumulation layers on n-type InSb with electrically excited submillimeter-wave lasers in order to compare their results with those obtained for inversion layers. They found a strong dependence of the surface cyclotron mass on the electric field and hence on the surface density of electrons. Furthermore, they concluded that within the quantizing range the effective potentials for accumulation and inversion in InSb differ. Cyclotron resonance with submillimeter-wave lasers was not only applied successfully to InSb but also to other semiconductors (Landwehr, 1977). Cyclotron resonance was observed in n-CdS by Button et al. (1970), as well as in Si (Landwehr, 1977) and in Ge (Kuchar, 1977; Otsuka et al., 1977). Furthermore, extensive studies of cyclotron resonance with 337- and 311-μm HCN lasers on bulk Te and on Te accumulation and inversion layers by von Ortenberg and Silbermann (1975, 1976).

Electrically excited submillimeter-wave lasers have also been used for the generation and the study of coherent and incoherent terahertz phonon pulses in solids. First attempts were made with the aid of a 337-μm HCN laser and Si crystals covered with an In film on one end and with a Sn superconducting bolometer on the other end (Schötzau et al., 1970a, b). Phonons were generated by mechanisms described by Abeles (1967) and by Halbritter (1970). Later, Grill and Weis (1975) excited coherent and incoherent terahertz phonon pulses in quartz using the electrically excited laser emissions of HCN at 337 μm and of H_2O at 28 and 118 μm. A survey on these experiments and calculations (Weis, 1975) on the excitation of short phonon pulses in the terahertz frequency was given the following year by Weis (1976).

B. Plasma Physics

Probably the first application of an electrically excited molecular sub-millimeter-wave laser to plasma diagnostics was the determination of the radial distribution of the electron density of a steady-state helium plasma with a maximum electron density of 1.5×10^{14} cm^{-3} by Kon et al. (1968). In the 1970s the research on tokamak plasmas resulted in an increase of the electron density and the plasma diameter. As a consequence, microwave interferometers for the determination of electron densities had to be replaced by submillimeter-wave interferometers. Therefore, French (Veron, 1974; Veron et al., 1977) and Japanese (Nishiziwa et al., 1976; Nishiziwa, 1978) research teams built HCN laser interferometers for the diagnostics of toka-mak plasmas. With these interferometers, spatial and temporal variations of electron densities between $10^{13}-10^{14}$ cm^{-3} in tokamak plasmas were measured. The first multichannel HCN interferometer has been operated since 1976 by Veron et al. (1977). At present, a submillimeter-wave polari-interferometer for simultaneous electron density and magnetic field measure-ments in tokamak plasmas is being developed by Dodel and Kund (1978).

According to Spalding (1979), specific applications for far-infrared diag-nostic lasers appear to lie in the following areas:

(i)   Far-infrared interferometry and correlation measurements at the relatively high densities inaccessible to microwave technique.

(ii)   Scattering under conditions where long wavelength fluctuations due to ion motions or instabilities can be probed at finite, i.e., experimentally feasible scattering angles with acceptable spatial resolution, i.e., $\alpha \geqq 1$ scattering when the Debye wavenumber is of the same order as $|k_o - k_s|$.

(iii)   Electron cyclotron absorption measurements, to complement present emission (ECE) measurements. It is noteworthy that ECE success-fully competes with scattering techniques in providing continuous measure-ments of T_e. The development of high-repetition-rate diagnostic lasers would clearly permit multiple-shot diagnostics, while retaining excellent spatial resolution with scattering techniques.

IX.   Conclusions

Fifteen years ago electrically excited molecular lasers were the only sources producing useful power levels of monochromatic submillimeter waves. In combination with high-frequency backward-wave oscillators (Carcinotrons) they closed the power gap between infrared and microwaves. Therefore, and since they were easy to build, many laboratories started to study their properties and mechanisms and applied them to solid-state spectroscopy, to the determination of atmospheric transmission, and to precise frequency and wavelength measurements. Simultaneously, the hunt for new laser emissions started. Not many molecules (HCN, DCN, ICN, H_2O, D_2O, SO_2, H_2S) were found to lase by electric excitation in the far infrared, yet the few emitted a reasonable number of relatively strong lines. Soon, most of the laser transitions were disclosed, but most mechanisms of population inversion remained hidden. On the other hand, because the mode separation surpasses Doppler width and pressure broadening in these low-pressure lasers, it allowed detailed studies of modes and the first experimental investigation and realization of waveguiding in laser resonators.

In 1970 two discoveries gave a new impetus to research on molecular lasers: the TEA laser and the optically pumped submillimeter-wave laser. The TEA laser produced strong pulses, while the optically pumped laser provided an immense number of lines. Therefore, laser studies concentrated on new emissions, mechanisms and waveguide resonators of optically pumped lasers, and on the extension of transverse electric excitation into the submillimeter-wave range. Since then most of the goals have been reached.

In recent years, tokamak plasmas increased in electron density and in plasma diameter. In consequence, submillimeter-wave lasers with specific properties are required for plasma diagnostics. Today, insiders notice that this is a good draft-horse for the future development of optically pumped as well as of electrically excited molecular submillimeter-wave lasers.

ACKNOWLEDGMENTS

Research on electrically excited submillimeter-wave lasers by the authors was and is supported by ETH, Zurich, and the Swiss National Science Foundation under various contracts.

The authors wish to thank B. Adam, Baden; H. J. Schötzau, Aarau; H. Steffen, Basel; J. Steffen, Thun; E. Affolter, W. Herrmann, K. Leutwyler, M. Rohr, H. R. Vogt, ETH, Zürich, for active help in their experimental and theoretical studies on longitudinally and transversely excited submillimeter-wave lasers.

For many valuable discussions in the course of their research and writing of this review the authors are greatly indebted to K Berger, ETH, Zurich; K. J. Button, MIT, Cambridge; G. W. Chantry, NPL, Teddington; P. D. Coleman, Urbana, Illinois; H. A. Gebbie, Appleton Lab., Slough; L. Genzel, MPI, Stuttgart; D. T. Hodges, Aerospace Corp., Los Angeles; F. Keilmann, MPI, Stuttgart; B. Lax, MIT, Cambridge; B. Z. Lengyel, Northridge, California; W. Lukosz, ETH, Zurich; P. L. Richards, Berkeley, California; W. Rieder, Wien; H. Schneider, Fribourg; S. Veprek, Zurich; and H. Yoshinaga, Osaka.

Furthermore, the authors are also very indebted to Miss Ch. Noll and Mrs S. Buchegger for typing the manuscript.

REFERENCES

Abeles, B. (1967). *Phys. Rev. Lett.* **19**, 1181-1183.
Adam, B., and Kneubühl, F. K. (1976). *Appl. Phys.* **8**, 281-291.
Adam, B., Schötzau, H. J., and Kneubühl, F. K. (1973). *Phys. Lett.* **45A**, 365-366.
Akitt, D. P., Jeffers, W. Q., and Coleman, O. D. (1966). *Proc. IEEE* **54**, 547-551.
Bahl, D. (1971). *Nature (London)* **234**, 332-334.
Barr, E. S. (1965). *Am. J, Phys.* **18**, 76.
Beaulieu, A. J. (1970). *Appl. Phys. Lett.* **16**, 504-505.
Becklacke, E. J., and Smith, M. A. (1970). *Rad. Electr. Eng.* **39**, 161.
Bekefi, G. (ed.) (1976). "Principles of Laser Plasmas." Wiley, New York.
Belland, P. (1975). Ph.D. Thesis, Univ. P. et M. Curie, Paris VI, Nov. 3, C.N.R.S.-A.O. 12715, and Report EUR-CEA-FC-806 (January 1976).
Belland, P., and Veron, D. (1973). *Opt. Commun.* **9**, 146-148.
Belland, P., Ciura, A. I., Veron, D., and Withbourn, L. B. (1974). *Digest Int. Conf. Submm Waves and Their Appl., 1st, Atlanta, Georgia* pp. 75-77.
Belland, P., Veron, D., and Withbourn, L. B. (1975). *J. Phys. D Appl. Phys.* **8**, 2113-2122.
Belland, P., Pigot, C., and Veron, D. (1976). *Phys. Lett.* **56A**, 21-22.
Benedict, W. S. (1968). *Appl. Phys. Lett.* **12**, 170-173.
Benedict, W. S., Pollack, M. A., and Tomlison, W. J. III (1969). *IEEE J. Quant. Electron.* **QE-5**, 108-124.
Bergstein, L., and Schachter, H. (1964). *J. Opt. Soc. Am.* **54**, 887-903.
Bicanic, D. D., and Dymanus, A. (1974). *Infrared Phys.* **14**, 153-163.
Bicanic, D. D., and Dymanus, A. (1975). *Opt. Commun.* **15**, 175-178.
Bletzinger, P., and Garscadden, A. (1971). *Proc. IEEE* **19**, 675.
Boettcher, J., Dransfeld, K., and Renk, K. F. (1968). *Phys. Lett.* **26A**, 146-148.
Born, G. K. (1968). *J. Appl. Phys.* **39**, 4479.
Brannen, E., Sochor, V., Sarjeant, W. J., and Froelich, H. R. (1967). *Proc. IEEE* **55**, 462-463.
Broida, H. P., Evenson, K. M., and Kikuchi, T. T. (1965). *J. Appl. Phys.* **36**, 3355.
Brossier, P., and Blanken, R. A. (1974). *IEEE Trans. MTT* **MTT-22**, 1053-1056.
Bruneau, J. L., Belland, P., and Veron, D. (1978). *Opt. Commun.* **24**, 259-264.
Burroughs, W. J., Pyatt, E. C., and Gebbie, H. A. (1966). *Nature (London)* **212**, 387-388.
Button, K. J. (1979). *Proc. Int. Conf. Infrared Phys., 2nd, (CIRP 2) Zürich* 31-49.
Button, K. J., Gebbie, H. A., and Lax, B. (1966). *IEEE J. Quant. Electron.* **QE-2**, 202-207.
Button, K. J., Lax, B., and Bradley, C. C. (1968). *Phys. Rev. Lett.* **21**, 350-352.
Button, K. J., Lax, B., and Cohn, D. R. (1970). *Phys. Rev. Lett.* **24**, 375-378.
Byer, R. L., Herbst, R. L., and Kildal, H. (1972). *Appl. Phys. Lett.* **20**, 463-466.
Chamberlain, J. E. *et al.* (1966) *Infrared Phys.* **6**, 195.
Chang, T. Y. (1974). *IEEE Trans. MTT* **MTT-22**, 983-988.
Chang, T. Y. (1977a). *Topics Appl. Phys.* **3**, (16), 215-274.
Chang, T. Y. (1977b). *IEEE J. Quant. Electron.* **13**, 937-942.
Chang, T. Y., and Bridges, T. J. (1970). *Opt. Commun.* **1**, 423-426.
Chang, T. Y., and McGee, J. D. (1976). *IEEE J. Quant. Electron.* **QE-12**, 62-65.
Chantry, G. W., Gebbie, H. A., and Chamberlain, J. E. (1965). *Nature (London)* **205**, 377.
Chantry, G. (1971). "Submillimeter Spectroscopy." Academic Press, New York.
Coleman, P. D. (1963). *IEEE Trans. MTT* **MTT-11**, 271-288.
Coleman, P. D. (1973). *IEEE J. Quant. Electron.* **QE-9**, 130-138.
Coleman, P. D. (1977). *J. Opt. Soc. Am.* **67**, 894-901.
Cotton, F. A., and Wilkinson, G. (1966). "Advances Inorganic Chemistry." Wiley, New York.

Crew, W. H., and Hulbart, E. D. (1927). *Phys. Rev.* **30**, 124–137.

Crocker, A., Gebbie, H. A., Kimmitt, M. F., and Mathias, L. E. S. (1964). *Nature* (*London*) **201**, 250–251.

Därr, A., Kotthaus, J. P., and Koch, J. F. (1975). *Solid State Commun.* **17**, 455–458.

DeTemple, T. A. (1977). *Proc. SPIE* **105**, 11–16.

Dodel, G., and Kunz, W. (1978). *Infrared Phys.* **18**, 773–776.

Downey, G. D., and Robinson, D. W. (1976). *J. Chem. Phys.* **64**, 2858–2862.

Duxbury, G., and Burroughs, W. J. (1970). *J. Phys. B* **3**, 98–111.

Duybko, S. F., Svich, V. A., and Valitov, R. A. (1968). *Sov. Phys.-JETP* **7**, 320.

Eckert, H. W., and Eckert-Reese, G. (1968). *Forschungsbericht Nordrhein-Westfalen*, No. 1554, April.

Engel, A. von, and Steenbeck, M. (1934). "Elektrische Gasentladungen." Springer-Verlag, Berlin and New York.

Evenson, K. M., Broida, H. P., Wells, J. S., Mahler, R. J., and Mizushima, M. (1968). *Phys. Rev. Lett.* **21**, 1038–1040.

Flesher, G. T., and Mueller, W. M. (1966). *Proc. IEEE* **54**, 543–546.

Francis, G. (1956). The glow discharge at low pressure. *In* "Handbuch der Physik," Vol. XXII. Springer-Verlag, Berlin and New York.

Franzen, D. L., and Collins, R. J. (1972). *IEEE J. Quant. Electron.* **8**, 400.

Frayne, P. G. (1969). *J. Phys. B.* (*A. Mol. Phys.*) **2**, 247.

Frenkel, L., Sullivan, T., Pollack, M. A., and Bridges, T. J. (1967). *Appl. Phys. Lett.* **11**, 344–345.

Gallagher, J. J., Blue, M. D., Bean, B., and Perkowitz, D. (1977). *Infrared Phys.* **17**, 43–55.

Garrett, C. G. B. (1965). *Int. Sci. Technol.* March, 39–44.

Garrett, C. G. B. (1966). "Physics of Quantum Electronics" (P. L. Kelley, B. Lax, and P. E. Tannenwald, eds.), pp. 557–566, McGraw-Hill, New York.

Gebbie, H. A., Stone, N. W. B., and Findlay, F. D. (1964a). *Nature* (*London*) **202**, 169–170.

Gebbie, H. A., Stone, N. W. B., and Findlay, F. D. (1964b). *Nature* (*London*) **202**, 685.

Gebbie, H. A., Stone, N. W. B., Findlay, F. D., and Pyatt, E. C. (1965). *Nature* (*London*) **205**, 377–378.

Gebbie, H. A., Stone, N. W. B., Slough, W., Chamberlain, J. E., and Sheraton, W. A. (1966). *Nature* (*London*) **211**, 62.

Gebbie, H. A., Stone, N. W. B., Putley, E. H., and Shaw, N. (1967). *Nature* (*London*) **214**, 165–166.

Glagolewa-Arkadiewa, A. (1924). *Z. Phys.* **24**, 153–165.

Grill, W., and Weis, O. (1975). *Phys. Rev. Lett.* **35**, 588–591.

Hadni, A., Thomas, R., and Weber, J. (1967). *J. Chim. Phys.* **64**, 71–79.

Hadni, A., Charlemagne, D., and Thomas, R. (1968). *C.R. Acad. Sci. Paris* **266B**, 1230–1233.

Halbritter, J. (1970). *Ext. Ber.*, *3/70-3*, Kernforschungszentrum Karlsruhe, *J. Appl. Phys.* **41**, 4581–4588.

Hard, T. M. (1969). *Appl. Phys. Lett.* **14**, 130.

Hartmann, B., and Kleman, B. (1968). *Appl. Phys. Lett.* **12**, 168–170.

Hassler, J. C., and Coleman, P. D. (1969). *Appl. Phys. Lett.* **14**, 135–136.

Hassler, J. C., Hubner, G., and Coleman, P. D. (1973). *J. Appl. Phys.* **44**, 795–801.

Herziger, G., and Holzapfel, Seelig, W. (1966). *Z. Phys.* **189**, 385.

Hocker, L. O., and Javan, A. (1967). *Phys. Lett.* **25A**, 489–490.

Hocker, L. O., and Javan, A. (1968a). *Phys. Lett.* **26A**, 255–256.

Hocker, L. O., and Javan, A. (1968b). *Appl. Phys. Lett.* **12**, 124–125.

Hocker, L. O., Javan, A., Rao, D. R., Frenkel, L., and Sullivan, T. (1967). *Appl. Phys. Lett.* **10**, 147–148.

Hodges, D. T. (1978). *Digest Int. Conf. Submm Waves Their Appl. 3rd Guildford* pp. 18–20.

Hodges, D. T., and Hartwick, T. S. (1973). *Appl. Phys. Lett.* **23**, 252–253.
Horiuchi, Y., and Murai, A. (1976). *IEEE J. Quant. Electron.* **QE-12**, 547–549.
Hubner, G., Hassler, J. C., and Coleman, P. D. (1970). *MRI Symp. Ser.* **20**, 69.
Hubner, G., Hassler, J. C., and Coleman, P. D. (1971). *Appl. Phys. Lett.* **18**, 511–512.
Jassby, D. L., Marhic, M. E., and Regan, P. R. (1973). *Appl. Opt.* **12**, 1403–1404.
Javan, A., Bennett, W. R., Jr., and Herriot, D. R. (1961). *Phys. Rev. Lett.* **6**, 106.
Jeffers, W. Q. (1967). *Appl. Phys. Lett.* **11**, 178–180.
Jeffers, W. Q., and Coleman, P. D. (1967). *Proc. IEEE Lett.* **55**, 1222–1223.
Jones, R. G., Bradley, C. C., Chamberlain, J., Gebbie, H. A., Stone, N. W. B., and Sixsmith, H. (1969). *Appl. Opt.* **8**, 701.
Kagan, Y. M., and Mitrofanov, N. K. (1971). *J. Tech. Phys.* **41**, 2065–2072.
Kasuya, T., Shimoda, K., Takeuchi, N., and Kobayashi, S. (1969). *Jpn. J. Appl. Phys.* **8**, 478.
Kenty, C., Easley, M. A., and Barnes, B. T. (1951). *J. Appl. Phys.* **22**, 1006–1011.
Kleen, W., and Müller, R. (1969). "Laser," p. 219. Springer-Verlag, Berlin and New York.
Kneubühl, F. K. (1977). *J. Opt. Soc. Am.* **67**, 959–963; *Anales Sci. Univ. Clermont* **64**, 117–127.
Kneubühl, F. K. (1978). *Digest Int. Conf. Submm Waves Their Appl. 3rd, Guildford* pp. 21–22.
Kneubühl, F. K., and Affolter, E. (1979). Infrared and submillimeter-waveguides. *In* "Infrared and Submillimeter Waves" (K. J. Button, ed.), Vol. 1. Academic Press, New York.
Kneubühl, F. K., Moser, J. F., Steffen, H., and Tandler, W. (1965). *J. Appl. Math. Phys.* **16**, 560–561.
Knight, D. J. E. (1978). *Nat. Phys. Lab. Rep.*, No. Qu 45.
Kon, Sh., Yamanaka, M., Yamamoto, J., and Yoshinaga, H. (1967). *Jpn. J. Appl. Phys.* **6**, 612–619.
Kon, Sh., Otsuka, M., Yamanaka, M., and Yoshinaga, H. (1968). *Jpn. J. Appl. Phys.* **7**, 434.
Kotthaus, J. P. (1968). Diploma Thesis, Tech. Univ., Munich.
Kozlov, Y. G., and Shukhtin, A. M. (1969). *Sov. Phys.-Tech. Phys.* **13**, 1197–1202.
Kuchar, F. (1977). *J. Opt. Soc. Am.* **67**, 935–938.
Kunstreich, S., and Lesieur, J. P. (1974). CERCEM-LOC, F-93'350 Le Bourget, Rapport interne No. 74/123.
Kunstreich, S., and Lesieur, J. P. (1975). *Opt. Commun.* **13**, 17–20.
Lam, M. F., Jassby, D. L., and Casperson, L. W. (1973). *IEEE J. Quant. Electron.* **QE-8**, 851–852.
Landwehr, G. (1977). *J. Opt. Soc. Am.* **67**, 922–928.
Li, T., and Zucker, H. (1967). *J. Opt. Soc. Am.* **57**, 984.
Lide, D. R., and Maki, A. G. (1967). *Appl. Phys. Lett.* **11**, 62–64.
Lowenstein, E. V., and Smith, D. R. (1971). *Appl. Opt.* **10**, 577–583.
Lüthi, H. R., and Seelig, W. (1972). *J. Appl. Math. Phys.* **23**, 665.
Maki, A. G. (1968). *Appl. Phys. Lett.* **12**, 122–124.
Maki, A. G., and Lide, D. R. (1967). *J. Chem. Phys.* **47**, 3206–3210.
Marcatili, E. A. J., and Schmeltzer, R. A. (1964). *Bell Syst. Tech. J.* **43**, 1783–1809.
Mathias, L. E. S., and Crocker, A. (1964). *Phys. Lett.* **13**, 35.
Mathias, L. E. S., Crocker, A., and Wills, M. S. (1965). *Electron. Lett.* **1**, 45–46.
McCaul, B. W. (1969). *Stanford Univ. Rep., No. 1750.*
McCaul, B. W. (1970). *Appl. Opt.* **9**, 653–663.
McCumber, D. E. (1965). *Bell Syst. Tech. J.* **44**, 333–363.
McFarlane, R. A., Faust, W. L., Patel, C. K. N., and Garrett, C. G. B. (1964a). *Quantum Electron.* **3**, 573–586.
McFarlane, R. A., Faust, W. L., Patel, C. K. N., and Garrett, C. G. B. (1964b). *Proc. IEEE* **52**, 318.
Minoh, A., Shimuzu, T., Kobayahi, Sh., and Shimoda, K. (1967). *Jpn. J. Appl. Phys.* **6**, 921–930.

Moran, T. F., and Hamill, W. M. (1963). *J. Chem. Phys.* **39**, 1413–1422.
Morimoto, M. (1979). *Int. Trans. Astron. Un.* **17A**, Part 3, 92–96.
Moser, J. F., Steffen, H., and Kneubühl, F. K. (1968). *Helv. Phys. Acta* **41**, 607–644 (in German).
Moser, J. F., Steffen, H., and Kneubühl, F. K. (1969). *Usp. Fiz. Nauk* **99**, 469 (in Russian).
Murotani, T., and Nisida, Y. (1970). *Solid State Commun.* **8**, 755–758.
Nichols, E. F., and Tear, J. D. (1923). *Phys. Rev.* **21**, 587–610.
Nichols, E. F., and Tear, J. D. (1925). *Astrophys. J.* **61**, 17–37.
Nishizawa, A. (1978). *Digest Int. Conf. Submm Waves Their Appl. 3rd, Guildford* pp, 270–271.
Nishizawa, A., Masuzaki, M., and Mohri, A. (1976). *Jpn. J. Appl. Phys.* **15**, 1753–1760.
Novgorodov, M. Z., Sviridov, A. G., and Sobolev, N. N. (1971). *IEEE J. Quant. Electron.* **7**, 508–512.
Oka, T., and Shimuzu, T. (1970). *Phys. Rev. A* **2**, 587–593.
Okajima, S., and Murai, A. (1972). *IEEE J. Quant. Electron.* **QE-8**, 677–679.
Palik, E. D. (1960). *J. Opt. Soc. Am.* **50**, 1329–1336.
Palik, E. D. (1971). In "Far Infrared Spectroscopy" (K. D. Möller and W. G. Rothschild, eds.), p. 679. Wiley (Interscience), New York.
Palik, E. D. (1977). *J. Opt. Soc. Am.* **67**, 857–865.
Patel, C. K. N., Faust, W. L., McFarlane, and R. A., Garrett, C. G. B. (1964). *Proc. IEEE* **52**, 713.
Petuchowski, S. J., Rosenberger, A. T., and DeTemple, T. A. (1977). *IEEE J. Quant. Electron.* **13**, 476–481.
Pichamuthu, J. P. (1974). *J. Phys. D.* **7**, 1096–1100.
Pichamuthu, J. P., Hassler, J. C., and Coleman, P. D. (1971). *Appl. Phys. Lett.* **19**, 510–512.
Planck, M. (1900). *Verhandl. Deutsch. Phys. Ges.* **2**, 237.
Planck, M. (1901). *Ann. Phys. (Leipzig)* **4**, 553.
Pollack, M. A. (1969). *IEEE J. Quant. Electron.* **QE-5**, 558–562.
Pollack, M. A., and Tomlinson, W. J. (1968). *Appl. Phys. Lett.* **12**, 173–176.
Pollack, M. A., Bridges, T. J., and Strand, A. R. (1967a). *Appl. Phys. Lett.* **10**, 182–183.
Pollack, M. A., Bridges, T. J., and Tomlinson, W. J. (1967b). *Appl. Phys. Lett.* **10**, 253–256.
Pressley, R. J. (1971). "Handbook of Lasers." The Chemical Rubber Co., Ohio.
Prettl, W., and Genzel, L. (1966). *Phys. Lett.* **23**, 443–444.
Robinson, B. J. (1976). *Int. Trans. Astron. Un.* **16A**, Part 3, 96–105.
Robinson, D. W. (1978). *Opt. Commun.* **27**, 281–286.
Robinson, L. C., and Whitbourn, K. B. (1971). *Proc. IRE* (Austr.) **32**, 355–360.
Röser, H. P., and Schultz, G. V. (1977). *Infrared Phys.* **17**, 531–536.
Rosenbluh, M., Temkin, R. J., and Button, K. J. (1976). *Appl. Opt.* **15**, 2635–2644.
Rubens, H. (1922a). *Z. Phys.* **19**, 377 (obituary). Rubens, H. (1922b). *Nature (London)* **110**, 740–741; *Naturwissenschaften* **48**, 1016 (obituary).
Rubens, H., and Kurlbaum, F. (1900). *Berliner Ber.* 929.
Rubens, H., and Kurlbaum, F. (1901). *Ann. Phys. (Leipzig) Folge* **4**, 649.
Rubens, H., and Michel, G. (1921). *Berliner Ber.* 590.
Sakai, K., Fukui, T., Tsunawaki, and Y. Yoshinaga, H. (1969). *Jpn. J. Appl. Phys.* **8**, 1046–1055.
Schötzau, H. J., and Kneubühl, F. K. (1974). *Phys. Lett.* **48A**, 205–206.
Schötzau, H. J., and Kneubühl, F. K. (1975a). *Appl. Phys.* **6**, 25–30.
Schötzau, H. J., and Kneubühl, F. K. (1975b). *IEEE J. Quant. Electron.* **QE-11**, 817–822.
Schötzau, H. J., and Veprek, S. (1975). *Appl. Phys.* **7**, 271–277.
Schötzau, H. J., Crettol, R., and Kneubühl, F. K. (1970a). *Helv. Phys. Acta* **43**, 507–508.
Schötzau, H. J., Crettol, R., Grieshaber, E., and Kneubühl, F. K. (1970b). *Proc. PIB Symp. Submillimeter Waves, New York* pp. 431–436.
Schötzau, H. J., Grathwohl, Ch., and Kneubühl, F. K. (1971). *J. Appl. Math. Phys.* **22**, 778–779.

Schötzau, H. J., Mahler, H. J., and Kneubühl, F. K. (1972a). *Phys. Lett.* **38A**, 286.
Schötzau, H. J., Mahler, H. J., Conti, F., and Kneubühl, F. K. (1972b). *J. Appl. Math. Phys.* **23**, 161–163.
Schottky, W. (1924). *Phys. Z.* **25**, 342, 635.
Schwaller, P., Steffen, H., Moser, J. F., and Kneubühl, F. K. (1967). *Appl. Opt.* **6**, 827–829.
Seligson, D., Ducloy, M., Leite, J. R. R., Sanchez, A., and Feld, M. (1977). *IEEE J. Quant. Electron.* **13**, 468–472.
Sharp, L. E., and Wetherell, A. T. (1972). *Appl. Opt.* **11**, 1737–1741.
Shimuzu, T., Shimoda, K., and Minoh, A. (1968). *J. Phys. Soc. Jpn.* **24**, 1185.
Smith, P. W., (1971). *Appl. Phys. Lett.* **19**, 132–133.
Sochor, V. (1968). *Czech. J. Phys.* **B18**, 910.
Spalding, I. J. (1979). *Proc. Int. Conf. Infrared Phys.*, *2nd*, *(CIRP 2)*, Zurich pp. 75–82.
Stafsudd, O. M., and Yeh, Y. C. (1969). *IEEE J. Quant. Electron.* **QE-5**, 377.
Stafsudd, O. M., Haak, F. A., and Radisavljevic, K. (1967). *IEEE J. Quant. Electron.* **QE-3**, 618–620.
Steffen, H., and Kneubühl, F. K. (1968). *IEEE J. of Quant. Electron.* **QE-4**, 922–1008.
Steffen, H., Steffen, J., Moser, J. F., and Kneubühl, F. K. (1966a). *Phys. Lett.* **20**, 20–21.
Steffen, H., Steffen, J., Moser, J. F., and Kneubühl, F. K. (1966b). *Phys. Lett.* **21**, 425–426.
Steffen, H., Schwaller, P., Moser, J. F., and Kneubühl, F. K. (1966c). *Phys. Lett.* **23**, 313–314.
Steffen, H., Keller, B., and Kneubühl, F. K. (1967). *Electron. Lett.* **3**, 561.
Stevenson, D. P. (1950). *J. Chem. Phys.* **18**, 1347–1351.
Sturzenegger, Ch., Adam, B., and Kneubühl, F. K. (1977). *IEEE J. Quant. Electron.* **QE-13**, 473–475.
Sturzenegger, Ch., Adam, B., Vetsch, H., and Kneubühl, F. K. (1978). *Digest Int. Conf. Subwm Waves and Their Appl. 3rd, Guildford* pp. 27–28.
Sturzenegger, Ch., Vetsch, H., and Kneubühl, F. K. (1979). *Infrared Phys.* **19**, 277–296.
Tait, G. D., Withbourn, L. B., and Robinson, L. C. (1973). *Phys. Lett. A*, **46**, 239–240.
Tait, G. D., Robinson, L. C., and Bartlett, D. V. (1974). *Digest Int. Conf. Submm Waves Their Appl. 1st Atlanta* pp. 79–82.
Temkin, R. J. (1977). *IEEE J. Quant. Electron.* **13**, 450–454.
Tomlinson, W. J., Pollack, M. A., and Fork, R. L. (1967). *Appl. Phys. Lett.* **11**, 150–153.
Turner, R. (1974). *Appl. Opt.* **13**, 968–973.
Turner, R. (1977). *Appl. Opt.* **16**, 1197–1203.
Turner, R., and Poehler, T. O. (1971). *J. Appl. Phys.* **42**, 3819–3826.
Turner, R., Hochberg, A. K., and Poehler, T. O. (1968). *Appl. Phys. Lett.* **12**, 104–106.
Ulrich, R. (1967). *Infrared Phys.*, **7**, 37–55.
Ulrich, R., Renk, K. F., and Genzel, L. (1963). *IEEE Trans. MTT* **MTT-11**, 363–371.
Ulrich, R., Bridges, T. J., and Pollack, M. A. (1970). *Appl. Opt.* **9**, 2511–2516.
Vanderkooy, J., and Kang, C. S. (1976). *Infrared Phys.* **16**, 627–637.
Veprek, S. (1972). Ph.D. Thesis, Univ. of Zurich.
Veprek, S., Brendel, C., and Schäfer, H. (1971). *J. Cryst. Growth* **9**, 266–272.
Véron, D. (1974). *Opt. Commun.* **10**, 95.
Véron, D., Certain, J., and Crenn, J. P. (1977). *J. Opt. Soc. Am.* **67**, 964–967.
Vogel, P., and Genzel, L. (1964). *Infrared Phys.* **4**, 257–262.
Volk, R. (1972). *Phys. Lett.* **42A**, 321.
von Ortenberg, M., and Silbermann, R. (1975). *Solid State Commun.* **17**, 617.
von Ortenberg, M., and Silberman, R. (1976). *Surf. Sci.* **58**, 202–206.
von Ortenberg, M., and Steigenberger, U. (1977). *J. Opt. Soc. Am.* **67**, 928–931.
Weis, O. (1975). *Z. Phys.* **B21**, 1–10.
Weis, O. (1976). *In* "Phonon Scattering in Solids" (L. J. Challis, ed.), pp. 416–420.

Weizel, W. (1958). "Lehrbuch der theoretischen Physik," Springer-Verlag, Berlin and New York.

Wells, W. H. (1966). *IEEE J. Quant. Electron.* **QE-2**, 94–102.

Wells, J. S., Evenson, K. M., Matarrese, L. M., Jennings, D. A., and Wichman, G. L. (1971). NBS Tech. Note No. 395.

Wiggins, J. D., Drozdowicz, Z., and Temkin, R. J. (1977). *IEEE J. Quant. Electron.* **14**, 23–30.

Withbourn, L. B., Robinson, L. C., and Tait, G. D. (1972). *Phys. Lett. A* **38**, 315–316.

Witteman, W. J., and Bleekrode, R. (1964). *Phys. Lett.* **13**, 126–127.

Wood, O. R. II. (1974). *Proc. IEEE* **62**, 355–397.

Wood, O. R., Burkhardt, E. G., Pollack, M. A., and Bridges, T. J. (1971). *Appl. Phys. Lett.* **18**, 261–262.

Woskoboinikow, P. (1974). MS Thesis, Rensselaer Polytechn. Inst., *NTIS Rep.* Nr. AFOSR-TR-74-060r, Catalog No. 778 478.

Woskoboinikow, P., Mulligan, W. J., Praddaude, H. C., and Cohn, D. R. (1978a). *Appl. Phys. Lett.* **32**, 527–529.

Woskoboinikow, P., Mulligan, W., Praddaude, H. C., Cohn, D. R., and Lax, B. (1978b). *Digest Int. Conf. Submm Waves Their Appl. 3rd, Guildford* pp. 31–32.

Yamada, Y., Mitsuishy, A., and Yoshinaga, H. (1962). *J. Opt. Soc. Am.* **52**, 17–19.

Yamanaka, M. (1976). *Rev. Las. Eng. (Jpn)*, **3**, 253–294.

Yamanaka, M. (1977). *J. Opt. Soc. Am.* **67**, 952–958.

Yamanaka, M., Yoshinaga, H., and Kon, Sh. (1968). *Jpn. J. Appl. Phys.* **7**, 250–256.

Yamanaka, M., Yamanacki, T., and Yoshinaga, H. (1971). *Jpn. J. Appl. Phys.* **10**, 1601–1603.

Yariv, A. (1971). "Introduction to Optical Electronics." Holt, New York.

Yoshinaga, H. (1965). *Jpn. J. Appl. Phys. Suppl. 1* **4**, 420–427.

Yoshinaga, H., Kon, Sh., Yamanaka, M., and Yamamoto, J. (1967). *Sci. Light* **16**, 50–63.

Zinman, W. G., (1960). *J. Am. Chem. Soc.* **82**, 1262.

CHAPTER 6

Submillimeter Magnetospectroscopy of Charge Carriers in Semiconductors by Use of the Strip-Line Technique

Michael von Ortenberg

Physical Institute of the University of Würzburg
Würzburg, Federal Republic of Germany

I. Introduction

During the last two decades the application of submillimeter waves, especially in combination with magnetospectroscopy, has become one of the most effective methods in the investigation of electronic energy levels in semiconductors. The outstanding success of this method started 25 years ago, not by application of submillimeter radiation but by application of microwaves. Dresselhaus *et al.* (1953) demonstrated that the measurement of cyclotron resonance is a useful tool in the investigation of the energy-band structure of semiconductors. There are many reasons for the still increasing applications of submillimeter magnetospectroscopy in semiconductor physics. The two most important are the versatility of this method in combination with other experimental techniques and the tremendous progress in high-magnetic-field and laser technology during the recent years.

The applications of submillimeter magnetospectroscopy have become so extensive, that it is no longer possible to cover the whole field in a single review. Too many specialized subtechniques have been developed: the measurement of photoconductivity using different cross-modulation techniques, the combination of ordinary transmission spectroscopy with application of high uniaxial stress, MOS-, MIS-surface spectroscopy, and the strip-line technique for highly conductive materials—to mention only some of them. Whereas the first three experimental methods are widely established, the last one is applied now by only a few groups, but it is raising increasing interest (von Ortenberg, 1974, 1976, 1978a, b; Ramage et al., 1974; Stradling, 1974; Kawamura et al., 1975; von Ortenberg et al., 1975; Ramage et al., 1975; Stradling, 1975; Kawamura et al., 1976; Schwarzbeck et al., 1976; Kawamura et al., 1978; Nishikawa, 1978; Ramage, 1978; Schwarzbeck, 1978; Schwarzbeck and von Ortenberg, 1978; Schwarzbeck et al., 1978). That is why we would like to concentrate here on the special prblems of this method using submillimeter-wave radiation. Despite the very special arrangement, which we will consider, the method of evaluating and analyzing the data is quite fundamental and applies in principle to any experiment in submillimeter magnetospectroscopy.

II. Submillimeter Magnetospectroscopy in Semiconductor Physics

A. Conditions for the Application

The range of submillimeter waves embedded between far-infrared and millimeter waves is of special interest: we have here a transition range of electromagnetic radiation. On the one side, submillimeter radiation reflects very definitely the single-photon energy in the experimental situation, thus representing the quasi-particle nature of radiation. On the other side, submillimeter waves can still be generated by classical resonance circuits as typical for wave phenomena. We will continue to meet this double aspect of wave and photon picture in the following considerations.

For a long time physicists met rather serious difficulties in the submillimeter range of radiation. The spectrum of shorter wavelengths was covered by powerful conventional sources and, rather early, even by lasers. The spectrum beyond the submillimeter range toward the longer wavelengths also had powerful sources in magnetrons, carcinotrons, and clystrons. Just the lack of such a powerful source in the submillimeter range kept physicists from being much attracted by this part of the spectrum. Only after the invention of the molecular-gas laser for this spectral-range did submillimeter waves become attractive to the semiconductor physicist in magnetospectroscopy (Gebbie et al., 1964). The molecular-gas laser made available

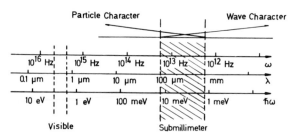

FIG. 1   The submillimeter range of radiation embedded between infrared and microwaves.

a powerful source that exceeded, by orders of magnitude, the spectral intensity of conventional sources like the mercury lamp.

The special interest of the semiconductor physicist in submillimeter radiation is caused by the following reason. From Fig. 1 we notice that the angular frequency ω of the submillimeter range is limited by about 2×10^{12} and 20×10^{12} Hz. This order of magnitude for the frequency, however, is typical for the classical angular frequency of electrons and holes in semiconductors. Or, if we apply the complementary photon picture, the energy of 1 and 10 meV corresponds to the energy differences of the electronic states within one energy band of a semiconductor. If the radiation energy equals such an energy difference of two states, we would expect resonance phenomena as a result of the interaction. By observation of such resonance, and with additional information on the radiation energy, we are able to give quantitative conclusions on the participating energy states.

The external magnetic field in submillimeter magnetospectroscopy changes the energy levels in such a way that for fixed radiation energy, the resonance interaction can be enforced, i.e., the magnetic field tunes the resonance. This phenomenon is well known and is applied in various other magnetospectroscopic methods, e.g., the Shubnikov–deHaas and magnetophonon effect of the electric resistance. Considering the latter two effects, the electromagnetic energy used to probe the energy levels in submillimeter magnetospectroscopy is replaced by the Fermi energy and the energy of optical phonons, respectively. Both energies, however, are much less exactly defined than the energy of the monochromatic laser radiation in submillimeter magnetospectroscopy.

In a very simple model the magnetic-field dependence of the energy levels, or the angular frequency of the electrons or holes, can be understood considering the Lorentz force, which is effective only in the plane perpendicular to the external field and enforces a circular motion within this plane. The particular shape of the motion depends naturally on the microscopic parameters of the considered charged particle within this plane. By this simple mechanism we can easily understand the second important reason for the application of a magnetic field in submillimeter magnetospectroscopy: any anisotropy of the electron parameters can be investigated by a definite variation of the crystal orientation relative to the external field because it causes an orientation-dependent resonance position on the magnetic field scale.

The resonance in the interaction of radiation and charged particles is clearly detectable only if it is not obscured by broadening due to damping effects. In terms of our simple model this means that the electrons in their periodic motion have to perform many revolutions before being scattered by some disturbance. Such strong disturbances are, e.g., thermal vibrations of the lattice constituents. To avoid them, we have to cool the investigated material to the temperatures of liquid helium. For many materials the cooling process improves the quality of the resonance considerably. There are still, however, a number of semiconductor materials that have not benefited from the cooling method because the origin of scattering disturbances is different from lattice vibrations.

In conclusion, we can state that powerful sources for submillimeter radiation and high-magnetic-field and low-temperature technology are necessary conditions for successful application of submillimeter magnetospectroscopy. From this point of view it is very clear that the invention of the molecular-gas laser by Gebbie et al. (1964), followed by the development of the optically pumped laser by Chang and Bridges (1970), had a very stimulating effect on the "optical" part of submillimeter magnetospectroscopy. Corresponding with the latter was the installation of high-magnetic-field facilities, e.g., the Francis Bitter National Laboratory in Cambridge, Mass., and the Clarendon Laboratory in Oxford, United Kingdom. Later on, the breathtaking development of the commercially produced, superconducting magnet systems gave nearly every laboratory the possibility of performing experiments in magnetic fields up to 10 T.

B.   Analysis of Method

Submillimeter spectroscopy is an experimental method. It provides the physicist with data that can be classified with respect to the measured quantity. In submillimeter magnetospectroscopy there are generally two possibilities in the choice of these quantities because there are two components, semiconductor and radiation, interacting with each other as indicated in Fig. 2. Therefore, the first kind of data is related to any change in the physical quantities of the radiation after interaction with the semiconductor. If we control the polarization of the electromagnetic radiation, we are performing a Faraday-type experiment. The measurement of the radiation intensity results mostly in a transmission or reflection experiment. In contrast to these measurements—where one of the radiation parameters is controlled—are those measurements related to the semiconductor material

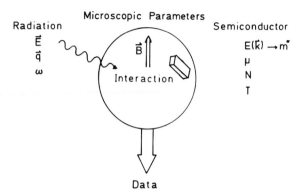

FIG. 2 The two components of submillimeter magnetospectroscopy, radiation and semi-conductor, are characterized by microscopic parameters. The experimental data obtained after interaction refer, however, to the macroscopic properties of the system.

itself. Changes in the carrier concentration, mobility, or temperature are easily detected by electrical measurements of the longitudinal or transverse voltage drop at the sample. The data of such an experiment are the numerical output of a definite causal sequence depending on definite values of the parameters. We are now confronted with the problem of extracting from the data the information concerning the microscopic structure of the charge-carrier system in which we are interested. For complete understanding and correct interpretation it is necessary to analyze the experiment step-by-step and then try to back into a step-by-step synthesizing simulation of the data. Only this closed-loop feedback process of analysis with subsequent synthesis guarantees an optimal understanding of the experiment. In the case of submillimeter magnetospectroscopy the analysis can mostly be divided into two steps as indicated in Fig. 3. The central physical quantity is the dielectric function. Deriving this function, we have already abstracted from the individual experimental situation, expressed in the particular optical arrangement and described by Maxwell's equations, including the dielectric boundary conditions. The connections between the dielectric function and the microscopic energy levels, in which we are interested, are rather complex and involve mostly quantum theoretical consideration of the problem [e.g., by the Kubo formalism (Kubo, 1957)], starting from an explicit model in linear-response theory and taking into account the Hamiltonian of the quasi-free electrons including scattering mechanisms.

Since most of the Hamiltonians in this field are "trial Hamiltonians," based on $\mathbf{k} \cdot \mathbf{p}$ approximation or the method of invariants (Kane, 1957; Luttinger, 1956), the particular model involved depends explicitly on a set of numerical coefficients, which can only be fitted from experimental data.

Microscopic Parameters

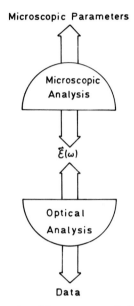

FIG. 3   The central quantity of the two-step analysis is the dielectric function.

To judge to what extent the applied model reflects the experimental situation, we have to simulate the experiment and to compare this result directly with experimental data. This procedure implies rather extended analytical and numerical calculations, which is, however, necessary for successful application of submillimeter magnetospectroscopy.

III.   The Application of the "Strip-Line" Technique in Submillimeter Magnetospectroscopy

The strip line in its original form was invented by Drew and Sievers (1967) for application in metal physics. It consisted of a repeated layer structure of metal tape and transparent dielectric material in a helical arrangement, as shown schematically in Fig. 4. The attenuation of the radiation transmitted through this arrangement reflects the dielectric properties of the two constituents. Actually the transmission is determined by the reflectivity of the metal surface and the transmission of the dielectric material. The special advantage of the strip-line application in comparison with an ordinary reflectivity experiment lies in the fact that by suitable choice of the strip-line length L, the attenuation can be adjusted to an optimum for the experimental situation. In this respect the strip-line arrangement is similar to an ordinary transmission measurement. Subsequently, the original concept of the strip

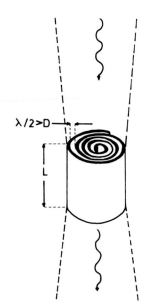

$\lambda/2 > D$

L

FIG. 4 The original concept of the strip line after Drew and Sievers (1967) uses a helical arrangement.

line was modified by Strom *et al.* (1971). The modified arrangement has the properties of a rectangular waveguide, whose width is generally large compared with the plate separation, as shown in Fig. 5. Depending on whether both plates are made of the semiconductor material to be investigated or one plate is a metallic mirror, we distinguish the symmetric and asymmetric strip-line configuration, respectively. Generally both configurations cannot be related only by symmetry arguments to each other.

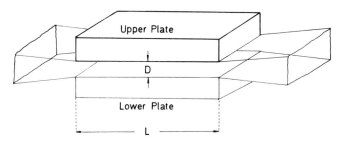

Upper Plate

D

Lower Plate

L

FIG. 5 The modified strip line after Strom *et al.* (1971) has the properties of a rectangular waveguide.

A. OPTICAL ANALYSIS OF THE STRIP-LINE PROPAGATION

1. *Classical Wave Equation of Electromagnetic Radiation in a Dielectric Medium*

The propagation of electromagnetic radiation in the strip line is completely determined by Maxwell's equations and the suitable boundary conditions at the dielectric interfaces are:

$$\nabla \times \mathbf{E} = -\dot{\mathbf{B}}, \tag{1}$$

$$\nabla \times \mathbf{H} = \dot{\mathbf{D}} + \mathbf{J}, \tag{2}$$

$$\nabla \mathbf{D} = \rho, \tag{3}$$

$$\nabla \mathbf{B} = 0. \tag{4}$$

Here $\mathbf{E}$ is the electric and $\mathbf{H}$ the magnetic field, $\mathbf{D}$ the displacement density, $\mathbf{J}$ the current, and ρ the space charge density. The current density $\mathbf{J}$ is related to the electric field by the conductivity tensor $\boldsymbol{\sigma}$ as

$$\mathbf{J} = \boldsymbol{\sigma}\mathbf{E}. \tag{5}$$

Introducing the dielectric tensor $\boldsymbol{\kappa}$, the corresponding relation for the displacement density reads

$$\mathbf{D} = \boldsymbol{\kappa}\mathbf{E}. \tag{6}$$

Since we confine our considerations to nonmagnetic materials, the magnetic field $\mathbf{H}$ and the induction $\mathbf{B}$ are collinear:

$$\mathbf{B} = \mu_0\mathbf{H}. \tag{7}$$

From Eqs. (1) and (2) we derive the wave equation describing the propagation in the dielectric medium:

$$\nabla \times (\nabla \times \mathbf{E}) - \mu_0 \boldsymbol{\kappa}\ddot{\mathbf{E}} - \mu_0\dot{\mathbf{j}} = 0. \tag{8}$$

For monochromatic radiation with a time dependence of the form

$$\mathbf{E}(\mathbf{r}, t) = \mathbf{E}_0(\mathbf{r})e^{-i\omega t}, \tag{9}$$

Eq. (8) reads

$$\nabla \times (\nabla \times \mathbf{E}) + \mu_0\omega^2\boldsymbol{\kappa}\mathbf{E} + i\omega\mu_0\mathbf{J} = 0. \tag{10}$$

There are two different approximations concerning the behavior of the above field and current vectors at a dielectric interface as indicated in Fig. 6. The usual Fresnel formulation considers the current density not as a separate

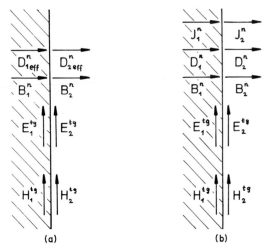

(a)   (b)

FIG. 6   The boundary at a dielectric interface after (a) Fresnel and (b) Sauter.

physical quantity, but relates it directly to the electric field and introduces an effective dielectric tensor

$$\varepsilon = \frac{\kappa + i\sigma/\omega}{\varepsilon_0},$$ (11)

defining an effective displacement density D_{eff} as

$$\mathbf{D}_{\mathrm{eff}} = \varepsilon\,\mathbf{E} = \mathbf{D} + i\mathbf{J}/\omega.$$ (12)

Only for this quantity does the continuity condition hold at the dielectric boundary for the components normal to the interface:

$$\mathbf{D}_{\mathrm{eff}\,1}\cdot\mathbf{n}_1 + \mathbf{D}_{\mathrm{eff}\,2}\cdot\mathbf{n}_2 = 0,$$ (13)

where $\mathbf{n}_1$ and $\mathbf{n}_2$ are the interface-normal vectors. In contrast to the above approximation, the formulation of Sauter (1967) considers the continuity of the displacement density $\mathbf{D}$ and the current density $\mathbf{J}$ separately:

$$\mathbf{D}_1\cdot\mathbf{n}_1 + \mathbf{D}_2\cdot\mathbf{n}_2 = 0,$$ (14)

$$\mathbf{J}_1\cdot\mathbf{n}_1 + \mathbf{J}_2\cdot\mathbf{n}_2 = 0.$$ (15)

The continuity conditions for the other quantities are the same in both models:

$$\mathbf{B}_1\cdot\mathbf{n}_1 + \mathbf{B}_2\cdot\mathbf{n}_2 = 0,$$ (16)

$$\mathbf{E}_1\times\mathbf{n}_1 + \mathbf{E}_2\times\mathbf{n}_2 = 0,$$ (17)

$$\mathbf{H}_1\times\mathbf{n}_1 + \mathbf{H}_2\times\mathbf{n}_2 = 0.$$ (18)

According to the Sauter formulation, the additional degree of freedom in the direct vicinity of the interface allows excitation of the longitudinal electromagnetic surface mode, which allows the normal component of the current density $\mathbf{J}$ to be continuous. The detailed investigation of the problem shows that the surface mode depends critically on the spatial structure of the nonperiodic mode, i.e., on nonlocal effects. In the following sections we will totally adopt the local approximation. In contrast to metal physics and to microwave spectroscopy using submillimeter radiation in semiconductor physics, nonlocal effects are mostly negligible (Perkowitz, 1969; Strom et al., 1973).

2. The Asymmetric Strip Line

The most often applied strip-line arrangement is the asymmetric strip line, where only one plate is made of sample material, as shown in Fig. 7. For simplicity we assume that the strip line is infinitely extended in x and z directions. From the physical point of view this means that there are no interfering reflections along these directions. The sample material is supposed to be extended all along the negative y axis. In the actual experimental situation this requires that the sample thickness be large compared with the penetration depth of the radiation. The dielectric properties of the three adjacent dielectric media are given by

$$\varepsilon_{\text{metal}} = -\infty \begin{pmatrix} 1 & 0 & 0 \\ 0 & 1 & 0 \\ 0 & 0 & 1 \end{pmatrix}, \tag{19}$$

$$\varepsilon_{\text{vac}} = 1 \begin{pmatrix} 1 & 0 & 0 \\ 0 & 1 & 0 \\ 0 & 0 & 1 \end{pmatrix}, \tag{20}$$

$$\varepsilon_{\text{sample}} = \begin{pmatrix} \varepsilon_{xx} & \varepsilon_{xy} & \varepsilon_{xz} \\ \varepsilon_{yx} & \varepsilon_{yy} & \varepsilon_{yz} \\ \varepsilon_{zx} & \varepsilon_{zy} & \varepsilon_{zz} \end{pmatrix}. \tag{21}$$

We admit explicitly the most general structure for the dielectric tensor in the semiconductor material. The large value of $\varepsilon_{\text{metal}}$ indicates the absence of any electric field in the metal. This is expressed by the special dielectric boundary condition for zero electric field in the metal parallel to the interface. Generally the strip-line mode consists in each dielectrically homogeneous part of a superposition of the two possible normal modes. Neglecting any

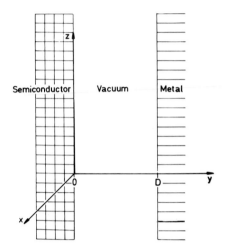

FIG. 7   The schematic of the asymmetric strip line.

mode structure along the x axis, we make the following *ansatz* for the electric field components of the strip-line mode:

vacuum:

$$E_x = A_v \sin[k_y(D - y)]\Big]$$ (22)

$$E_y = B_v \cos[k_y(D - y)]\Big\} e^{i(hz - \omega t)}$$ (23)

$$E_z = C_v \sin[k_y(D - y)]\Big]$$ (24)

semiconductor:

$$E_x = A_1 e^{-ik_{y1}y} + A_2 e^{-ik_{y2}y}\Big]$$ (25)

$$E_y = B_1 e^{-ik_{y1}y} + B_2 e^{-ik_{y2}y}\Big\} e^{i(hz - \omega t)}$$ (26)

$$E_z = C_1 e^{-ik_{y1}y} + C_2 e^{-ik_{y2}y}\Big]$$ (27)

Here the subscripts represent the two normal modes resulting from the above *ansatz* and the wave equation (8). The sign of the wave vectors in the semiconductor material has been chosen so that the energy flow decays along the negative y axis. The actual quantity of interest is the strip-line propagation constant h, because the imaginary part of this quantity determines the strip-line attenuation.

The fact that both the strip-line arrangement and the field *ansatz* are completely translationally invariant in x, reduces the generally six independent boundary conditions at the dielectric interface to four nonequivalent equations:

$$E_{x\,\text{vac}} = E_{x\,\text{sem}} \qquad \Leftrightarrow \qquad B_{y\,\text{vac}} = B_{y\,\text{sem}}, \tag{28}$$

$$E_{y\,\text{vac}} = D_{y\,\text{sem}} \qquad \Leftrightarrow \qquad H_{x\,\text{vac}} = H_{x\,\text{sem}}, \tag{29}$$

$$E_{z\,\text{vac}} = E_{z\,\text{sem}}, \tag{30}$$

$$H_{z\,\text{vac}} = H_{z\,\text{sem}}. \tag{31}$$

Applying Maxwell's equation, especially the wave equation (9) and the four boundary conditions (Eqs. (28)–(31)), we obtain with the *ansatz* of Eqs. (22)–(26), after elementary but tedious calculation, the following set of coupled equations for four parameters (von Ortenberg, 1978):

$$h^2 + k_y^2 = k_0^2 = (2\pi/\lambda)^2, \tag{32}$$

$$
\begin{aligned}
(h^2 &+ k_{y1,2}^2 - k_0^2\varepsilon_{xx})(h^2 - k_0^2\varepsilon_{yy})(k_{y1,2}^2 - k_0^2\varepsilon_{zz}) \\
&+ k_0^4\varepsilon_{xz}\varepsilon_{yx})hk_{y1,2} - k_0^2\varepsilon_{yz}) + k_0^4\varepsilon_{zx}\varepsilon_{xy}(hk_{y1,2} - k_0^2\varepsilon_{yz}) \\
&- k_0^2\varepsilon_{zx}\varepsilon_{xz}(h^2 - k_0^2\varepsilon_{yy}) \\
&- (hk_{y1,2} - k_0^2\varepsilon_{zy})(hk_{y1,2} - k_0^2\varepsilon_{yz})(h^2 + k_{y1,2}^2 - k_0^2\varepsilon_{xx}) \\
&- k_0^4\varepsilon_{xy}\varepsilon_{yx}(k_{y1,2}^2 - k_0^2\varepsilon_{zz}) \\
&= 0.
\end{aligned}
\tag{33}
$$

$$
\begin{aligned}
[G_1 k_0^2 \cos k_y D &- ik_y \sin k_y D(G_1 k_{y1} + hB_1)][k_y \cos k_y D - ik_{y2} \sin k_y D] \\
&- [G_2 k_0^2 \cos k_y D - ik_y \sin k_y D(G_2 k_{y2} + hB_2)] \\
&\times [k_y \cos k_y D - ik_{y1} \sin k_y D] = 0,
\end{aligned}
\tag{34}
$$

$$B_{1,2} = \frac{k_0^4\varepsilon_{yx}\varepsilon_{xz} - (h^2 + k_{y1,2}^2 - k_0^2\varepsilon_{xx})(hk_{y1,2} - k_0^2\varepsilon_{yz})}{(h^2 - k_0^2\varepsilon_{yy})\varepsilon_{xz}k_0^2 - k_0^2\varepsilon_{xy}(hk_{y1,2} - k_0^2\varepsilon_{yz})} \tag{35}$$

$$G_{1,2} = \frac{[k_0^2\varepsilon_{yy} - h^2][k_0^4\varepsilon_{yx}\varepsilon_{xz} - (k_{y1,2}^2 + h^2 - k_0^2\varepsilon_{xx})(hk_{y1,2} - k_0^2\varepsilon_{yz})]}{\{[hk_{y1,2} - k_0^2\varepsilon_{yz}][(h^2 - k_0^2\varepsilon_{yy})\varepsilon_{xz}k_0^2 - (hk_{y1,2} - k_0^2\varepsilon_{yz})\varepsilon_{xy}k_0^2]\}}$$
$$+ \frac{k_0^2\varepsilon_{yx}}{hk_{y1,2} - k_0^2\varepsilon_{yz}} \tag{36}$$

Actually, Eqs. (32)–(36) represent a system of eight independent equations determining the eight parameters, h, k_y, k_{y1}, k_{y2}, B_1, B_2, G_1, G_2. From the

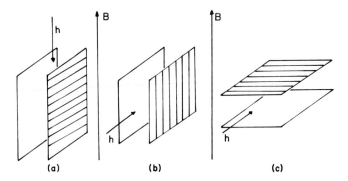

(a)   **(b)**   **(c)**

FIG. 8 Depending on the relative orientation of magnetic field **B** and the propagation vector **h** in the strip line, three different high-symmetry configurations can be defined: (a) parallel, (b) perpendicular, and (c) surface.

above equations the dispersion relation $h(\omega)$ can be calculated directly. For the most general structure of the ε tensor, as assumed, it is hardly possible to get any insight into the physical structure of the solution. It is very useful, however, to discuss the problem for the more restricted "extended gyrotropic" form of the dielectric tensor, where only two off-diagonal components are nonzero. The extended gyrotropic form can be realized in three different ways as shown in Fig. 8.

Each of the three forms represents a highly symmetric configuration, where the external magnetic field is parallel to one of the favored axes of the strip-line system. Under these restrictions, the symmetric strip line was theoretically investigated by Kanada *et al.* (1976) and the asymmetric strip line by von Ortenberg (1976) and Nakayama and Tsuji (1977). Due to the special application of the different configurations we have named them appropriately to the experimental situation. In the following we shall discuss the three favored configurations separately.

a. *The Parallel Configuration.* From the experimental point of view, the parallel configuration is the one most often applied. The magnetic field is parallel to the propagation constant h and results in an ε tensor of the form (Palik and Furdyna, 1970)

$$\varepsilon_{\text{par}} = \begin{pmatrix} \varepsilon_{xx} & \varepsilon_{xy} & 0 \\ -\varepsilon_{xy} & \varepsilon_{yy} & 0 \\ 0 & 0 & \varepsilon_{zz} \end{pmatrix}. \tag{37}$$

Applying Eq. (37) leads to a considerable simplification of Eqs. (32)–(36):

$$k_0^2 = h^2 + k_y^2, \tag{38}$$

$$[\varepsilon_{yy}(k_0^2 \varepsilon_{zz} - k_{y1,2}^2) - h^2 \varepsilon_{zz}][\varepsilon_{xx} k_0^2 - h^2 - k_{y1,2}^2] = -\varepsilon_{xy}^2 k_0^2 (k_0^2 \varepsilon_{zz} - k_{y1,2}^2), \tag{39}$$

$$(k_y \sin k_y D + i k_{y2} \cos k_y D)(k_y \cos k_y D - i k_{y1} \sin k_y D)$$

$$\times \frac{-\varepsilon_{xy}^2 k_0^2 (k_0^2 \varepsilon_{zz} - k_{y1}^2)}{[\varepsilon_{yy}(k_0^2 \varepsilon_{zz} - k_{y2}^2) - h^2 \varepsilon_{zz}][\varepsilon_{xx} k_0^2 - h^2 - k_{y1}^2]}$$

$$- \left(k_y \sin k_y D + \frac{i k_{y1}}{\varepsilon_{zz}} \cos k_y D \right)(k_y \cos k_y D - i k_{y2} \sin k_y D)$$

$$= 0. \tag{40}$$

From Eqs. (39) and (40) it becomes quite evident that there is a coupling of the two normal modes in the semiconductor material caused by the Hall component ε_{xy}. This means that for nonzero ε_{xy}, and hence for finite magnetic fields, there is no pure TE or TM mode propagating in the strip line, but only "quasi-TE" and "quasi-TM" modes having small admixtures of the TM and TE mode, respectively. For zero magnetic fields, however, no mode coupling occurs. The most important strip-line parameter is the plate separation D. The ratio D/λ is the critical determinant of the modes that are able to propagate in the strip line. For real values of the dielectric tensor components, that means, in the lossless case, the conditions for the propagating strip-line modes were discussed in detail by Nakayama and Tsuji (1977). However, since the imaginary part of the dielectric function cannot be neglected in the experimental situation for strongly degenerate materials, we will discuss the lossy case in detail.

For the lowest "quasi-TM" mode the spatial variation of the electric field between the plates is very small, so that the following approximation holds:

$$k_y^2 \ll k_0^2. \tag{41}$$

For large absolute values of the dielectric function of the different tensor components,

$$|\varepsilon_i| \gg 1, \tag{42}$$

the dispersion relation of Eq. (39) can be factorized in the following form:

$$[k_0^2 \varepsilon_{zz} - k_{y1,2}^2][(\varepsilon_{xx} k_0^2 - k_{y1,2}^2) + (\varepsilon_{xy}^2/\varepsilon_{yy})k_0^2] = 0. \tag{43}$$

From the physical point of view this means, that we have decoupled the TM and TE modes. Using the same approximation, the boundary equation Eq. (40) reads

$$k_y \sin k_y D + (i k_{y1}/\varepsilon_{zz}) \cos k_y D = 0. \tag{44}$$

With the approximations

$$D < \lambda \tag{45}$$

and

$$|\varepsilon_{zz}| \gg 1, \tag{46}$$

the dispersion relation of the propagation constant h of the lowest "quasi-TM" mode can be written as

$$h \approx k_0 (1 + i/2k_0 D\sqrt{\varepsilon_{zz}}). \tag{47}$$

The intensity attenuation coefficient is then given by

$$\kappa = 2\mathrm{Im}(h) = \frac{1}{D} \, Re\left\{\frac{1}{\sqrt{\varepsilon_{zz}}}\right\}. \tag{48}$$

With decreasing plate separation D, the interaction becomes stronger and the attenuation increases.

We would like to emphasize that the above result depends necessarily on the condition of Eq. (46). Kanada *et al.* (1976) have pointed out that this approximation breaks down in the vicinity of the dielectric anomaly in ε_{zz}. This statement, however, holds only in the lossless case. The investigation of both the real and imaginary part of the dielectric function in the lossy case shows, however, that the imaginary part of the dielectric function in highly degenerate materials is always sufficiently large so that Eq. (46) is fulfilled.

The result of Eq. (47) can easily be interpreted with reference to Fig. 9. The plane wave of the "quasi-TM" mode incident nearly parallel to the dielectric interface is strongly refracted by the large value of ε_{zz}, so that the normal mode in the semiconductor material is well approximated by the ordinary Voigt mode. This means that the field component perpendicular to the dielectric interface is nearly completely screened and is much less important. This fact is typical for the strip-line experiments and will be described later.

The result of Eq. (47), however, holds only as long as there is no mode coupling as assumed in Eq. (39). In the presence of mode coupling in this configuration, the anomaly of the extraordinary Voigt mode becomes effective and seriously influences the attenuation (von Ortenberg, 1978a,b). This effect will be discussed in detail using numerical values for the dielectric function.

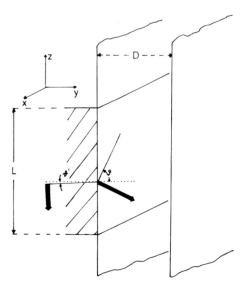

FIG. 9   Because of the large refraction, the electric-field vector of the quasi-TM mode in the semiconductor is well polarized parallel to the interface, so that the attenuation is mainly determined by ε_{zz} ($k \approx \mathrm{Re}\{1/D\sqrt{\varepsilon_{zz}}\}$).

The above approximation can be easily extended for higher "quasi-TM" modes as long as:

$$\varepsilon_{yy}k_0^2 \gg h^2 = k_0^2 - k_y^2, \tag{49}$$

where k_y is determined by

$$k_y D = n\pi + \delta, \qquad |\delta| \ll 1. \tag{50}$$

Equation (49) determines the upper limit for the mode index n depending on ε_{yy}. The propagation constant h_n is then given by

$$n \geq 1: \quad h_n^2 = k_0^2 - \left(\frac{n\pi}{D} - \frac{ik_0}{n\pi\sqrt{\varepsilon_{zz}}}\right)^2. \tag{51}$$

From Eq. (49) it follows that the formula for h_n holds essentially, as long as:

$$\varepsilon_{yy} \gg n^2\pi^2/k_0 D. \tag{52}$$

For the "quasi-TE" mode the condition of Eq. (41) does not hold any longer. Nevertheless, the dispersion relation can be factorized as long as

$$|\varepsilon_{xx}k_0^2|, |\varepsilon_{yy}k_0^2| \gg h^2 = k_0^2 - k_y^2. \tag{53}$$

The boundary condition of Eq. (40) is simply reduced to

$$k_y \cos k_y D - iky_2 \sin k_y D = 0. \tag{54}$$

The corresponding approximations as used for the higher "quasi-TM" modes yield

$$n \geq 1: \quad h^2 = k_0^2 - \left(\frac{n\pi}{D} - \frac{in\pi}{k_0 D^2 \sqrt{\varepsilon_\perp}}\right)^2; \quad \varepsilon_\perp = \frac{\varepsilon_{zz}\varepsilon_{yy} + \varepsilon_{yz}^2}{\varepsilon_{yy}}. \tag{55}$$

The influence of the plate separation D is much more pronounced and results in a considerably stronger attenuation than for the "quasi-TM" modes.

b. *The Perpendicular Configuration.* For the perpendicular configuration the magnetic field is oriented along the x axis and the corresponding ε tensor is given by (Palik and Furdyna, 1970)

$$\varepsilon = \begin{pmatrix} \varepsilon_{xx} & 0 & 0 \\ 0 & \varepsilon_{yy} & \varepsilon_{yz} \\ 0 & -\varepsilon_{yz} & \varepsilon_{zz} \end{pmatrix}. \tag{56}$$

The dispersion equation decouples exactly for pure TM and TE modes:

$$\text{TM:} \quad k_0^2(\varepsilon_{zz}\varepsilon_{yy} + \varepsilon_{yz}^2) - k_{y1}^2\varepsilon_{yy} - h^2\varepsilon_{zz} = 0, \tag{57}$$

$$\text{TE:} \quad k_0^2\varepsilon_{xx} - k_{y2}^2 - h^2 = 0. \tag{58}$$

The boundary condition, too, can be factorized exactly in two terms:

$$\text{TM:} \quad (hk_{y1} + k_0^2\varepsilon_{yz})\cos k_y D - ik_y(h\varepsilon_{zz} + k_{y1}\varepsilon_{yz})\sin k_y D = 0, \tag{59}$$

$$\text{TE:} \quad k_y \cos k_y D - ik_{y2} \sin k_y D = 0. \tag{60}$$

It should be noted that the component ε_{yz} enters linearly in the boundary condition of Eq. (59). This means that any magnetic field reversal will affect the dispersion of the propagation constant h. Such behavior corresponds to the experimental data and will be discussed in detail later.

From Eqs. (57) and (59) it follows for the lowest pure TM-mode in the same approximation as in the parallel configuration that:

$$h^2 = k_0^2 + \frac{ik_0^2}{k_0 D} \times \frac{\varepsilon_{yz} + \sqrt{\varepsilon_\perp}}{\varepsilon_{yz}\sqrt{\varepsilon_\perp} + \varepsilon_{zz}}; \quad \varepsilon_\perp = \frac{\varepsilon_{yy}\varepsilon_{zz} + \varepsilon_{yz}^2}{\varepsilon_{yy}}. \tag{61}$$

Here the propagation constant h depends definitely on the sign of ε_{yz} and thus on the polarity of the magnetic field. Only for large values of ε_{yz}, where

$$\varepsilon_{yz}^2 \gg \varepsilon_{zz} \tag{62}$$

does it follow that h is determined only by the extraordinary Voigt mode and the polarity of the magnetic field can be neglected.

The situation for the TE mode is much simpler and the propagation constant is determined by the ordinary Voigt mode:

$$n \geq 1: \qquad h^2 = k_0^2 - \left(\frac{n\pi}{D} - \frac{in\pi}{k_0 D \sqrt{\varepsilon_{xx} D}}\right)^2. \tag{63}$$

Equation (63) holds as long as:

$$|\varepsilon_{xx} k_0^2| \gg h^2. \tag{64}$$

c. *The Surface Configuration.* In the surface configuration the magnetic field is oriented along the y axis and the corresponding ε tensor is given by (Palik and Furdyna, 1970):

$$\varepsilon = \begin{pmatrix} \varepsilon_{xx} & 0 & \varepsilon_{xz} \\ 0 & \varepsilon_{yy} & 0 \\ -\varepsilon_{xz} & 0 & \varepsilon_{zz} \end{pmatrix}. \tag{65}$$

The dispersion relation now reads

$$(k_0^2 \varepsilon_{xx} - k_{y1,2}^2)(k_0^2 \varepsilon_{zz} - k_{y1,2}^2) + k_0^4 \varepsilon_{xz}^2$$
$$= \frac{h^2}{\varepsilon_{yy}} [k_0^2 \varepsilon_{xz}^2 + \varepsilon_{zz}(k_0^2 \varepsilon_{xx} - k_{y1,2}^2) + \varepsilon_{yy}(k_0^2 \varepsilon_{zz} - k_{y1,2}^2)] \tag{66}$$

The boundary conditions of Eq. (34) are now rather complex, because there is a strong mode coupling depending on the anisotropy of ε_{xx}, ε_{zz}.

$$\left[\frac{k_{y1}^2 \varepsilon_{yy} k_y}{k_0^2 \varepsilon_{yy} - h^2} \sin k_y D + ik_{y1} \cos k_y D\right][k_y \cos k_y D - ik_{y2} \sin k_y D]$$
$$\times \frac{k_{y2}}{k_{y1}} \times \frac{-k_0^2 \varepsilon_{xz}(k_0^2 \varepsilon_{yy} - h^2)}{[k_0^2 \varepsilon_{xx} - h^2 - k_{y2}^2][\varepsilon_{yy}(k_0^2 \varepsilon_{zz} - k_{y1}^2) - h^2 \varepsilon_{zz}]} = 0. \tag{67}$$

With the usual approximations used for the other configurations we obtain:

$$h^2 = k_0^2 + \frac{ik_0^2}{k_0 D}$$
$$\times \frac{\varepsilon_{xz}^2(1 - \sqrt{\varepsilon_-} k_0 D) + (\varepsilon_{xx} - \varepsilon_-)(\varepsilon_{zz} - \varepsilon_+)(1 - \sqrt{\varepsilon_+} k_0 D)}{\varepsilon_{xz}^2 \sqrt{\varepsilon_+}(1 - i\sqrt{\varepsilon_-} k_0 D) + \sqrt{\varepsilon_-}(\varepsilon_{xx} - \varepsilon_-)(\varepsilon_{zz} - \varepsilon_+)(1 - \sqrt{\varepsilon_+} k_0 D)} \tag{68}$$

where

$$\varepsilon_{\pm} = \frac{\varepsilon_{xx} + \varepsilon_{zz}}{2} \pm \sqrt{\frac{(\varepsilon_{xx} - \varepsilon_{zz})^2}{4} - \varepsilon_{xz}^2}. \tag{69}$$

For isotropic ε in the sense that

$$\varepsilon_{xx} = \varepsilon_{zz}, \tag{70}$$

Eq. (68) can be considerably simplified and the propagation constant depends only on the normal modes in Faraday configuration $\varepsilon_{\pm}$ (Palik and Furdyna, 1970):

$$h^2 = k_0^2 + \frac{ik_0^2}{k_0 D} \frac{(1 - i\sqrt{\varepsilon_-}\,k_0 D) + (1 - i\sqrt{\varepsilon_+}k_0 D)}{\sqrt{\varepsilon_+}(1 - i\sqrt{\varepsilon_-}\,k_0 D) + \sqrt{\varepsilon_-}(1 - i\sqrt{\varepsilon_+}\,k_0 D)}. \tag{71}$$

We have now to consider the two limits:

$$1 \gg |\sqrt{\varepsilon_{\pm}}\,k_0 D|: \qquad h^2 = k_0^2 + \frac{ik_0^2}{k_0 D} \frac{2}{(\sqrt{\varepsilon_+} + \sqrt{\varepsilon_-})}, \tag{72}$$

and

$$1 \ll |\sqrt{\varepsilon_{\pm}}\,k_0 D|: \qquad h^2 = k_0^2 + \frac{ik_0^2}{k_0 D} \left(\frac{1}{\sqrt{\varepsilon_+}} + \frac{1}{\sqrt{\varepsilon_-}} \right). \tag{73}$$

The results of Eqs. (72) and (73) are considerably different. In the experiments, however, mostly the condition of Eq. (73) is fulfilled.

Since we have in the surface configuration an extremely strong coupling of the modes, it is very difficult to attribute even a "quasi-mode" character of TM or TE to the higher modes. Actually the dispersion of the propagation constant for the higher modes can be approximated for the isotropic case by:

$$h_n^2 = k_0^2 - \left[\frac{n\pi}{D} + \frac{2in\pi}{D} \frac{1}{k_0 D(\sqrt{\varepsilon_+} + \sqrt{\varepsilon_-})} \right]^2, \tag{74}$$

and

$$h_n^2 = k_0^2 - \left[\frac{n\pi}{D} + \frac{ik_0 D}{2n\pi D} \left(\frac{1}{\sqrt{\varepsilon_+}} + \frac{1}{\sqrt{\varepsilon_-}} \right) \right]^2. \tag{75}$$

The above formulas for the three favored configurations are quite useful for rough estimation of which tensor component should be effective in the considered case. Nishikawa (1978) has given in his thesis a quite general formula for the approximate dispersion relation of the propagation constant:

$$h \approx k_0 - ik_0 Z/2D, \tag{76}$$

where Z is defined as the surface impedance of the sample for normal incidence and is given by:

$$Z = \frac{E_{z1}/k_{y1} - E_{z2}/k_{y2}}{E_{z1} - E_{z2}}. \tag{77}$$

Here $k_{y1,2}$ are the wave vectors of the normal modes of the magnetoplasma in the sample, and $E_{z1,2}$ the electric field components of the modes parallel to the propagation direction of h.

3. The Triplet Strip Line

In some experiments a more complex form of the strip line had to be used as shown in Fig. 10 (Kamgar et al., 1974; Kneschaurek et al., 1976). The waveguide part of the strip line has a multilayer structure and is embedded between two metal plates. The different dielectric media are characterized as

$$\boldsymbol{\varepsilon}_1 = \begin{pmatrix} \varepsilon_1 & 0 & 0 \\ 0 & \varepsilon_1 & 0 \\ 0 & 0 & \varepsilon_1 \end{pmatrix}, \tag{78}$$

$$\boldsymbol{\varepsilon}_2 = \begin{pmatrix} \varepsilon_{xx} & 0 & 0 \\ 0 & \varepsilon_{yy} & 0 \\ 0 & 0 & \varepsilon_{zz} \end{pmatrix}, \tag{79}$$

$$\boldsymbol{\varepsilon}_3 = \begin{pmatrix} \varepsilon_3 & 0 & 0 \\ 0 & \varepsilon_3 & 0 \\ 0 & 0 & \varepsilon_3 \end{pmatrix}. \tag{80}$$

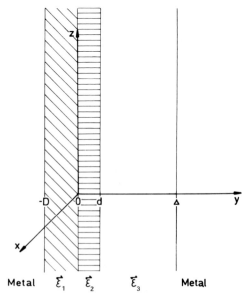

FIG. 10   The schematic of the triple strip line.

For the above choice of the dielectric tensors, the strip-line modes can be classified in pure TM or pure TE modes. The propagation constant h is determined by the following set of equations:

$$h^2 + k_{y1}^2 = \varepsilon_1 k_0^2, \tag{81}$$

$$h^2 + k_{y3}^2 = \varepsilon_3 k_0^2, \tag{82}$$

$$h^2 \varepsilon_{zz} + k_{y2}^2 \varepsilon_{yy} = k_0^2 \varepsilon_{yy} \varepsilon_{zz}. \tag{83}$$

TM: $(k_{y1}/\varepsilon_1)\mathrm{tg}\, k_{y1} D + (k_{y3}/\varepsilon_3)\, \mathrm{tg}\, k_{y3}(\Delta - d) + (k_{y2}/\varepsilon_2)\, \mathrm{tg}\, k_{y2} d$

$$= (k_{y1}/\varepsilon_1)\, \mathrm{tg}\, k_{y1} D(\varepsilon_{zz}/k_{y2})\, \mathrm{tg}\, k_{y2} d(k_{y3}/\varepsilon_3)\, \mathrm{tg}\, k_{y3}(\Delta - d), \tag{84}$$

TE: $k_{y1} k_{y2}\, \mathrm{tg}\, k_{y3}(\Delta - d) + k_{y2} k_{y3}\, \mathrm{tg}\, k_{y1} D + k_{y1} k_{y3}\, \mathrm{tg}\, k_{y2} d$

$$= k_{y2}^2\, \mathrm{tg}\, k_{y1} D\, \mathrm{tg}\, k_{y2} d\, \mathrm{tg}\, k_{y3}(\Delta - d). \tag{85}$$

Setting one or two of the separation parameters equal to zero, the formulas hold as well for the twin and single strip line.

4. The Symmetric Strip Line

The schematic of the symmetric strip line is shown in Fig. 11. Only for those configurations and modes where the electric field in the symmetry axis of the strip line is normal to the plates, can the results of the asymmetric strip line be applied (Kanada *et al.*, 1976). After suitable modification of the effective plate separation for these cases—actually the pure TM-mode in the perpendicular and one mode of the surface configuration—the above formulas, applying

$$D_{\mathrm{eff}} = D/2, \tag{86}$$

can be used. This means that the attenuation coefficient is actually doubled in the symmetric strip line with the same plate separation, because the radiation interacts now with two semiconductor surfaces instead of one.

5. Reflection Losses of Strip-Line Modes

So far we have investigated the dispersion relation of the propagating modes in an infinitely extended strip line. In the experiment, however, we have to consider also the finite length of the interaction region. In Fig. 12 we have shown the actual arrangement. The radiation is fed through a narrow waveguide, which actually acts under suitable geometric conditions as a polarizer for a pure TEM mode, to the entrance of the interaction part. Because of the different mode structure in the interaction part, we expect reflection losses at the entrance of the strip line. The same argument holds

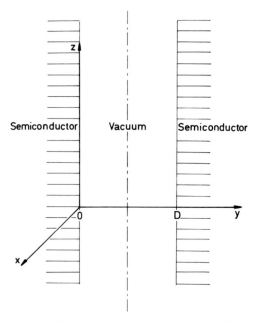

FIG. 11   The schematic of the symmetric strip line.

for the exit. In principle we not only have to consider the losses due to reflections but also interference effects of reflected modes within the interaction part of the strip line. From the mathematical point of view, to solve the boundary problem along the y axis as shown in Fig. 13, we have to investigate which modes are actually excited in the transmitted and reflected wave. We assume that the incident wave has the following mode structure:

$$\mathbf{E}^i = \sum_{n=0}^{\infty} c_n^i \mathbf{E}_n^i. \tag{87}$$

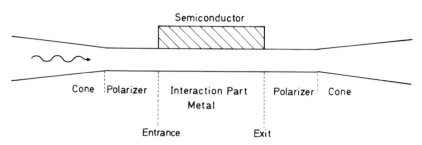

FIG. 12   The strip line in the actual experiment is a complex waveguide system of cones, polarizers, and interaction part with different impedance.

In Eq. (87) we have summed over all possible waveguide modes n, which are supposed to be explicitly known analytically. The reflected and transmitted waves are of the form:

$$\mathbf{E}^r = \sum_{n=0}^{\infty} c_n^r \mathbf{E}_n^r, \tag{88}$$

$$\mathbf{E}^t = \sum_{n=0}^{\infty} c_n^t \mathbf{E}_n^t. \tag{89}$$

Because of the translational invariance along the x axis, we again have only four independent boundary conditions:

$$E_x^i(y) + E_x^r(y) = E_x^t(y), \tag{90}$$

$$D_z^i(y) + D_z^r(y) = D_z^t(y), \tag{91}$$

$$E_y^i(y) + E_y^r(y) = E_y^t(y), \tag{92}$$

$$H_y^i(y) + H_y^r(y) = H_y^t(y). \tag{93}$$

The rather complex problem reduces considerably if we assume that only a pure TEM mode is incident and excites only TM modes in the strip line. For this case

$$E_x^i = E_z^i = 0, \qquad H_y^i = H_z^i = 0, \tag{94}$$

holds, and the boundary conditions are

$$E_x^{r,t}(y) = 0, \tag{95}$$

$$D_z^r(y) = D_z^t(y), \tag{96}$$

$$E_y^i(y) + E_y^r(y) = E_y^t(y), \tag{97}$$

$$H_y^{r,t}(y) = 0. \tag{98}$$

By expansion of $E_j^i(y)$ and $E_j^r(y)$ in series of strip-line modes, the problem can be solved by straightforward calculation. If we assume that in the strip line only the lowest mode, $n = 0$, really propagates and that the higher modes are exponentially damped, then the interesting intensity transmission coefficient T of the boundary is given by the expression

$$T = \int |\mathbf{E}_0^t|^2 \, dy / \int |\mathbf{E}_0^i|^2 \, dy. \tag{99}$$

The corresponding situation is met at the lower dielectric interface of Fig. 13. The actual effect on the numerical results of the total strip-line transmission, including reflection effects, will be discussed later.

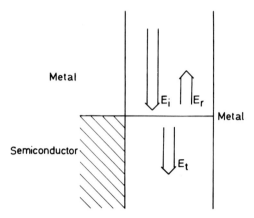

FIG. 13   Because of the dielectric boundary at the entrance of the interaction part of the strip line, the incident radiation is partially reflected.

B.   MICROSCOPIC ANALYSIS OF THE DIELECTRIC TENSOR

So far we have discussed the "optical analysis" of the strip line as used in submillimeter magnetospectroscopy. We have analyzed the different optical configurations and have related the strip-line attenuation directly to the different components of the ε tensor. This means actually that we have abstracted from the particular experimental situation and have extracted the relevant information of the experiment in form of the dielectric tensor. The essential reason for performing an experiment in submillimeter magneto-spectroscopy is to obtain detailed information on the microscopic parameters of the charge-carrier system. The link between the dielectric function and these parameters is the linear response theory of high-frequency electric conductivity. We do not want here to give a detailed and comprehensive discussion of this subject, but present the essential features necessary to understand the magnetic-field dependence of the dielectric function. For a detailed review of magnetoplasma effects we refer the reader to the article by Palik and Furdyna (1970).

There are essentially two approaches to the linear-response theory of electric conductivity. First the classical or "Drude" theory starts from the classical equation of motion of the charged carriers in the presence of an external electric field. This theory has been very successful, especially for the investigation of anisotropy effects of a homogeneous gas of charged carriers with parabolic energy-dispersion relation. The special feature of this treatment is that all carriers are equal and contribute the same response. If we want, however, to investigate the resonance transition between individual electron states, we have to apply quantum theory. The most elegant formulation in this field is Kubo's response theory using the density-operator formalism (Kubo, 1957). The quantum theoretical formulation includes another important feature—the spin of the electron. We shall show that there are special configurations of the strip-line technique that are very sensitive to spin–flip transitions.

1. Classical Formulation

a. *Single-Valley System.* The "one-electron Drude model" assumes that all electrons are stationary without external electric field. The response of the system to an external radiation field is completely determined by the classical equation of motion:

$$\frac{d}{dt}(\mathbf{mv}) = e(\mathbf{E} + \mathbf{v} \times \mathbf{B}) - \frac{\mathbf{mv}}{\tau}. \tag{100}$$

Here $\mathbf{E}$ and $\mathbf{B}$ represent the total fields, $\mathbf{v}$ is the velocity vector of the carrier characterized by an effective-mass tensor $\mathbf{m}$ and manifests the presence of the lattice, and the parameter τ describes the relaxation time of the carrier system. Usually the magnetic field of the radiation field can be neglected, because the effective force due to the magnetic field interaction is reduced by a factor v/c in comparison with the dielectric interaction, c being the velocity of light. In this approximation the total magnetic field in Eq. (100) is given by a constant external field $\mathbf{B}_0$:

$$\mathbf{B} = \mathbf{B}_0, \tag{101}$$

and the total electric field only by the radiation field

$$\mathbf{E} = \mathbf{E}_{rad}(\omega) = \tilde{E}e^{i\omega t}. \tag{102}$$

The response of the system as represented by the electric current density is then given by (Palik and Furdyna, 1970)

$$\mathbf{j}(\omega) = \frac{Ne^2[e\mathbf{E}_{rad}(\omega) \times (\mathbf{mB}_0) + ie^2\mathbf{B}_0(\mathbf{E}_{rad}(\omega)\mathbf{B}_0)/(\omega + i/\tau) - i\|\mathbf{m}\|(\mathbf{m})^{-1}\mathbf{E}_{rad}(\omega)(\omega + i/\tau)]}{e^2\mathbf{B}_0\mathbf{mB}_0 - (\omega + i/\tau)^2\|\mathbf{m}\|}. \tag{103}$$

Here $\|\mathbf{m}\|$ is defined as the determinant, and $(\mathbf{m})^{-1}$ as the inverse tensor of the effective mass tensor $\mathbf{m}$. N is the carrier concentration of the system. By use of Eq. (11) all components of the ε tensor can be obtained from Eq. (103). There is a definite resonance structure in the denominator of Eq. (103). The resonance frequency is given by

$$\omega_c^2 = \frac{e^2\mathbf{B}_0\mathbf{mB}_0}{\|\mathbf{m}\|}, \tag{104}$$

and yields, for a mass tensor exhibiting rotational symmetry with respect to the magnetic field orientation, the usual expression for the cyclotron frequency:

$$\mathbf{m} = \begin{pmatrix} m_T & 0 & 0 \\ 0 & m_T & 0 \\ 0 & 0 & m_L \end{pmatrix} \qquad \Rightarrow \qquad \omega_c = \frac{eB_0}{m_T}. \qquad (105)$$

In Eq. (105) m_T is the transverse mass component determining the motion perpendicular to the magnetic field orientation.

Taking into account the lattice part of the dielectric function and having chosen the z axis parallel to the magnetic field, the different components of the ε tensor are given in the gyrotropic form as follows:

$$\varepsilon_{xx} = \varepsilon_{yy} = \varepsilon_l - \frac{Ne^2}{m_T \varepsilon_0 \omega 2} \left[\frac{1}{(\omega + \omega_c) + i/\tau} + \frac{1}{(\omega - \omega_c) + i/\tau} \right], \qquad (106)$$

$$\varepsilon_{xy} = -\varepsilon_{yx} = -\frac{Ne^2}{m_T \varepsilon_0 \omega 2} \left[\frac{1}{(\omega + \omega_c) + i/\tau} - \frac{1}{(\omega - \omega_c) + i/\tau} \right], \qquad (107)$$

$$\varepsilon_{zz} = \varepsilon_l - \frac{Ne^2}{m_L \varepsilon_0 \omega} \frac{1}{\omega + i/\tau}, \qquad (108)$$

$$\varepsilon_{xz} = \varepsilon_{zx} = \varepsilon_{yz} = \varepsilon_{zy} = 0. \qquad (109)$$

It should be noted, that the component ε_{zz} is strictly magnetic-field independent and only the other components exhibit the field-dependent resonance structure. This is the direct consequence of the Lorentz force, which becomes effective only in the plane perpendicular to the magnetic field and does not affect the motion parallel to the field orientation.

This argument, however, does not hold as soon as the magnetic field is tilted with respect to the symmetry axis of the mass tensor as shown in Fig. 14.

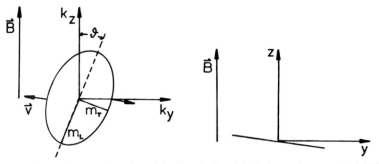

FIG. 14   For arbitrary orientation of the Fermi-ellipsoid relative to the magnetic field **B**, as shown in the upper part, the orbit of the charge carriers in coordinate space is tilted so that a component parallel to the field becomes effective, as shown in the lower part.

For this case the components of the ε tensor are given by

$$\varepsilon_{xx} = \varepsilon_l - \frac{Ne^2}{m_T \varepsilon_0 \omega} \frac{(\omega + i/\tau)}{(\omega + i/\tau)^2 - \omega_c^2}, \tag{110}$$

$$\varepsilon_{yy} = \varepsilon_l - \frac{Ne^2}{m_T \varepsilon_0 \omega} \frac{(\omega + i/\tau)}{(\omega + i/\tau)^2 - \omega_c^2} \frac{(m_L \cos^2 \vartheta + m_T \sin^2 \vartheta)}{m_L}, \tag{111}$$

$$\varepsilon_{zz} = \varepsilon_l - \frac{Ne^2}{m_T \varepsilon_0 \omega} \left(\frac{(\omega + i/\tau)}{(\omega + i/\tau)^2 - \omega_c^2} \frac{(m_T \cos^2 \vartheta + m_L \sin^2 \vartheta)}{m_L} \right.$$
$$\left. - \frac{e^2 B_0^2 / m_T m_L (\omega + i/\tau)}{(\omega^2 + i/\tau)^2 - \omega_c^2} \right), \tag{112}$$

$$\varepsilon_{xy} = -\varepsilon_{yx} = - \frac{iNe^2}{m_T \varepsilon_0 \omega} \frac{[eB_0/m_T][(m_L \cos^2 \vartheta + m_T \sin^2 \vartheta)/m_L]}{(\omega + i/\tau)^2 - \omega_c^2}, \tag{113}$$

$$\varepsilon_{yz} = +\varepsilon_{zy} = - \frac{Ne^2}{m_T \varepsilon_0 \omega} \frac{(\omega + i/\tau)}{(\omega + i/\tau)^2 - \omega_c^2} \frac{(m_T - m_L) \sin \vartheta \cos \vartheta}{m_L}, \tag{114}$$

$$\varepsilon_{zx} = -\varepsilon_{xz} = - \frac{iNe^2}{m_T \varepsilon_0 \omega} \frac{eB_0/m_T}{(\omega + i/\tau)^2 - \omega_c^2} \frac{(m_T - m_L) \sin \vartheta \cos \vartheta}{m_L}, \tag{115}$$

where the cyclotron frequency is given by

$$\omega_c^2 = \frac{e^2 B_0^2}{m_T^2} \frac{m_L \cos^2 \vartheta + m_T \sin^2 \vartheta}{m_L}. \tag{116}$$

Since the cyclotron frequency ω_c depends on the relative orientation of the magnetic field and the symmetry axis of the mass tensor, the measurement of the cyclotron resonance is a perfect tool to investigate any anisotropy effects of the carrier system.

For arbitrary tilt angle, as long as there is no coincidence of one of the principal axes of the mass tensor and the magnetic field orientation, all tensor components including ε_{zz} are resonant. Since there is now a strong coupling of the motion parallel and perpendicular to the magnetic field, the Lorentz force affects the velocity component of the electron parallel to the field, resulting in a "tilted orbit" motion in coordinate space as indicated in Fig. 14. The "tilted orbit" is a direct consequence of the nondiagonal components of the mass tensor.

b. *Multivalley Systems.* If the charge carriers of a semiconductor can be characterized by an ellipsoidal mass tensor, as discussed so far, the corresponding ellipsoid in k-space is usually only part of the total Fermi surface within the Brillouin zone. Depending on the crystal symmetry, there are mostly several ellipsoids that differ in their position and relative orientation,

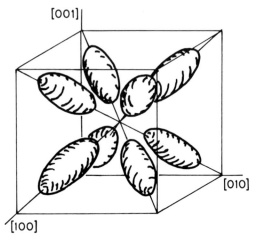

FIG. 15   Fermi ellipsoids centered along the [111] axes of a cubic system, e.g., in PbSe and PbTe.

but are otherwise equivalent. For the semiconductors PbSe and PbTe, for example, the ellipsoids are centered along the [111] axes as shown in Fig. 15. The carriers of each ellipsoid will contribute to the total current and hence to the dielectric function, so that the total dielectric tensor for such a system is given by

$$\varepsilon_{tot} = \sum_j^{j_{max}} \varepsilon_j, \tag{117}$$

where j runs over all different valleys in the Brillouin zone. For the system of [111] ellipsoids and the magnetic field within the (010) plane the components of ε are given by (von Ortenberg et al., 1975):

$$\varepsilon_{xx} = \varepsilon_l - \frac{Ne^2}{2\varepsilon_0 \omega}\left(\omega + \frac{i}{\tau}\right)\left[\frac{(2/m_T + 1/m_L)/3 + \sin 2\vartheta[(1/m_L - 2/m_T)/6]}{\omega_+^2 - (\omega + i/\tau)^2}\right.$$

$$\left. \times \frac{(2/m_T + 1/m_L)/3 - \sin 2\vartheta[(1/m_L - 2/m_T)/6]}{\omega_-^2 - (\omega + i/\tau)^2}\right], \tag{118}$$

$$\varepsilon_{yy} = \varepsilon_l - \frac{Ne^2}{2\varepsilon_0 \omega}\left(\omega + \frac{i}{\tau}\right)\left[\frac{(2/m_T + 1/m_L)/3}{\omega_+^2 - (\omega + i/\tau)^2}\right.$$

$$\left. + \frac{(2/m_T + 1/m_L)/3}{\omega_-^2 - (\omega + i/\tau)^2}\right], \tag{119}$$

$$\varepsilon_{zz} = \varepsilon_1 - \frac{Ne^2}{2\varepsilon_0 \omega}\left(\omega + \frac{i}{\tau}\right)$$

$$\times \left[\frac{-e^2 B_0^2/m_T^2 m_L(\omega + i/\tau)^2 + (2/m_T + 1/m_L)/3 + \sin 2\vartheta[(1/m_L - 1/m_T)/6]}{\omega_+^2 - (\omega + i/\tau)^2} \right.$$

$$\left. + \frac{-e^2 B_0^2/m_T^2 m_L(\omega + i/\tau)^2 + (2/m_T + 1/m_L)/3 - \sin 2\vartheta[(1/m_L - 1/m_T)/6]}{\omega_-^2 - (\omega + i/\tau)^2} \right]$$

$$(120)$$

$$\varepsilon_{xy} = +\varepsilon_{yx} = \frac{iNe^2}{2\varepsilon_0 \omega}\frac{eB_0}{m_T^2 m_L}\left[\frac{(2m_T + m_L)/3 + \sin 2\vartheta[(m_L m_T)/6]}{\omega_+^2 - (\omega + i/\tau)^2} \right.$$

$$\left. + \frac{(2m_T + m_L)/3 - \sin 2\vartheta[(m_L - m_T)/6]}{\omega_-^2 - (\omega + i/\tau)^2} \right], \qquad (121)$$

$$\varepsilon_{xz} = -\varepsilon_{zx} = \frac{Ne^2}{2\varepsilon_0 \omega}\left(\omega + \frac{i}{\tau}\right)$$

$$\times \left[\frac{\cos 2\vartheta[(1/m_L - 1/m_T)/3]}{\omega_+^2 - (\omega + i/\tau)^2} - \frac{\cos 2\vartheta[(1/m_L - 1/m_T)/3]}{\omega_-^2 - (\omega + i/\tau)^2} \right], \qquad (122)$$

$$\varepsilon_{yz} = -\varepsilon_{zy} = \frac{iNe^2}{2\varepsilon_0 \omega}\frac{eB_0}{m_T^2 m_L}$$

$$\times \left\{ \frac{\cos 2\vartheta[(m_L - m_T)/3]}{\omega_+^2 - (\omega + i/\tau)^2} - \frac{\cos 2\vartheta[(m_L - m_T)/3]}{\omega_-^2 - (\omega + i/\tau)^2} \right\}, \qquad (123)$$

where $\omega_\pm$ is defined by

$$\omega_\pm = \frac{e^2 B_0^2}{m_T^2 m_L}\left[\frac{2m_T + m_L}{3} \pm \frac{\sin 2\vartheta(m_L - m_T)}{6} \right]. \qquad (124)$$

In the above formulas ϑ is the angle between the magnetic field orientation and the [001] axis. For an orientation of the magnetic field different from a symmetry axis of the crystal, there are two sets of ellipsoids in the Brillouin zone with different tilt angles ϑ, resulting in two different values for the cyclotron frequency ω_+ and ω_-. It should be noted that even for nonzero magnetic field the nondiagonal components ε_{xz} and ε_{yz} vanish for

$$\vartheta = \tfrac{1}{2}m\pi, \qquad m = 0, 1, 2, \ldots, \qquad (125)$$

despite the actual "tilted orbit" motion of the carriers represented by the different ellipsoids. Even though the contribution of the carriers of one ellipsoid is nonzero, the superposition averages the contributions out to zero because of the different sign.

2. Quantum Mechanical Formulation

The quantum theoretical formulation of the dielectric response exceeds the classical theory in many aspects. Not only is the purely quantum theoretical quantity of the electron spin included, but also the temperature dependence of the resonance. This effect manifests not only in the generally increasing resonance width with increasing temperature, but also in a dramatic intensity variation of the different quantum transitions due to population effects. The most transparent formulation of the dielectric quantum response in the local approximation uses the density-operator formalism after Kubo (1957). Using this concept and neglecting the magnetic field of the radiation, Kubo has shown that the high-frequency conductivity is given by the expression

$$\sigma_{\mu\nu}(\omega) = \frac{1}{V} \int_0^\infty dt \, e^{-i\omega t} \int_0^{1/kT} d\lambda \, \langle J_\nu(-i\hbar\lambda)J_\mu(t)\rangle. \tag{126}$$

This formula gives the exact amplitude and phase of the induced electric current as a response to an applied electric field oscillating with the frequency ω. Here kT has the usual meaning, and J_μ is the many-particle current along the μ direction in the volume V:

$$J_\nu = \int_V \psi^+(\mathbf{r})j_\nu\psi(\mathbf{r}) \, d\mathbf{r}^3. \tag{127}$$

With use of the one-electron current operator

$$j_\nu = -ev_\nu, \tag{128}$$

$\psi(\mathbf{r})$ and $\psi^+(\mathbf{r})$ are the normalized wave functions. The Heisenberg operator $j_\nu(t)$ is determined by the natural motion of the current in the absence of the external electric field:

$$j_\nu(t) = e^{i\mathscr{H}_0 t/\hbar}j_\nu e^{-i\mathscr{H}_0 t/\hbar}, \tag{129}$$

where $\mathscr{H}_0$ is the Hamiltonian of the carrier system including scattering interactions. The brackets in Eq. (126) denote the average with the equilibrium-density operator

$$\langle A \rangle = \mathrm{Tr}(\rho A) \triangleq \mathrm{trace}(\rho A), \tag{130}$$

where the density operator ρ is given by

$$\rho = e^{-(\mathscr{H}_0 - \mu N)/kT}/\mathrm{Tr}(\rho). \tag{131}$$

Here μ represents the chemical potential of the system and N is the many-particle number operator. After reduction of the many-particle trace to a sum over single particle states, the high-frequency conductivity is given by the expression

$$\sigma_{\nu\mu}(\omega) = \frac{\hbar}{V} \int_{-\infty}^{+\infty} dE \ \mathrm{Tr}\{\delta(\hbar - E)j_\nu(E - \hbar)^{-1}[f(\hbar) - f(E)]$$

$$\times \ [\pi\delta(\hbar - E - \hbar\omega) + i\mathscr{P}(\hbar - E - \hbar\omega)^{-1}]j_\mu\}. \tag{132}$$

In the above equation $\hbar$ is the single-particle Hamiltonian of the system including scattering interactions, $\mathscr{P}$ denotes the principle value of the argument, and f indicates the Fermi–Dirac distribution function

$$f(E) = \frac{1}{1 + e^{(E - \mu)/kT}}. \tag{133}$$

Assuming that the eigenstates of $\hbar$ are known, the trace in Eq. (132) can easily be performed:

$$\sigma_{\nu\mu}(\omega) = \sum_n \sum_m \frac{\pi\hbar}{V} \frac{f(E_n) - f(E_m)}{\hbar\omega} j_{nm}^\nu j_{mn}^\mu \delta(E_m - E_n - \hbar\omega)$$

$$+ \frac{i\hbar}{V} \sum_n \sum_m \frac{f(E_n) - f(E_m)}{E_m - E_n} \mathscr{P} \frac{1}{E_m - E_n - \hbar\omega} j_{nm}^\nu j_{mn}^\mu. \tag{134}$$

The subscripts in the above equation denote the complete set of quantum numbers of the eigenstates of h. Equation (134) is an exact result, in the sense that it includes in the second term the diamagnetic part, which becomes important at high frequencies. In the limit of

$$\hbar\omega \gg (E_n - E_m), \tag{135}$$

the second term after purely analytical transformation can be written as $-ie^2N/m\omega$, where N is the carrier concentration and m the effective mass of the energy band considered. Near the resonance, however, this approximation does not hold.

The intensity of the conductivity is determined not only by the matrix elements of the current operators but also by the thermodynamic population of the initial and final states. Thus, by variation of the temperature it is possible to probe the contribution of different energy levels.

The formulation of the high-frequency conductivity as a trace in Eq. (132) has the special property of being independent of the set of functions we are using for the explicit calculation of the trace. This fact is of special advantage, because mostly the eigenfunctions of the electron–scatterer system are not known. Performing the trace in Eq. (132) with eigenfunctions of the system without scatterers, the operators involving h are nondiagonal and the δ functions are smeared out. The detailed investigation of this problem leads to the damping theoretical treatment of the conductivity, and instead of the δ function introduces the "spectral function," which determines the energy broadening of the individual microscopic quantum transition. It depends on nearly every quantum number, the magnetic field, and the temperature. The superposition of all the microscopic transitions yields the line shape of the microscopic resonance. The central quantity determining the resonance structure in Eq. (134) is the single particle Hamiltonian

$$h = h_0 + V_{\text{scat}}. \tag{136}$$

Whereas the scattering interaction V_{scat} modifies the line shape, the resonance position itself is in good approximation, determined only by the single particle Hamiltonian h_0 of the electron in the presence of the lattice potential and an external magnetic field.

In its general form, including relativistic especially spin effects, h_0 can be written as:

$$h_0 = \frac{(\mathbf{p} + e\mathbf{A})^2}{2m_0} + V_{\text{per}} - \frac{\hbar^2}{2m_0 c^2}(\nabla V_{\text{per}})\nabla + \frac{\hbar}{4m_0 c^2}\boldsymbol{\sigma}(\nabla V_{\text{per}} \times \mathbf{p}). \tag{137}$$

Here $\mathbf{p}$ is the canonical momentum, $\mathbf{A}$ the vector potential associated with the external magnetic field, V_{per} the periodic part of the effective lattice potential, and $\boldsymbol{\sigma}$ represents the Pauli spin operators. The external magnetic field, as represented by the vector potential $\mathbf{A}$, permits the energy levels to be changed in a rather arbitrary, although well-defined way.

The handling of Eq. (136) in this form is hardly possible, but requires fundamental approximations. Actually by an analysis of only the symmetry properties of the lattice, an effective Hamiltonian in matrix formulation can be established. The numerical parameters of this matrix, which are related to interband matrix elements of the momentum operator and the relativistic correction terms of Eq. (137), have to be fitted from the experiment, e.g., from data of submillimeter magnetospectroscopy. The parameters determining the energy-band structure of the considered solid are actually the final aim of most experiments in submillimeter magnetospectroscopy. They represent all the features of the coupling mechanism of the energy levels.

Since the models established in this way by fitting procedures are essentially "trial Hamiltonians," the energy levels calculated in this trial model are much better represented than the eigenfunctions. This is a typical feature of any "trial procedure" in quantum theory.

The second important contribution to the dielectric function to be considered, in addition to the response of the charged carrier system in the energy band, is ε_l. This is due to the response of all remaining charges, electrons, and lattice constituents, and is supposed to be independent of the magnetic field. The numerical values for both the real and imaginary parts of ε_l, however, are crucial to the determination of the attenuation constant of the strip line. This means that a considerable knowledge of the other dielectric interactions is also necessary for the interpretation of strip-line data. Therefore, ε_l cannot always be approximated by the static dielectric function, but has to take lattice resonances fully into account.

Thus far we have accumulated the necessary tools to perform the twofold analysis of any experiment in submillimeter magnetospectroscopy. However, in the optical part we have laid emphasis on the strip-line technique.

C. THE EXPERIMENTAL SETUP FOR THE STRIP-LINE TECHNIQUE IN SUBMILLIMETER MAGNETOSPECTROSCOPY

The wavelength of submillimeter radiation ranges from 100 μm to 1 mm. This order of magnitude for the geometrical dimensions determines the frame for the actual construction of the strip line. As already discussed, single-mode excitation of the strip line is possible only if the plate separation is smaller than half the radiation wavelength. Therefore, the usual plate separation is about 50–100 μm. Since the strip-line transmission for constant dielectric properties depends strongly on the wavelength, in the optimal experimental setup the radiation frequency is fixed and the magnetic field swept. This fact makes it possible to use powerful lasers as a radiation source. Usually, the magnetic field is provided by a superconducting solenoid magnet. Because of the limited bore of the magnet, the "parallel configuration" is most easily applied. By use of split-coil magnets, however, the perpendicular or surface configuration is preferable.

In Fig. 16 we have plotted the schematic of a single strip line for the parallel configuration. The sample, of nearly arbitrary shape, is glued on top of the 100-μm-deep channel in the brass body of the lower strip-line part. For low-temperature glue we used cellulose tridecanoate, which is supposed to remain plastic even at helium temperatures (Apel *et al.*, 1971). Special care has to be given to the finish of the waveguide and interaction part of the strip line, as well as to the sample surface. For the sample it was generally found that a well-reflecting, optically perfect surface had worse properties in the strip line than a chemically etched surface exhibiting cracks and grooves because as a result of the polishing procedure all the inevitable cracks are smeared out by the polishing paste and act as optical inhomogeneities. The sensitivity of the strip-line arrangement depends mainly on the radiation intensity transferred from the waveguide into the strip-line mode. Because of the very unfavorable ratio of the waveguide to the strip-line cross section—which is about 500!—mode matching is a problem. Sufficient energy transfer requires a very small aperture of the matching cone; otherwise the reflection losses are too large.

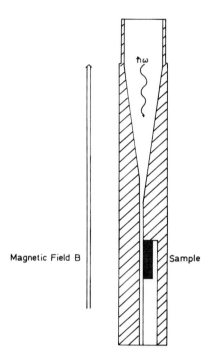

FIG. 16   In the simple experimental realization of the strip line for the parallel configuration, the sample is glued on top of a 100-μm-deep channel machined into the brass body of the sample holder.

For the investigation of anisotropy effects a special strip line with rotating sample has been introduced by Ramage *et al.* (1974) as shown in Fig. 17. The relatively large, circular-shaped sample can be rotated while pressed on distance blocks made of Teflon. Only the middle part of the sample is exposed to the radiation. Because of the circularly shaped dielectric boundary at the entrance and exit of the interaction part, we can no longer expect the strip-line mode to be constant over the width in x direction. This implies that the propagation constant h is no longer determined solely by ε_{zz}, as derived in Eq. (47), but that there are admixtures of the other components also. The influence of ε_{zz}, however, should be dominant. This kind of strip line is now being applied by different groups investigating anisotropy effects (Ramage *et al.*, 1974; von Ortenberg, 1974; Kawamura *et al.*, 1975). Despite the limited space in the bore of an axial superconducting magnet, the principle of the rotating sample has been adopted in the perpendicular and surface configuration using sophisticated light-pipe systems. The schematics are shown in Figs. 18 and 19, respectively.

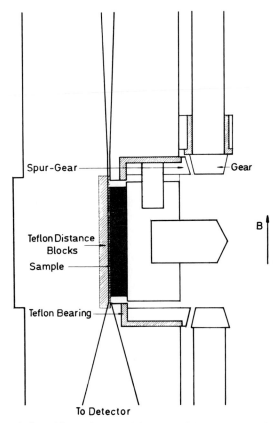

FIG. 17   The strip line with rotating sample in the parallel configuration for the investigation of anisotropy effects.

In our experimental setup, the strip line is part of the integrated sample-holder–detector system, which is kept in a low-pressure helium atmosphere as exchange gas for better thermal coupling. The schematic of the arrangement is shown in Fig. 20. The long tail of the integrated sample-holder–detector system is in direct contact with the liquid helium. The temperature of the sample is measured by a carbon thermometer. By variation of the exchange-gas pressure and by additional heating, the temperature can be increased up to 60 K. For temperatures below 4.2 K, the pressure above the helium bath is reduced by pumping. Thus, temperatures down to the lower limit of about 1.6 K are available.

As detectors we tried different kinds of bolometers: Ge, C, SnO_2. Whereas the sensitivity of Ge and C depends strongly on the magnetic field, this effect

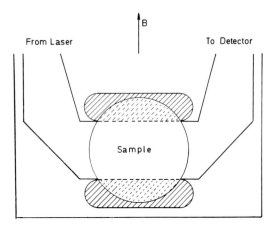

FIG. 18   The strip line with rotating sample in the perpendicular configuration uses a complex system of cones and mirrors because the space in the bore of an axial superconducting magnet is rather limited.

can be neglected for the SnO_2 detector (von Ortenberg *et al.*, 1977). Therefore, this kind of detector can be placed directly below the strip line in the center of the magnet, thus avoiding additional light pipes.

As the source for the submillimeter radiation we used molecular-gas lasers, both of the optically and electrically pumped type. Thus, quite a number of different radiation wavelengths are available, ranging from 28 up to 496 μm. The lasers are installed in a radial arrangement with respect

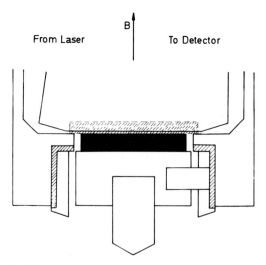

FIG. 19   The strip line with rotating sample in the surface configuration.

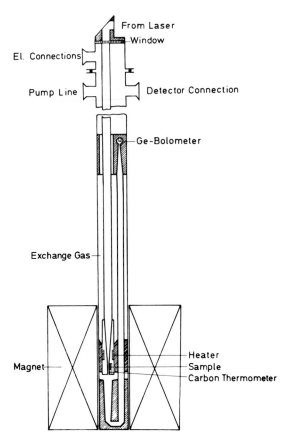

FIG. 20   The integrated sample-holder–detector system using a Ge bolometer outside the stray field of the magnet.

to the magnet, so that changing frequency by switching to the next laser is quite easy. In Fig. 21 we show in the photograph the three output ends of the DCN-, H_2O-, and HCN-laser aiming at the head of the sample holder sitting in a 12 T superconducting magnet.

The signal of the detector in response to the chopped laser beam transmitted through the strip line is processed by conventional lock-in technique. To enhance weak structures on a sufficiently stable background signal, we applied magnetic-field modulation in connection with continuously incident laser radiation.

In Fig. 22 we give the schematic of the overall electronic system. We like to emphasize that most of our samples were provided with electrical contact leads suitable for the simultaneous measurement of optical and dc magneto

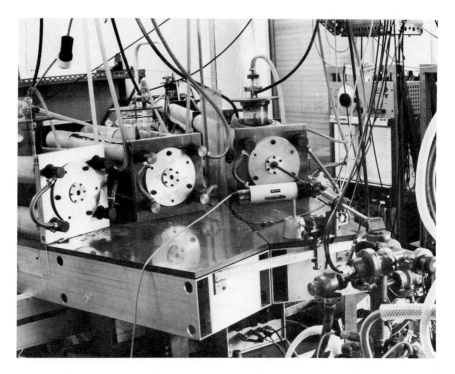

FIG. 21 The output ends of three, conventional, molecular-gas lasers aiming at the head of the sample holder.

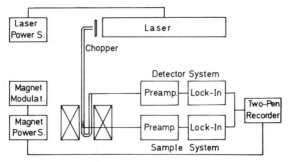

FIG. 22 The electronic system has a detector and a sample channel for simultaneous measurement of submillimeter and dc properties.

effects. The experience has proved that this combination of optical and electrical measurements is necessary for an unambiguous interpretation of the data.

D. EXPERIMENTAL DATA AND SIMULATION OF THE STRIP-LINE TRANSMISSION

1. Investigation of the Effective-Mass Anisotropy

By suitable choice of the strip-line configuration, different combinations of the dielectric tensor components can be investigated separately. In this sense, the configuration implies the polarization of the submillimeter radiation. Mode-coupling effects, however, can seriously obscure the spectra. Considering an isotropic charge-carrier model, the components of the dielectric tensor in the parallel configuration are given by Eqs. (106)–(109). In Fig. 23 we have plotted the calculated strip-line transmission spectra for 195- and 337-μm wavelength radiation as functions of the magnetic field. The numerical values of the parameters are $\varepsilon_l = 25.6$, $N = 10^{17}$ cm^{-3}, $m_T = m_L = 0.03\, m_0$, $\omega\tau = 10$. The length of the interaction part of the strip line is set to $L = 10$ mm. The configuration is indicated at the top of the figure and is, from left to right, perpendicular, surface, and parallel configuration.

At zero magnetic field the relative transmission of the strip line is 80%. In the magnetic-field dependence the spectra differ considerably. All spectra are calculated by the full set of equations determining the propagation constant h. The resonance positions are indicated by arrows.

For the perpendicular configurations the spectra depend on the polarity of the magnetic field involving some kind of "strip-line Hall-effect" (solid and broken curves). The decrease in transmission at high field is caused by the anomaly in $\varepsilon_\perp$ which determines the spectrum after Eq. (61). The nondiagonal component has, however, a nonnegligible influence, as demonstrated by the polarity dependence.

The spectra for the surface configuration in the middle part of Fig. 23 decrease more rapidly with increasing field. These spectra and the spectra for the parallel configuration are independent of the field polarity.

Whereas for 337-μm wavelength radiation the strip-line transmission in the parallel configuration is independent of the magnetic field up to 10 T, there is a pronounced structure at about 6 T for 195-μm wavelength radiation. This line is caused by the anomaly in $\varepsilon_\perp$, which demonstrates that in this magnetic-field range there is a very strong mode coupling, so that the influence of the $\varepsilon_\perp$ mode becomes nearly dominant. However, as long as there is no anomaly in any of the modes for the parallel configuration, the spectrum

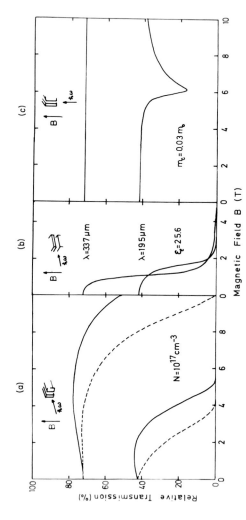

FIG. 23 The simulated strip-line transmission for the three favorite configurations [(a) perpendicular, (b) surface, (c) parallel] as calculated for an isotropic Drude model. Only in the perpendicular configuration does the spectra depend on the magnetic field polarity, as shown by the solid and broken curves.

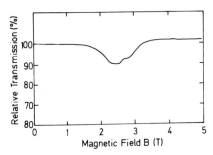

FIG. 24 The experimental strip line transmission for p-type PbSe in the parallel configuration exhibits two "tilted-orbit" resonances for the magnetic field oriented 22.5° off the [001] axis.

is well determined by ε_{zz}, which is in the isotropic model independent of the magnetic field.

For all of the three different spectra, there is no evident relationship to the resonance in which we are interested. Nevertheless, all configurations are suitable for the detection of weak quantum transitions, as we shall prove in detail. In cases where there is no anomaly in $\varepsilon_\perp$ over the considered field range, the spectra of the parallel configuration reflect perfectly the field dependence of ε_{zz}. As shown by Eq. (112), without considering quantum effects, any structure in ε_{zz} is caused by the tilted-orbit phenomena. Therefore, the parallel configuration is well suited for investigation of the anisotropy effects.

a. *The Parallel Configuration.* In Fig. 24 we have reproduced the experimental spectra of the strip-line transmission for the parallel configuration, using a p-type PbSe sample having a carrier concentration of about 10^{18} cm^{-3}. The sample surface lies in the (010) plane. The [001] axis has a tilt angle of $\vartheta = 22.5°$ relative to the magnetic-field direction. We observe a pronounced structure at about 2.3 T. Actually, the observed resonance is split into two separate lines, representing the two different sets of ellipsoids.

In Fig. 25 we have plotted the results of the corresponding simulation. The parameters are given directly in the figure. For the calculation we used the approximation of Eq. (47). In the upper part of the figure the real and imaginary parts of ε_{zz} as functions of the magnetic field are shown. The resonance position is most clearly determined by Im(ε_{zz}). In the lower part of the figure the imaginary part of the propagation constant h and the simulated spectrum are plotted. For comparison, we have reproduced the experimental data by the dashed curve and find a very good agreement of both curves.

We like to point out that the minima in the strip-line transmission curve do not indicate the resonance position, which actually is much more related

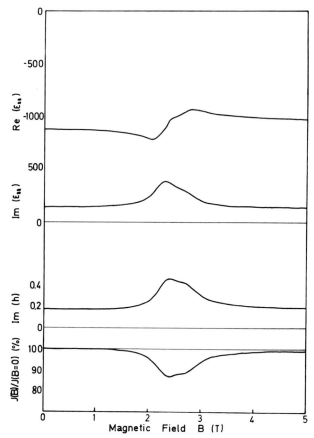

FIG. 25   The tilted-orbit resonance for the magnetic field oriented 22.5° off the [100] axis as expressed in both the real and imaginary parts of ε_{zz} determines the imaginary part of the propagation constant h and thus the strip line transmission for PbSe (von Ortenberg *et al.* 1975). ($m_1 = m_2 = 0.0626\ m_0$, $m_3 = 0.11\ m_0$, $\varepsilon_l = 203$, $p = 1 \times 10^{18}\ \text{cm}^{-3}$ $\hbar\omega = 3.68$ meV, $D = 110$ μm, $L = 0.48$ cm, $\omega\tau = 92$).

to the turning point in the slope before the minimum. This fact has actually caused some confusion in the interpretation of experimental data.

In Fig. 26 we have plotted the sequence of resonance curves of the same crystal for the orientation relative to the magnetic field varied in steps of 4.5°. The symmetry configuration of the Fermi ellipsoids in PbSe becomes quite obvious: for $\vartheta = 0$, all ellipsoids contribute to the same resonance, so that a single, strong line is observed. Tilting the field, two different sets of ellipsoids become effective, so that the resonance is split. For $\vartheta = 45°$, we still have two nonequivalent sets of ellipsoids. Those lying in the $(\bar{1}01)$ plane, however,

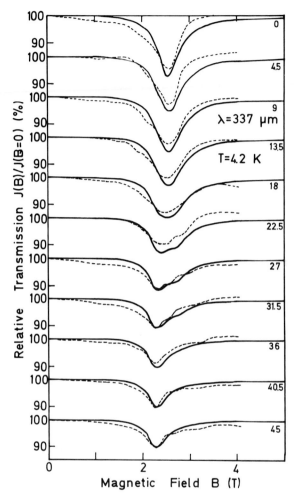

FIG. 26 The simulated strip-line spectra agree well with the experimental data for *p*-type PbSe. The parameters in the right part of the figure indicate the tilt angle of the [001] axis relative to the magnetic-field orientation for rotation of the sample in the (010) plane (——, theory; ——, experimental) (von Ortenberg *et al.* 1975).

do not contribute to the tilted-orbit resonance because the principal axis of the ellipsoids is parallel to the field direction. Therefore, at angles larger than 30° this resonance fades.

In Fig. 27 we have summarized the various measurements in a plot of the minimum position versus tilt angle. As expected, there is a fourfold symmetry. For general orientation the resonance splits. In Fig. 28 we have plotted the resonance spectra of the same sample for the high-symmetry configuration

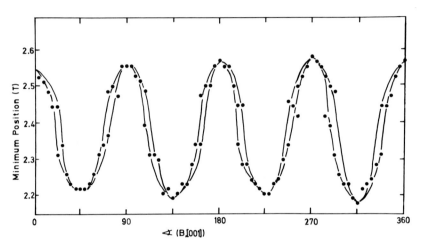

FIG. 27   The minimum position versus orientation exhibits only a single resonance for PbSe (010) for the high-symmetry configurations. For the other orientations the resonance splits because of the two nonequivalent sets of Fermi ellipsoids (von Ortenberg and Schwarzbeck, 1974).

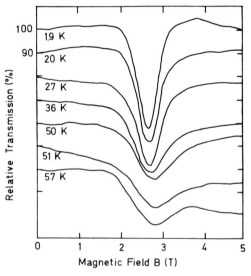

FIG. 28   The experimental strip line transmission for PbSe with the magnetic field oriented parallel to the [001] axis is strongly temperature dependent (von Ortenberg et al. 1975).

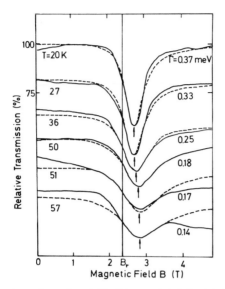

FIG. 29   The comparison of experimental (solid curves) and simulated (broken curves) strip-line spectra for PbSe shows that the temperature affects only the damping parameter Γ. The resonance position is related to the turning point of the slope before the minimum as indicated by the vertical line. The minimum, however, shifts with increasing temperature to higher magnetic-field intensities as indicated by the arrows (von Ortenberg and Schwarzbeck, 1974).

$B\|(001)$ for different parameters of the temperature. All adjacent curves have a relative offset of 10% on the intensity axis. The narrow line at low temperature broadens asymmetrically with increasing temperature. There is a definite shift of the transmission minima to higher magnetic field intensities at higher temperatures. The simulation of Fig. 29, however, demonstrates that both features are easily understood by an increasing damping parameter $\Gamma = \hbar/\tau$, with growing temperature and no change in the effective mass values. We like to emphasize that the damping parameter Γ represents the linewidth of the actual resonance in the dielectric function, and is not defined as the half-width of the experimental data. In this respect Γ is a much more basic quantity.

In Fig. 30 we have summarized the temperature dependence of the damping parameter. The functional dependence can be approximated by a T^2 law indicating an electron–electron scattering mechanism.

For the discussed (001) sample configuration we want to demonstrate how sensitive the strip-line method is with respect to sample preparation and waveguide alignment. In Fig. 31 we compare the transmission curve of a symmetric strip-line arrangement for different finishes of the sample

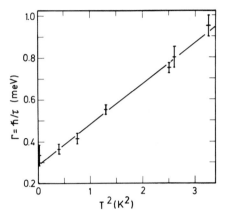

FIG. 30   The damping parameter Γ is a linear function of the square of the temperature, indicating an electron–electron scattering mechanism (von Ortenberg and Schwarzbeck, 1974).

surfaces. The broken curve was obtained for a couple of samples with mechanically polished surfaces, without subsequent etching. The surfaces did not show any defects and were highly reflective in the visible region. Evidently, the surface has a low carrier mobility due to the mechanical preparation technique. After etching, the sample surfaces were of minor quality by eye inspection, because grooves and cracks could be observed. The resultant strip-line spectra show, however, that the remaining sample material is of much higher quality and the damaged parts have been removed by the etching process.

In Fig. 32 the transmission spectra of a PbSe sandwich in the symmetric strip line is shown for $\vartheta = 45°$. Because of the sandwich structure, the attenuation is increased by a factor of two. For $\vartheta = 45°$ there are two nonequivalent sets of Fermi ellipsoids. One set, however, is oriented within the ($\bar{1}01$) plane, so that it should not contribute to any tilted-orbit resonance. Actually

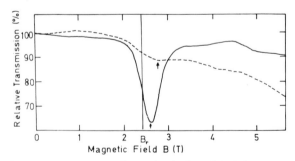

FIG. 31   The two strip-line spectra of a mechanically polished (broken curve) and chemically etched PbSe sample (solid curve) demonstrate the influence of surface preparation.

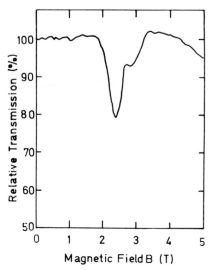

FIG. 32   Despite the high-symmetry configuration in the (010) plane, the tilted resonance for the magnetic field oriented 45° off the [100] axis is split, indicating that both nonequivalent sets of Fermi ellipsoids contribute to the strip line spectrum for PbSe.

this fact holds only as long as the propagation constant h is strictly parallel to the field direction. Depending on the alignment of the waveguide system, a mode structure parallel to the strip-line faces can be excited. This can be effectively described by tilted propagation vectors, as schematically shown in Fig. 33. For tilted vectors h, however, the normal mode in the semiconductor material can no longer be approximated by ε_{zz}, but couples also

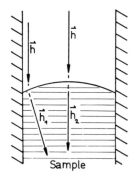

FIG. 33   In the sample holder with rotating sample, refraction effects become effective at the entrance of the interaction part of the strip line.

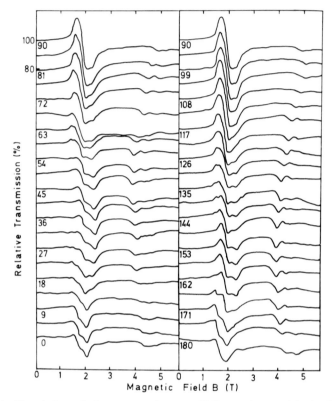

FIG. 34   The relative strip-line spectra of a p-type PbSe sample rotated in the (100) plane. The parameters indicate the tilt angle of the [011] axes and the magnetic field (von Ortenberg *et al.*, 1975).

with the other components of the dielectric tensor. This effect of additional transverse mode excitation is extremely sensitive to the waveguide alignment.

For a PbSe sample with (011) surface the strip-line spectra exhibit a twofold symmetry as shown in Fig. 34. Whereas the low-field structure is definitely related to the tilted-orbit resonance, the high-field resonance is still in discussion. Similar additional lines have been observed by Ramage *et al.* (1975) in PbTe. One possible mechanism to explain the additional lines is based on mode-coupling effects, as demonstrated by the numerical simulation for $B \| [111]$ shown in Fig. 35. In the upper part of the figure we have plotted the calculated strip-line spectra for different parameters of the carrier concentration. The solid curve was calculated for $p = 5 \times 10^{17}$ cm^{-3} and a damping parameter of $\omega\tau = 10$. For the broken curve, the damping parameter was decreased to $\omega\tau = 2$. The two dashed curves were calculated

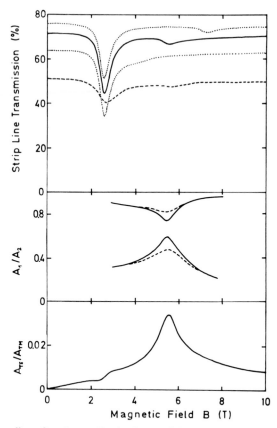

FIG. 35 The effect of mode coupling in the parallel configuration is clearly demonstrated by the concentration-dependent position of the weak high-field minimum. The solid and broken curves are simulations for a carrier concentration of $p = 5 \times 10^{17}$ cm^{-3}, whereas the lower and upper dotted curves correspond to $p = 2 \times 10^{17}$ cm^{-3} and $p = 1 \times 10^{18}$ cm^{-3} respectively (von Ortenberg, 1978b).

for carrier concentration at $p = 10^{18}$ cm^{-3} and $p = 2 \times 10^{17}$ cm^{-3} in the upper and lower parts of the figure, respectively. Depending on the carrier concentration, a weak additional minimum is found in the spectrum. This line is caused by the dielectric anomaly in $\varepsilon_\perp$. At the anomaly there is a strong mode coupling. Since the field position of the anomaly depends directly on the carrier concentration, the additional minimum shifts with increasing p toward higher magnetic field intensities.

In the middle part of Fig. 35 we have plotted the amplitudes of the two normal modes in the semiconductor material contributing to the strip-line mode. The upper curves indicate the amplitude of the "quasi-ordinary

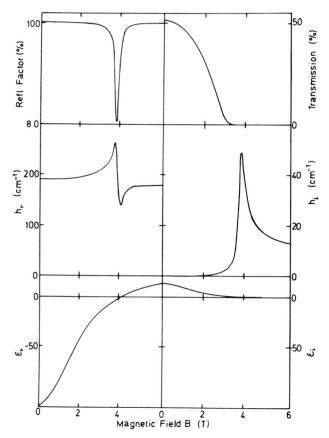

FIG. 36   The dielectric anomaly in ε determines the strip-line transmission in the perpendicular configuration, not only via the imaginary part of the propagation constant h, but also by
the reflection losses at the entrance and exit of the interaction part. The influence of the latter contribution is characterized by the reflection factor in the upper-left part of the figure ($\lambda = 337\,\mu$m,
$m = 0.03\,m_0$, $N = 4 \times 10^{16}$ cm^{-3}, $\omega\tau = 10$) (von Ortenberg, 1978a).

Voigt mode," while the lower curves represent the amplitude of the "quasi-
extraordinary Voigt mode." Near the anomaly the latter clearly peaks up.
In the lower part the same situation is reflected by the amplitude ratio of
the TE/TM modes in the vacuum part of the strip line. The mode-coupling
phenomenon also influences the spatial variation of the electric field perpendicular to the plates of the strip line.

   b.   *Mode-Matching Effects.*   The influence of the spatial field distribution
on the effective strip-line transmission is demonstrated for pure-mode
excitation in the perpendicular configuration in Fig. 36. In the lower part of

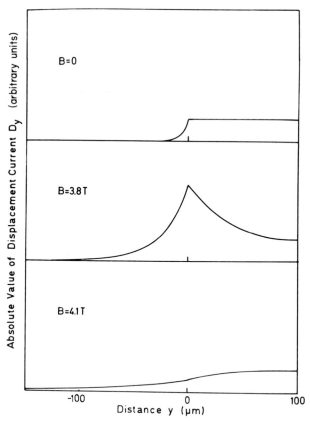

FIG. 37 The character of the propagating strip-line TM mode changes dramatically near the dielectric anomaly ($\lambda = 337 \, \mu$m, $N = 4 \times 10^{16} \, \text{cm}^{-3}$, $m_c = 0.03 \, m_0$, $\omega\tau = 10$, $\varepsilon_l = 25.6 \, \varepsilon_0$) (von Ortenberg, 1978a).

the figure we have plotted the magnetic-field dependence of both the real and imaginary parts of the dielectric function. Near the anomaly there is a strong dispersion in the propagation constant h, as can be seen from the curves in the middle part at the figure. The effect of the anomaly on h is so strong that even the real part clearly exhibits a magnetic-field dependence (von Ortenberg, 1978a,b). Near the dielectric anomaly the spatial variation of the mode changes dramatically.

In Fig. 37 we have plotted the displacement current normal to the interface as a function of the transverse coordinate for three different values of the magnetic-field intensity. For zero magnetic field, the displacement current is nearly constant in the vacuum part and decays exponentially in the semi-conductor material. Near the anomaly, however, for small negative values of

ε_r (B = 3.8 T), the strip-line mode peaks up at the interface and adopts the character of a real surface mode. The opposite behavior of a decreasing displacement current at the interface results for small positive values of ε_r. Both kinds of distortion from the constant-field distribution give rise to reflection losses at the entrance and exit of the strip line due to mode matching. In Section A.5 we calculated the correction factor to account for the reflection losses in the effective strip-line transmission. This factor is plotted in the upper part of Fig. 36. Only near the anomaly are the reflection losses nonnegligible and enhance the effect of the attenuation.

In our simulation we have chosen a low carrier concentration, because in this case the imaginary part of the dielectric function is very small near the anomaly and the zero in ε_r becomes very effective. For higher carrier con-

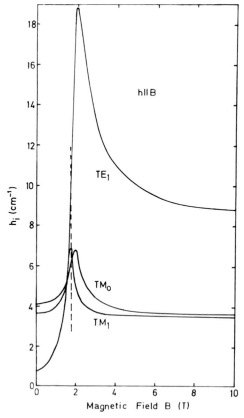

FIG. 38   Changing the radiation wavelength to λ = 118.6 μm in a strip-line with D = 100-μm plate-separated multimode excitation becomes possible. The position of the dielectric anomaly is indicated by the vertical broken line.

centrations, however, the strip-line modes are less disturbed and the influence of mode matching can be neglected.

c. *Multimode Excitation.* An essential criterion for the mode excitation in the strip line is the ratio of wavelength over plate separation λ/D. Therefore, in the experiment a change of frequency is often accompanied by a change of mode excitation. For $D = 100 \, \mu m$ and a radiation wavelength of $\lambda = 118 \, \mu m$, not only is the "quasi-TM_0" mode able to propagate in the "parallel configuration" but also the "quasi"-TM_1 and -TE_1 modes. In Fig. 38 we have plotted the imaginary part of the propagation constant h for such a situation. The corresponding dielectric functions of the carrier system in the semiconductor are reproduced in Fig. 39. The peaks in the "quasi-TM" modes are completely due to the strong mode coupling with the quasi-TE mode near the dielectric anomaly. The maximum attenuation of the quasi-TE mode is about three times as large as those of the quasi-TM modes, so that in most parts of the spectrum the TE-mode attenuation is dominant. Only at low magnetic-field intensities are the TE modes less attenuated than the TM mode.

In Fig. 40 we have plotted the corresponding real part of the propagation constant. Whereas the dispersion of the TM mode is negligible, for the TE mode there is a pronounced variation of h_r near the anomaly. This fact should give rise to mode-matching effects as discussed before. If all the propagating modes are excited with equal intensity, the strip-line transmission should be mainly due to the TM modes, because the TE mode is nearly completely attenuated. This argument should hold also for smaller values of λ/D. In the latter case, however, the TM modes are nearly unattenuated and there is still sufficient transmission of the TE mode, so that the structure in the total transmission is caused by the latter.

This experimental situation was encountered by Ramage et al. (1974). These authors actually applied the parallel configuration with a plate separation larger than half the wavelength. The advantage of large overall transmission in such a setup is, however, compensated for by the rather complex structure of the extraordinary Voigt mode, which dominates the attenuation. Since screening effects occur in the extraordinary Voigt mode, a knowledge of plasma frequency is necessary for a quantitative interpretation of the data (Palik and Furdyna, 1970). Nevertheless Ramage et al. (1974) succeeded, in a series of excellent experiments on PbTe and PbSnTe, in using the TE-moded parallel strip-line configuration.

d. *The Perpendicular Configuration.* At first inspection the perpendicular configuration seems to win, because only pure TM or TE modes are excited and no mode coupling obscures the results. The TM mode, however, is not only determined by $\varepsilon_\perp$ but also by the other components of the

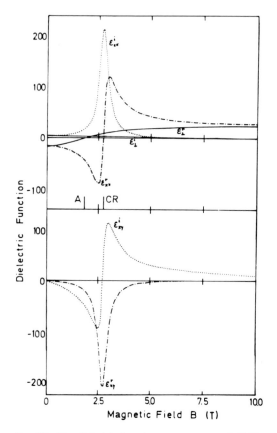

FIG. 39   The dielectric functions versus magnetic field.

dielectric tensor, as can be seen from Eq. (61). This implies that not only does the same complex structure of $\varepsilon_\perp$ as discussed in the preceding paragraph have to be envisaged, but also the additional admixture of ε_{yz}, which depends on the field polarity. In Fig. 41 we demonstrate this fact by experimental results for HgSe. This semiconductor material has an isotropic energy band in good approximation. The influence of ε_{yz} is clearly visible by the difference in the transmission curves for reversed magnetic field. The weak resonance lines are caused by quantum effects and will be discussed in detail later.

The perpendicular configuration has been applied successfully by the Japanese group for the investigation of the anisotropy of the Fermi surface of $Pb_{1-x}Sn_xTe$ and $Pb_{1-x}Ge_xTe$ with prolate ellipsoids along the [111] direction (Kawamura et al., 1978). The excellent experimental data are interpreted, however, only in terms of $\varepsilon_\perp$ and the admixture of the other com-

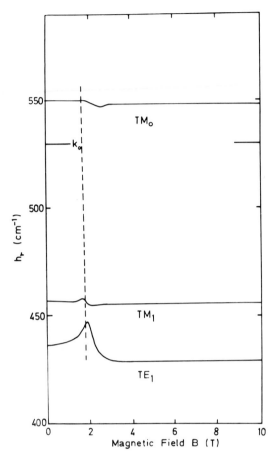

FIG. 40 For multimode excitation the real part of the propagation constant is strongly affected for the quasi-TE$_1$ mode near the dielectric anomaly, whose position is indicated by the broken vertical line ($\lambda = 118.6 \ \mu$m, $D = 100 \ \mu$m, $N = 10^{17} \ \mathrm{cm}^{-3}$, $\Gamma = 0.1$, $m_c = 0.03 \ m_0$, $\varepsilon_l = 25.6$).

ponents is neglected. The novel feature introduced by Kawamura *et al.* (1978) in the interpretation of the strip-line data is the fact that not only does the real part of ε_1 have to be considered, but also the imaginary part, as the latter affects considerably the strip-line transmission for PbTe as shown in Fig. 42. The spectra were calculated by Eq. (61) in the limit for large ε_{yz}. Depending on the imaginary part of ε_l, there is not only a considerable broadening of all the lines in the spectrum but also a shift of the transmission minima. Therefore, any line shift in a strip-line experiment is not necessarily connected with a shift in the resonance position. As can be seen from Fig. 42

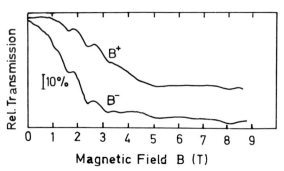

Fig. 41   The experimental strip-line spectra in the perpendicular configuration for HgSe depend definitely on the magnetic-field polarity.

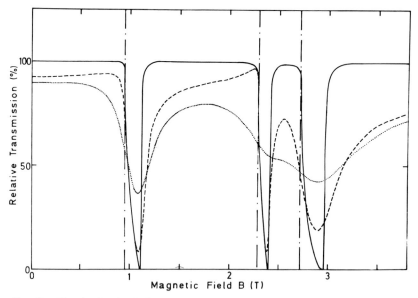

Fig. 42.   The simulated strip-line transmission in the perpendicular configuration for PbTe calculated for different damping contributions. For the solid curve ε_1 has been assumed to have only a real part $\varepsilon_1 = -40$, whereas the broken curves consider also an imaginary part $\varepsilon_1 = -40 + 8i$. For the dotted curve the additional influence of the finite damping $\omega\tau = 10$ of the carrier system becomes obvious. The vertical lines indicate the resonance positions, which are well approximated by the turning point in the slope before the transmission minimum.

the resonance position, which is marked by vertical lines, is very well approximated by the turning point in the descending slope before the transmission minimum. This fact holds also for increased damping, as shown by the dotted curve.

e. *The Surface Configuration.* The surface configuration is not as often applied in the experiment because of the generally strong mode coupling, which compensates for the generally simple structure of the dielectric function in Faraday configuration. Nishikawa (1978) has performed some of his measurements in this configuration. Schwarzbeck (1978) applied this configuration for some measurements of quantum effects in HgSe.

2. *Investigation of Quantum Effects*

a. *Quantum Resonances.* So far we have discussed resonance effects which are well described by classical theory. This implies, from a quantum mechanical point of view, that only cyclotron transitions contribute to the dielectric response and that all cyclotron transitions from the individual quantum states have the same energy. This assumption, however, does not hold for materials having a strongly nonparabolic energy–wave number relation for the charge carriers. There are basically two effects that modify the classical type of transmission spectrum. First, because of the nonparabolicity, the cyclotron resonance may split into lines of different intensity. Second, in addition to cyclotron transitions, electric-field-induced combined spin–flip transitions may occur, which are resonant in the ordinary Voigt mode (Schwarzbeck *et al.*, 1976, 1978). For this transition the cyclotron resonance is accompanied by a spin–flip process, so that the resonance energy of this process subtracts or adds to the cyclotron energy. Schematically, this situation is shown in Fig. 43 where we have plotted the nonparabolic, spin-split Landau-levels of the zero-gap material HgSe. The different transitions for the same radiation energy are indicated by arrows. HgSe has, with good approximation, a spherical Fermi surface, so that no tilted-orbit effects are expected. Any resonance structure in the parallel configuration should be due to mode coupling or spin–flip transitions.

In Fig. 44 we elucidate the situation by a simulated spectrum in the parallel configuration. Despite the strong cyclotron resonance in the dielectric function ε_+ in the right part of the figure (dotted curves), the strip-line transmission in the left part is constant within this field range (broken line) because the dominant component ε_{zz} is independent of the magnetic field. If we admit, however, that in ε_{zz} a weak resonance of only 1 % of the cyclotron intensity is present, the spectrum is completely changed (solid curves). At the position of the additional resonance, a strong line has modified the transmission spectrum. The simulation demonstrates how extremely sensitive

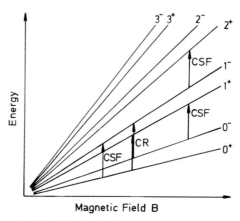

FIG. 43   The schematic of combined spin–flip and cyclotron resonance as transitions between different energy levels.

the strip-line transmission is with respect to small changes in the dielectric function, so that very weak transitions can be also detected by this method.

In Fig. 45 we have plotted the corresponding results for the perpendicular configuration, which is dominated by $\varepsilon_\perp$. Because of the coupling of plasma oscillations and cyclotron motion in the electron gas, there is no resonance in the dielectric function for the chosen set of parameters and only the dielectric anomaly becomes effective. If we admit an additional weak resonance, however, the latter especially will be observable in the corresponding strip-line spectrum. The weak resonance modifies not only the dielectric function but is also clearly visible in the imaginary part of the propagation constant. In the transmission spectrum the structure is directly related to the resonance position, which is marked by arrows in the figure. This fact actually permits us to identify the peak position in the transmission spectra to a good approximation with the resonance position in the dielectric function. The fact that any weak resonance in the strip-line spectrum is favored makes this experimental method a very effective tool for the investigation of spin and combined spin–flip transitions, which are generally much weaker in intensity than the ordinary cyclotron resonance. In Fig. 46 we have plotted the strip-line transmission of HgSe using 337-μm wavelength radiation for the perpendicular configuration. Whereas the strong decrease is due to the anomaly, the contributions of different quantum transitions are clearly visible in the spectrum (Schwarzbeck, 1978).

For the parallel configuration small quantum effects in the ordinary Voigt mode are experimentally manifested in two different ways. First, as already described, the combined spin–flip is resonant in ε_{zz}. The corresponding spectra for different orientations of the magnetic field relative to

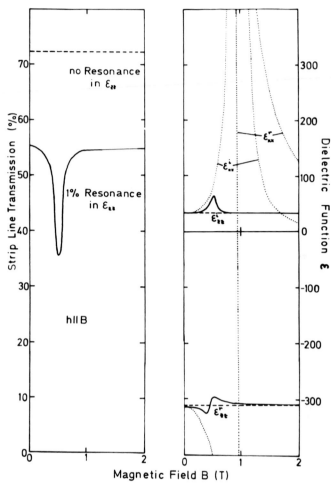

FIG. 44   In the parallel configuration the strip-line transmission is determined by ε_{zz}, which is independent of the magnetic field in the Drude model. A weak resonance in ε_{zz} produces in the strip-line (TM mode) transmission a strong line directly at the resonance position ($\lambda = 337$ μm, $D = 100$ μm) (von Ortenberg, 1978a).

the crystal axes of the HgSe sample within the (110) plane are shown in Fig. 47. The structures are relatively weak and only about 5% of the total strip-line transmission despite a rather large sample length of 10 mm.

The Fermi surface of HgSe is rather spherical and the warping terms are very small. The effect of the weak distortion from sphericity, however, is clearly visible in the sequence of spectra for different orientations in Fig. 47. Besides the two spin–flips, a cyclotron-resonance transition is resolved. This

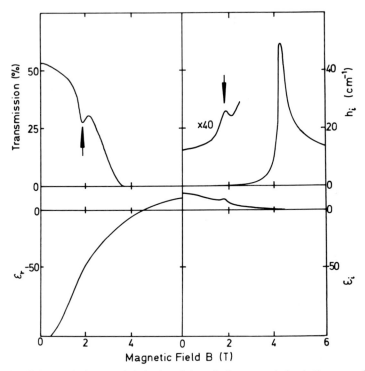

FIG. 45 Whereas the large-scale behavior of the strip-line transmission in the perpendicular configuration depends on the dielectric anomaly, the fine structure is caused by weak resonances whose position is the same in the dielectric function and the strip-line transmission ($\lambda = 337$ μm, $m_1 = 0.03$ m_0, $N = 4 \times 10^{16}$ cm^{-3}, $\omega\tau = 10$, $m_2 = 0.06$ m_0, $N_2 = 2 \times 10^{15}$ cm^{-3}) (von Ortenberg 1978a).

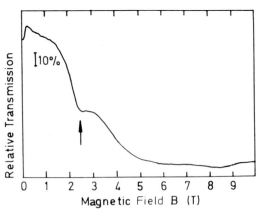

FIG. 46 The effect of a weak quantum transition in the experimental strip-line spectrum of HgSe in the perpendicular configuration using 118.6-μm wavelength radiation.

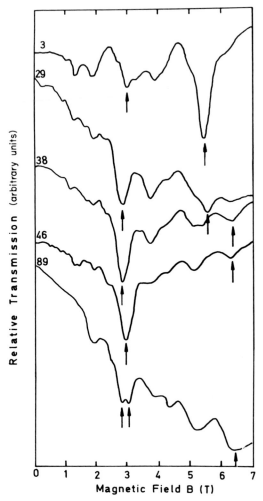

FIG. 47 The combined spin–flip transitions for HgSe depend on the angle between the [001] axes and the magnetic field in the (101) plane as given by the parameter (Schwarzbeck, 1978).

latter is of considerably smaller intensity and should actually not be observable in a purely parallel configuration. However, since in this measurement we applied a circular-shaped sample, the propagation constant is not exactly parallel to the magnetic field, and mixing effects occur.

In Fig. 48 we summarize the results in a plot of the total angular dependence of the resonance position for the different lines. The dots indicate the experimental values, and the solid curves represent the theoretical results based

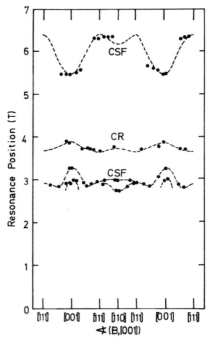

FIG. 48   The angular dependence of the experimentally determined resonance position in HgSe (dots) is in good agreement with theoretical fit represented by the broken curves (Schwarzbeck, 1978).

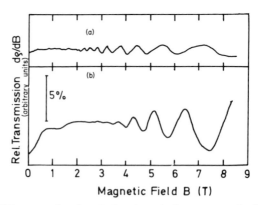

FIG. 49   For 337-μm-wavelength radiation the strip-line spectrum for HgSe is modulated by Shubnikov–deHaas oscillations as confirmed by the direct comparison with the dc magnetoresistance in (a) $n = 6.7 \times 10^{17}$ cm^{-3}, $T = 4.2$k, in (b) $D = 100$ μm, $T = 1.9$k (Schwarzbeck, 1978).

on the **k · p** model for the energy levels in HgSe (Schwarzbeck and von Ortenberg, 1978).

b. *Nonresonant Quantum Effects.* By changing the radiation wavelength from $\lambda = 118.6$ μm to $\lambda = 337$ μm, the strip-line-transmission spectrum of the same sample with a carrier concentration of $n = 7.6 \times 10^{17}$ cm^{-3} is completely changed, and the second type of quantum effect is observable as shown in Fig. 49. Due to the smaller radiation energy, all resonance lines are shifted to smaller magnetic-field intensities. Because of the increased level, which broadens by a factor of 3 due to the frequency change, the resonances are hardly resolved. Only a weak indication is observed at about 1.3 T. In the high-field range, however, Shubnikov–deHaas-like oscillations are observed. This interpretation of the oscillations is confirmed by direct comparison with the simultaneously measured dc magnetoresistance in the upper part of Fig. 49 (Schwarzbeck and von Ortenberg, 1978).

Figures 47 and 49 demonstrate how universal the strip-line technique actually is: By only a change of frequency two quite different physical phenomena of one and the same sample can be detected and combined for an unambiguous interpretation. This interpretation of the shown data depends entirely on the scheme of energy levels in the presence of an external magnetic field. As discussed in Section B.2, a resonance occurs if, for an allowed transition, the radiation energy fits the level separation. In Fig. 50

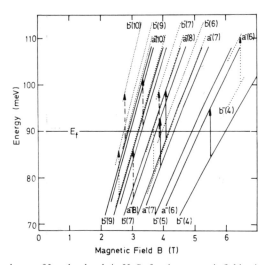

FIG. 50 The scheme of Landau levels in HgSe for the magnetic field oriented parallel to the [001] and [111] axes is given by the solid and dotted curves, respectively. The possible transitions near the Fermi energy are indicated by arrows ↑↑, strong transitions; ↑↑, weak transitions (Schwarzbeck and von Ortenberg, 1978).

we have plotted the scheme of Landau levels in HgSe for $B\|[001]$ and $B[111]$, by solid and broken curves respectively (Schwarzbeck and von Ortenberg, 1978). The arrows indicate the possible transitions near the Fermi energy. The value of the Fermi energy E_F, however, has been determined from the Shubnikov–deHaas-like oscillations in Fig. 49.

The parameters of the theoretical model involved have been fitted in such a way that for all different orientations an optimal agreement of experimental and theoretical resonance positions is obtained. The comparison of the experimental and theoretical resonance positions is summarized for all orientations in Fig. 48.

c. *Bound-Carrier Effects.* In the experimental strip-line spectra additional lines are observed, which cannot be explained by the resonances of quasi-free carriers as discussed so far. In Fig. 51 these additional structures for HgSe are indicated by arrows (Schwarzbeck *et al.*, 1978). In spite of the high Fermi energy, we interpret these lines as caused by internal impurity transitions between the bound states of a shallow attractive potential, as shown schematically in Fig. 52. Due to the fact that each of the impurities can bind only one electron, the higher impurity states are unoccupied, even if they lie far below the Fermi energy. For the quantitative understanding of the impurity resonances, the detailed problem of an electron bound to an

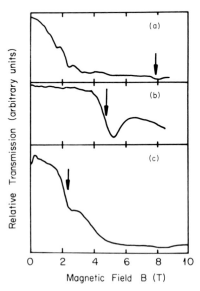

FIG. 51   The resonances indicated by arrows in the strip-line spectra of HgSe cannot be explained by quasi-free-carrier transitions (a) 118 μm, (b) 195 μm, (c) 337 μm (Schwarzbeck *et al.*, 1978).

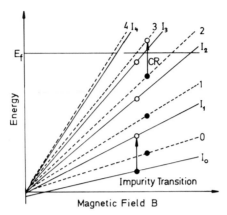

FIG. 52 Since the impurity potential can bind only one electron, the excited impurity states are unoccupied; therefore, impurity transitions are possible even far below the Fermi level, in constrast to the cyclotron resonance, which requires a final state above the Fermi energy.

attractive potential in the presence of an external magnetic field, with its kinetic energy described by the matrix of the $\mathbf{k} \cdot \mathbf{p}$ model, has to be solved. For a multiband Hamiltonian as necessary for zero-gap semiconductors, this problem is rather complex. Nevertheless, by comparing numerical results and experimental data for the resonance position in Fig. 53, the agreement is quite satisfactory for the shown case of HgSe (Schwarzbeck et al., 1978).

3. The Strip Line in Surface Spectroscopy

In addition to magnetospectroscopy the strip-line technique has been successfully applied in surface spectroscopy (Kamgar et al., 1974; Kneschaurek et al., 1976). Applying electric-field modulation, the electrons in the space-charge region of a quantizing surface potential have been investigated. This problem was treated by Nakayama (1977) considering nonlocal effects. In the local approximation the schematic of the situation is shown in Fig. 54. The external electric field is applied by the gate voltage between the metallic-gate electrode and the semiconductor–substrate material. Both are separated by a thin insulating layer, consisting either of natural oxide or externally applied varnish. Inside the semiconductor the electric field is screened by the space-charge of an inversion or accumulation layer, depending on the electric-field polarity and the majority carriers of the substrate. Because the carriers are confined into a narrow space-charge region of about only 50 Å thickness, a pronounced energy quantization results. The corresponding eigenfunctions extend perpendicular to the surface. Therefore, any optical transition between the subbands associated with the quantized levels requires an electric-field polarization perpendicular

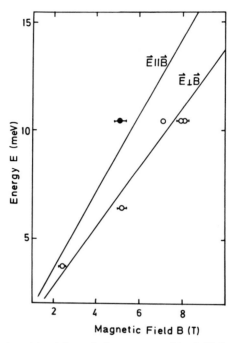

Fig. 53   The experimental and theoretical resonance position in HgSe are in good agreement (Schwarzbeck *et al.*, 1978).

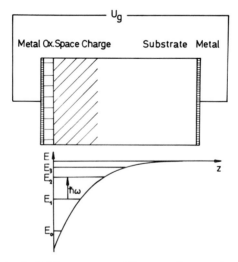

Fig. 54   Schematic of a MOS arrangement. The space charge at the surface gives rise to a quantizing potential.

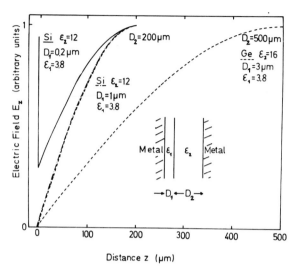

Distance z (μm)

FIG. 55 The electric field distribution in a MOS arrangement in the strip line shows that only for very thin insulation layers is the electric field at the interface nonzero ($\lambda = 100$ μm) (von Ortenberg, 1978b).

to the surface. A suitable experimental arrangement for this polarization is shown by the triple strip line in Fig. 10. The three dielectric media of oxide, space-charge region, and substrate are embedded between the metallic-gate electrode and a metallic-mirror plate. Both experiment and simulation have shown that the intensity of the electric-field vector of the radiation in the space-charge region depends critically on the thickness of the insulating layer (von Ortenberg, 1978).

For the simulation we applied Eqs. (81)–(84) using parameters close to the experimental situation. In Fig. 55 we have plotted the calculated mode structure of the electric-field component perpendicular to the surface for vanishing space-charge region, e.g., for zero gate voltage. For the lowest strip-line mode in a Si–MOS–FET, the effective radiation field decreases at the interface to about only 20% of its peak value for a thin oxide layer of 2000 Å thickness. For increased thickness of the oxide layer to 1 μm, the effective field is much more reduced. The same result holds for a Ge substrate with a 3-μm-thick varnish layer for insulation. The theoretical results are in excellent agreement with the experiments (von Ortenberg, 1978). They demonstrate that for surface–intersubband spectroscopy only Si–MOS–FETs with a rather thin oxide layer are applicable, because the radiation field at the interface is otherwise too small to excite intersubband transitions. The situation does not change significantly considering higher strip-line modes as plotted in Figs. 56 and 57.

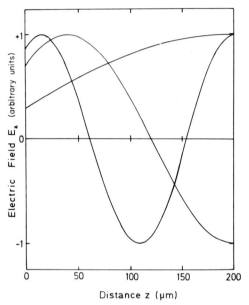

FIG. 56   The higher TM modes in the double strip line of a Si–MOS–FET ($\lambda = 100$ μm, $\varepsilon_2 = 12$, $\varepsilon_1 = 3.8$, $D_1 = 0.2$ μm) (von Ortenberg, 1978b).

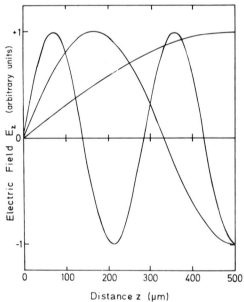

FIG. 57   The higher TE modes in the double strip line of a Si–MOS–FET ($\lambda = 100$ μm, $\varepsilon_2 = 16$, $\varepsilon_1 = 3.8$, $D_1 = 3$ μm) (von Ortenberg, 1978b).

We include now in our simulation the effect of a space-charge layer, approximated by a three-dimensional homogeneous medium of thickness D_1. The nonresonant dielectric tensor components parallel to the surface are given by

$$\varepsilon_{xx} = \varepsilon_{zz} = \varepsilon_l - \frac{N_s e^2}{Dm\varepsilon_0} \frac{1}{\omega(\omega + i/\tau)}. \tag{138}$$

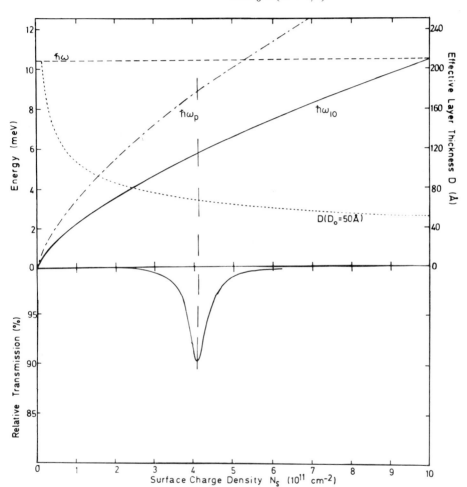

FIG. 58 The relative transmission of the triple strip line in the lower part is essentially dominated by the screened resonance in the space-charge region. For the applied classical model of homogeneous media, this means that not only the resonance energy $\hbar\omega_{10}$ but also the plasmon energy $\hbar\omega_p = N_s e^2/m\varepsilon_0 D$ determines the position of the observable line. Both quantities, as well as the effective layer thickness D, depend in the applied model on the surface density N_s, as shown in the upper part.

The resonant intersubband component has the following form:

$$\varepsilon_{zz} = \varepsilon_l - \frac{N_s e^2}{Dm\varepsilon_0} \frac{1}{\omega^2 - \omega_{10}^2 + i\omega/\tau}. \tag{139}$$

Here N_s is the surface–carrier density. Using these tensor components for characterizing the dielectric properties of the space-charge layer, we calculated directly the strip-line transmission for 118.6-μm-wavelength radiation as shown in Fig. 58. The parameters are close to the experimental situation for the Si–MOS–FET and consider explicitly the dependence of the resonance and layer thickness as functions of the gate voltage in the triangular-potential model. In agreement with other results, the position of the transmission line is considerably shifted by screening effects with respect to the bare resonance (Chen et al., 1976; Allen et al., 1976).

IV. Summary

The strip-line technique is a useful tool for the investigation of highly conductive materials using submillimeter radiation. Because of the long radiation wavelength, single-mode excitation in the strip line is possible and admits a definite polarization for probing the dielectric tensor of the sample material. By variation of the strip-line configuration, the polarization can be changed arbitrarily so that the different components of the dielectric tensor can be investigated almost separately. Due to mode-coupling effects, the interpretation of the strip-line transmission data can sometimes be rather complex. A general feature of strip-line data is that the resonance position of weak structures agrees well with the transmission minimum, whereas strong lines are due to dielectric anomalies.

ACKNOWLEDGMENTS

The author would like to express his sincere gratitude to Dr. Karl Schwarzbeck, whose skill and enthusiasm in the experimental techniques were fundamental for this paper. The financial support of the Deutsche Forschungsgemeinschaft is gratefully acknowledged.

REFERENCES

Allen, J. S., Tsui, D. C., and Vinter, B. (1976). *Solid State Commun.* **20**, 425–428.
Apel, J. R., Poehler, C. R., Westgate, C. R., and Joseph, R. I. (1971). *Phys. Rev. B* **4**, 436–451.
Chang, T. Y., and Bridges, T. J. (1970). *Opt. Commun.* **1**, 423–426.
Chen, W. P., Chen, Y. J., and Burstein, E. (1976). *Proc. Int. Conf. Electron. Properties Quasi-Two-Dimensional Syst. Providence* pp. 263–265.
Dresselhaus, G., Kip, A. F., and Kittel, C. (1953). *Phys. Rev.* **92**, 827.

Drew, H. D., and Sievers, A. J. (1967). *Phys. Rev. Lett.* **19**, 697–699.
Gebbie, H. A., Findlay, F. D., Stone, N. W. B., and Ross, J. A. (1964). *Nature (London)* **202**, 169–170.
Kamgar, A., Kneschaurek, P., Dorda, G., and Koch, J. F. (1974). *Phys. Rev. Lett.* **32**, 1251–1254.
Kanada, S., Nakayama, M., and Tsuji, M. (1976). *J. Phys. Jpn.* **41**, 1954–1961.
Kane, E. O., (1957). *J. Phys. Chem. Solids* **1**, 249–261.
Kawamura, H., Murase, K., Nishikawa, S., Nishi, S., and Katayama, S. (1975). *Solid State Commun.* **17**, 341–344.
Kawamura, H., Nishikawa, S., and Nishi, S. (1976). *Proc. Int. Conf. Phys. Semicond. 13th, Rome* pp. 310–313.
Kawamura, H., Nishikawa, S., and Murase, K. (1978). *Lecture Notes Int. Conf. Appl. High Magn. Fields Semicond, Phys., Oxford* (J. F. Ryan, ed.), pp. 170–186. Clarendon Laboratory, Oxford.
Kneschaurek, P., Kamgar, A., and Koch, J. F. (1976). *Phys. Rev.* B **14**, 1610–1622.
Kubo, R. (1957). *J. Phys. Soc. Jpn.* **12**, 570–586.
Luttinger, J. M. (1956). *Phys. Rev.* **102**, 1030–1041.
Nakayama, M., (1977). *Solid State Commun.* **21**, 587–589.
Nakayama, M., and Tsuji, M. (1977). *J. Phys. Soc. Jpn.* **43**, 164–172.
Nishikawa, S. (1978). Thesis, Osaka Univ.
Palik, E. D., and Furdyna, J. K. (1970). *Rep. Prog. Phys.* **33**, 1193–1322.
Perkowitz, S. (1969). *Phys. Rev.* **182**, 828–837.
Ramage, J. C. (1978). *Lecture Notes Int. Conf. Appl. High Magn. Fields Semicond. Phys.* (J. F. Ryan, ed.), pp. 370–383. Clarendon Lab., Oxford.
Ramage, J. C., Stradling, R. A., and Tidey, R. J. (1974). *Proc. Int. Conf. Phys. Semicond., Stuttgart* pp. 531–535.
Ramage, J. C., Stradling, R. A., Aziza, A., and Balkanski, M. (1975). *J. Phys. C Solid State Phys.* **8**, 1731–1736.
Sauter, F. (1967). *Z. Phys.* **203**, 488–494.
Schwarzbeck, K. (1978). Thesis, Univ. of Würzburg.
Schwarzbeck, K., von Ortenberg, M., Landwehr, G., and Galazka, R. R. (1976). *Proc. Int. Conf. Phys. Semicond., Rome* pp. 435–438.
Schwarzbeck, K., Bangert, E., and von Ortenberg, M. (1978). *Proc. Int. Conf. Solids Plasmas High Magn. Fields, Cambridge* pp. 119–121.
Stradling, R. A. (1974). *Proc. Int. Conf. Appl. High Magn. Fields Semicond. Phys., Würzburg* (G. Landwehr, ed.), pp. 257–271.
Stradling, R. A. (1975). *Colloq. Int. C.N.R.S. No. 242–Phys. sous Champs Magn. Intenses, Grenoble* pp. 317–326.
Strom, U., Drew, H. D., and Koch, J. F. (1971). *Phys. Rev. Lett.* **26**, 1110–1114.
Strom, U., Kamgar, A., and Koch, F. J. (1973). *Phys. Rev.* **7**, 2435–2450.
von Ortenberg, M. (1974). *Lecture Notes Int. Conf. Appl. High Magn. Fields Semicond. Phys., Würzburg* (G. Landwehr, ed.) pp. 656–671. Univ. of Würzburg.
von Ortenberg, M. (1976). *Lecture Notes Int. Conf. Appl. High Magn. Fields Semicond. Phys., Würzburg* (G. Landwehr, ed.) pp. 654–662. Univ. of Würzburg.
von Ortenberg, M., (1978a). *Lecture Notes Int. Conf. Appl. High Magn. Fields Semicond. Phys., Oxford* (J. F. Ryan, ed) pp. 346–359. Clarendon Lab. Oxford.
von Ortenberg, M. (1978b). *Infrared Phys.* **18**, 735–740.
von Ortenberg, M., and Schwarzbeck, K. (1974). Unpublished.
von Ortenberg, M., Schwarzbeck, K., and Landwehr, G. (1975). *Colloq. Int. C.N.R.S. No. 242– Phys. sous Champs Magn. Intenses, Grenoble* pp. 305–309.
von Ortenberg, M., Link, J., and Helbig, R. (1977). *J. Opt. Soc. Am.* **67**, 968–971.

CHAPTER 7

Cyclotron Resonance and Related Studies of Semiconductors in Off-Thermal Equilibrium

Eizo Otsuka

Department of Physics
College of General Education
Osaka University
Toyonaka, Osaka, Japan

I. Introduction

The glorious achievement of the microwave cyclotron resonance on germanium and silicon (Lax *et al.*, 1954; Dresselhaus *et al.*, 1955) is quite unparalled in the history of semiconductor physics in the past 25 years. The most direct determination of the band parameters, in fact, solved many of the enigmas and uncertainties observed in the transport works. One should,

347

however, note that the success is largely owed to the art of purifying materials. Otherwise, the resonance absorption line is broadened by carrier scatterings at the residual impurity centers. Difficulty in purifying materials other than germanium and silicon caused a considerable delay for observing cyclotron resonance in a third material. Meanwhile, technical progress was remarkable in the far-infrared cyclotron resonance. The first several years were spent in getting signals with the help of conventional far-infrared spectrometers. It is of interest to note that the first material taken for investigation by far infrared was indium antimonide (Burstein et al., 1956). The fundamental merit of raising frequency lies in increased resolution, which, in turn, can be done at the cost of achieving a high magnetic field. Electrons in indium antimonide would perhaps have been the easiest to handle, since their effective mass was the lightest of all the carriers in the semiconductors then available. Then came, finally, the laser era. In 1966 the first data of far-infrared cyclotron resonance were presented by Button et al. (1966). The experiment was made not on indium antimonide but on p-type germanium, through the wavelength 337 μm from the HCN laser. Measurement at an elevated temperature was also made, showing the usefulness of the infrared for extending the scope of investigation. Many more new materials were then attacked, one after the other by the MIT group, and the wavelengths employed were extended with use of DCN, H_2O, and D_2O lasers. The primary interest of the investigators, however, remained in determining the band mass parameters. That is, of course, the original motive of cyclotron resonance and its importance naturally comes first. But the technology of cyclotron resonance offers us further strong means of physical investigation. Studies of carrier scattering mechanisms through linewidth measurement and of carrier lifetimes through intensity measurement are the next comers. Such types of use, transport or kinetics, have been made even for millimeter-wave cyclotron resonance (Kawamura et al., 1962, 1964; Fukai et al., 1964; Sekido et al., 1964; Otsuka et al., 1966a, b, 1968; Otsuka and Yamaguchi, 1967; Murase and Otsuka, 1968, 1969; Ohyama et al., 1970). The employed semiconductors naturally remained silicon and germanium. Indium antimonide now makes a similar and convenient object of investigation at far infrared because of its small electronic effective mass and the relative progress in purification technology. Specifically, the effective mass value of 0.014 m_0, where m_0 is the free electron mass, requires a magnetic field only slightly over 1.3 T for the wavelength of 118.7 μm from a water-vapor laser.

The cyclotron resonance technology can be further extended to the study of degenerate semiconductors. A clear resonance peak in the classical sense will no longer be obtainable and the entire signal contains various features of plasma or magnetoplasma effects. A special form of degeneracy can be found in the so-called electron–hole drop, most commonly observed in germanium and silicon. This is a kind of Fermi liquid in which both electrons and holes exist for a finite length of time after an intense illumination by the bandgap light. The response of the electron–hole drop to far infrared is quite manifold, especially in the presence of magnetic field. The concept of cyclotron resonance still has a definite meaning in analyzing the peaks due to the drop. Naturally the main part of the drop has been investigated by luminescence, which is the most direct way of catching the behavior of the drop. The aspects of luminescence spectra, however, are rather similar to each other even under different physical conditions, since the wavelength region investigated is always the same. The far-infrared magnetoplasma resonance, on the other hand, gives rise to entirely different spectra, depending on the wavelength employed. One has to face a complexity in analysis—and that certainly causes a drawback— to see the global nature of the drop. But the same complexity can also be exploited to detect subtle changes in the property of the drop under different physical conditions.

The far-infrared technology can also be applied to the study of impurity states in semiconductors. It is well known that at high magnetic fields the impurity states can no longer be considered as good quantum states corresponding to a hydrogenic or similar Coulomb-like potential. When the inequality $\hbar\omega_c/2E_b \gg 1$ holds, where ω_c is the cyclotron frequency of the carrier of the same kind as that bound to the impurity and E_b the binding energy, each of the impurity states is more influenced by the magnetic field than by the Coulomb field, and the Landau quantum number should be taken into account for describing the impurity states (Wallis and Bowlden, 1958; Hasegawa and Howard, 1961). As a matter of fact, the impurity state in indium antimonide, which becomes important in the study of hot electrons, displays a quite well-defined magneto-optical absorption peak on the low magnetic-field side of the ordinary cyclotron resonance for conduction electrons (Kaplan, 1969). We commonly call this impurity-associated absorption peak "impurity cyclotron resonance." The idea remains as one deals with impurity states in other semiconductors, including the excitonic states. A somewhat puzzling notion "cyclotron resonance of an exciton" can then be justified with proper explanation.

The vast application of far infrared to cyclotron resonance and related fields can in no way be covered in a limited review. This article, accordingly, will restrict its treatment chiefly to the resonance phenomena under off-thermal equilibrium. Thus, the topics of hot electrons and excitonic effect will certainly constitute the main trend of recent application.

II. Far-Infrared Cyclotron Resonance of the Hot-Electron System in InSb

A. Historical Background

The experimental study of hot electrons in InSb at high magnetic fields has its historic roots most prominently in the works by Komatsubara and

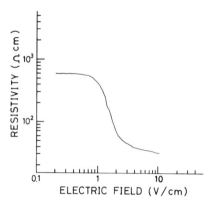

FIG. 1   A typical resistivity versus electric-field relation in the presence of a magnetic field obtained for a high-purity InSb samples (from data taken by K. L. I. Kobayashi).

Yamada (1966), Miyazawa and Ikoma (1967), and Kotera *et al.* (1972). These people dealt with the problem by the techniques of dc transport and the employed samples were of rather high purity, in other words, not degenerate even at zero magnetic field. The common observation was that at liquid helium temperatures, the magnetoresistivity remains ohmic only under very low applied electric field, even less than 1 V cm^{-1}, and it suddenly drops to much lower value as typically illustrated in Fig. 1. The feature of the drastic change in resistivity is all the more pronounced in the presence of a high magnetic field. Komatsubara and his collaborators, or the Hitachi group, interpreted the behavior in terms of the mobility change of the conduction electrons through strong coupling with optical phonons. The explanation was based on the theoretical work by Kazarinov and Skobov (1962). Miyazawa and Ikoma, or the Toshiba group, on the other hand, suggested another possible interpretation, after the model of two-band impurity conduction. According to them, the electric field applied to the specimen corresponds to temperature, through the change of which the two-band conduction is characterized. If the mobility of electrons in the impurity band is much less than that in the conduction band, one should then be able to observe a maximum in Hall coefficient as the electric field is changed. Miyazawa and Ikoma actually observed such a peak in Hall coefficient. They thus interpreted the drop in resistivity as the transfer of electrons from the impurity states to the conduction band. This model requires a definite separation of the impurity states off the bottom of the conduction band, which is well guaranteed for purest InSb samples at high magnetic fields. At zero or rather low magnetic fields of the order of several tenths of a tesla, however, the guarantee becomes perilous even for samples with $N_D - N_A \sim$ 10^{14} cm^{-3}, where N_D and N_A are the densities of donors and acceptors, respectively. In any case, one had to face twofold interpretation concerning the nonlinear transport in InSb. So long as one stuck to dc transport, it would have been difficult to draw a definite conclusion about the right process of the phenomenon. A kind of differential method was recently applied to cyclotron resonance, making use of a modulation of applied electric field (Kobayashi and Otsuka, 1974). A merit of cyclotron resonance is the ability to distinguish donor and conduction electrons. On top of that, the differential technique was able to tell the number of electrons transferred from donor to conduction band and, further, to show the change in mobility as well. The essential feature of the result will be presented in Section II.D. Another merit of cyclotron resonance for InSb is exploiting the nonparabolicity of the conduction band. This leads to observation of the so-called quantum lines of electron cyclotron resonance. The topic of the quantum lines was first mentioned for the valence bands of germanium. The warping nature of the valence bands, however, makes the analysis extremely complicated. In comparison with the valence bands, the quantum effect due to nonparabolicity of the conduction band, especially that of InSb, is quite simple because of the absence of warping. One can readily identify the transition corresponding to each line of the cyclotron resonance. Then, the possible next step in analysis would be to determine the electron distribution in the conduction band. Under the application of an electric field, the distribution of electrons becomes different from that under thermal equilibrium. In that case, one can assign a temperature to the electron system different from that for the lattice. We call this new temperature, which is inherent only to the electron system, "electron temperature." The concept of electron temperature fails whenever the form of the distribution function becomes fundamentally different from the Maxwell–Boltzmann form. In the hot-electron cyclotron resonance work by Kobayashi and Otsuka, in which the wavelength of 119 μm from a water-vapor laser was used, determination of the electron temperature was made up to 30 K at the lattice temperature of 4.2 K, and compared with the result by Miyazawa (1969). One has to keep in mind that both dc and cyclotron resonance work until that time had treated only the case $E \perp H$; i.e., the applied electric and magnetic fields are perpendicular to each other. Choice of this geometry was encouraged by technical and theoretical easiness. Quite recently, as will be described in Section II.E treatment for $E \| H$ was also carried out (Matsuda and Otsuka, 1978). A striking difference in electron distribution between the two different geometries is demonstrated with proper choice of laser wavelength, in particular 84 μm, where the profile of electron distribution can be seen even more plausibly by direct absorption than by modulation spectroscopy.

In the course of studying the distribution of hot electrons in a semiconductor, one should not forget the theoretical contributions. Perhaps special mention should be made of the work by Kurosawa (1965). This idea was based on the diffusion process of the electrons within the energy space under the application of an electric field. Later Yamada and Kurosawa (1973) extended the idea to the case of magnetic-field application. The experimental group of Hitachi (Kotera *et al.*, 1972) further carried out dc measurements and assigned different electron distributions for the cases $E \perp H$ and $E \| H$, in accordance with Kurosawa's prediction. Basically, Kurosawa's idea was built on classical statistics, and treatment of the electron scatterings was limited to those by phonons and ionized impurities. Later on, more elaborate calculations appeared treating the problem rather quantitatively. Nowadays, theory (Partl *et al.*, 1978) includes the effect of electron–electron scattering which was ignored by Kurosawa. The elaboration of theoretical calculation was made in analyzing the results of cyclotron emission—the inverse process of cyclotron resonance absorption that will mainly be treated here. The phenomenon of cyclotron emission was independently discovered by Gornik (1972) and by Kobayashi *et al.* (1973). Gornik and his group call the effect the "Landau emission." The study of emission also yields information of electron temperature, but not to such an advanced degree as achieved by laser cyclotron resonance. The phenomenon of emission, however, makes people dream of the possibility of constructing a powerful far-infrared source, tunable by magnetic field, and perhaps brought to lasing. Unfortunately, the dream has not yet come true.

Since the present article does not intend to go into detailed theoretical aspects, we will simply indicate in later sections the qualitative prediction by Kurosawa and the Hitachi group in order to help in understanding the result of cyclotron resonance.

For those who are interested in more quantitative fitting between theory and experiment, the relevant part of the bibliography should be consulted.

B. Electron Distribution in the Presence of a Strong Magnetic Field

The magnetic quantization of electron energy within the conduction band gives rise to a sharp peak of state density at the bottom of each Landau subband. In the extreme quantum limit, or under the condition $\hbar\omega_c \gg k_B T_L$, where T_L is the lattice temperature, electrons populate, for practical purposes, only the lowest subband. This is actually the case for InSb below 4.2 K and under the magnetic field over 1 T. The upper subbands will also be populated only at elevated lattice temperature or on application of an electric field.

On looking at the whole view of the electron distribution, one comes across the presence of donor electrons, which reside in the levels below the bottom of the conduction band, or at the lowest Landau subband. The separation between the donor states and the conduction band becomes quite obscure at zero magnetic field even for the purest InSb material available. Under the application of a strong magnetic field, the donor states come down with respect to the conduction band edge and more electrons tend to reside there. This effect is called magnetic freezeout and reflects the shrinkage of the wave function of the donor electron, as predicted by the variational calculations (Yafet *et al.*, 1956). As mentioned in the preceding section, the donor electrons are transferred to the conduction band on application of an electric field. From redistribution of electrons one can tell the electron temperature in accordance with the fashion prescribed.

The most direct way to see the electron distribution is to watch the cyclotron resonance, especially when the cyclotron transitions are manifold, as in the case for InSb. In Fig. 2, we present cyclotron resonance traces by the laser wavelength of 84 μm for three lattice temperatures. At 4.8 K, one observes two definite peaks. The one on the lower magnetic field (I_1) is the so-called impurity cyclotron resonance, while the other on the higher magnetic field (C_1) corresponds to the cyclotron transition from the lowest Landau subband. If one denotes the Landau subbands by $N^{\pm}$, where N is the Landau quantum number and $\pm$ indicate the spin states, the latter transition can be written $0^+ \rightarrow 1^+$. One may be rather amazed at the similarity between the impurity and conduction band transitions. In fact, it is this similarity that has caused the name "impurity cyclotron resonance." The final state of the impurity cyclotron resonance transition is swallowed up by the conduction band. In other words, after transition the impurity electron is no longer bound. After Yafet *et al.*, the eigenvalue problem concerning the impurity state was further treated, both by variational and by perturbation methods at the

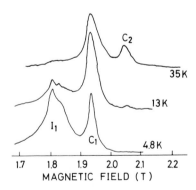

FIG. 2 Cyclotron resonance traces from an *n*-type InSb sample at different lattice temperatures obtained through a D$_2$O laser wavelength of 84 μm. The peak indicated I_1, most conspicuous at 4.8 K, is commonly known as impurity cyclotron resonance, while C_1 and C_2 are associated with cyclotron transitions within the conduction band (from data taken by O. Matsuda).

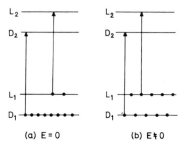

(a) E = 0   (b) E ≠ 0

FIG. 3   A simplified two-state scheme for donor and Landau-level residing electrons with and without electric field. On application of electric field, intensity of the transition arising from D_1 (the impurity ground state) decreases, while that from L_1 (the lowest Landau level) increases.

quantum limit (Wallis and Bowlden, 1958; Hasegawa and Howard, 1961). It was found that each of the discrete states was associated with the Landau subband. In this respect, it would not be surprising if the eigenstates of the impurity under a high magnetic field contain the Landau quantum number in their designation. The transition corresponding to the lower field peak at 4.2 K in Fig. 3 is described as $(000) \to (110)$ after the manner of Hasegawa and Howard, with the first numbers, 0 and 1, showing, respectively, the quantum numbers of the Landau subbands to which the initial and final impurity states belong. Only the rest of the numbers are relevant to the impurity states.

As seen in Fig. 2, the impurity cyclotron resonance—henceforth we shall abbreviate it ICR—diminishes on increase of temperature. This is not surprising considering thermal exhaustion of the donor electrons. The ordinary cyclotron resonance at the higher field, which we shall call conduction band cyclotron resonance (or CCR, to distinguish it from ICR), is also affected by temperature. Its signal intensity first increases when temperature is raised; then it saturates, and at higher temperatures declines somewhat. Meanwhile, the linewidth broadens with temperature and a small asymmetry also appears. One more important thing to be noted is the emergence of a new peak (indicated C_2) on an even higher magnetic field side. This peak corresponds to the cyclotron transistion $0^- \to 1^-$. If the conduction band were really parabolic, one could never expect appearance of this transition at a magnetic field different from that corresponding to the transition $0^+ \to 1^+$. The emergence of the $0^- \to 1^-$ transition indicates the moving up of the conduction electrons to the higher energy states and is partially responsible for the eventual decline of the C_1 line. It is with the help of this new peak, as well as with the analysis of the line shape and linewidth of the CCR, that we are led to the information of the electron distribution in the conduction band. The distribution is certainly expected to be Boltzmann-like in thermal equilibrium, corresponding to the case presented in Fig. 2. But nobody can expect the same when the electron system is perturbed by an electric field. As stated in Section A, applying an electric field to the sample corresponds, in a sense, to raising temperature. Redistribution of electrons by the use of electric field will be examined by watching the behavior of cyclotron resonance.

C. PULSED ELECTRIC-FIELD-MODULATED CYCLOTRON RESONANCE

The idea of displacing the electron distribution with the help of electric field in the experiment of cyclotron resonance is not new. The electron temperature can be elevated either by the rf field needed to see cyclotron resonance or by externally applying a dc field. The effect of the rf heating was first observed for Ge by the use of microwaves (Kawamura et al., 1962), and then for InSb by 337 μm from the HCN laser (Murotani and Nisida, 1972). The use of dc electric field is also seen both in microwaves (Kawamura et al., 1961) and in far infrared (Oka and Narita, 1970). For the purpose of exploring the form of electron distribution, in the author's opinion, none of the trials cited so far could make a decisive breakthrough. The experimental piece of work introduced here would perhaps be the first to give a plausible analysis, in particular for InSb, concerning the electron distribution problem.

In order to raise resolution and to identify the origin of an observed transition, use is made of a differential method in which one is able to see the change in population at a given level or subband between, with and without application of electric field (Kobayashi et al., 1971; Kobayashi 1973; Kobayashi and Otsuka 1974). As may be seen in Fig. 2, the most important energy levels for electrons in InSb are the donor ground state and the lowest Landau subband in the conduction band, since they are the only states populated by electrons at liquid helium temperatures in thermal equilibrium.

With a simple scheme we shall give an explanation of the method. Figure 3 contains a pair of two level systems: one can take L_1 and L_2 to be the initial and final states of the cyclotron transition, while D_1 and D_2 are those of the impurity transition. The D_1 and D_2 states are considered to belong to the first and second Landau levels, respectively, in accordance with the statement in Section B. Suppose, in thermal equilibrium or under zero electric field ($E = 0$), that the electrons will mainly be found in D_1 and L_1. On application of an electric field, a part of the electrons in D_1 will be shifted to L_1. The cyclotron resonance transitions, including ICR, are indicated by arrows. As one sees, the $D_1 \rightarrow D_2$ transition rate decreases if one applies an electric field. On the other hand, the $L_1 \rightarrow L_2$ transition increases on application of the electric field. If we set our operation so as to measure the amount of change in resonance intensity, the signal corresponding to the $D_1 \rightarrow D_2$ transition will appear negative. For determining the electron population at each level, the direct absorption measurement, of course, excels in comparison with the differential method. But if one is more concerned with the transfer of electrons from D_1 to L_1 and also with identification of the initial state of the transition—in other words, if the initial state is the bottom state or not for electrons—the differential method offers us a definite advantage. Identifications of the transitions originating from the bottom state becomes particularly important when one performs a resonance experiment at an intermediate magnetic field. The reason is that at an extremely strong magnetic field, the selection rules permit only one transition from the bottom state, and that is the commonly known ICR. The selection rule relaxes at weaker magnetic fields and then one can expect more transitions from the bottom state. Even with a laser wavelength of 119 μm, one meets harmonic resonances that appear at lower magnetic fields than those for CCR or ICR, where one has to yield to the conditions inherent to the intermediate field case.

For performing the differential method, use of a cw far-infrared laser is preferable. The laser beam is to be chopped by a mechanical chopper, say at 30 Hz, and the electric field is applied to the sample in pulses having the width of 400 μs each at the frequency of 15 Hz, or half the chopping frequency (Fig. 4). We thus become ready to observe resonance absorption alternately with and without application of an electric field. The two kinds of signal are led to a two-channel boxcar integrator, and we obtain the change in absorption that reflects the role of electric field.

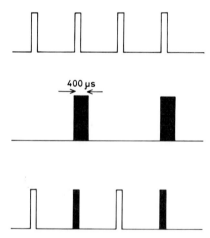

FIG. 4   Essence of the pulsed electric-field modulation. (a) Repetition of the chopped far-infrared laser beam. (b) Pulsed electric fields in synchronization with laser chopping but at half the frequency. (c) Combination of the above two, or laser probes with and without application of electric field.

To be more precise, one is measuring the amount of transmission of the laser beam through the sample. If one writes the amounts of transmission $T(0)$ and $T(E)$ for $E = 0$ and $E \neq 0$, respectively, the change in absorption coefficient can be expressed by

$$\Delta\kappa(E) = \ln\{[T(E) - (I_0'/2)]/[T(0) - (I_0'/2)]\}, \tag{1}$$

with

$$I_0' = I_0(1 - R). \tag{2}$$

Here I_0 is the intensity of the incident laser beam and R is the reflectivity. The factor $\frac{1}{2}$ multiplying I_0' is because of the linearly polarized nature of the incident beam; in other words, electrons can absorb at most half the beam corresponding to the proper component of circular polarization. The basic setup of the method is given in Fig. 5. Further application of the method would be found in the study of impurity conduction, general magneto-optical or transport study of degenerate semiconductors, and so on, in which grasping the electron transfer process from the bottom states becomes quite important.

In the present case of InSb, the merit of the differential method can be more than exploited in the investigation of the ICR and related transitions. In Fig. 6, the relevant transitions are indicated through the arrows I_1, I_2, and I_3, where I_1 is the standard ICR transition, only which is expected to remain at the high magnetic-field limit. The corresponding resonance traces are given in Fig. 7 for direct transmission $T(E)$, for $\Delta\kappa(E)$ described above, and also for photoconductances. There is no doubt about the signals I_1 and C_1. The signal

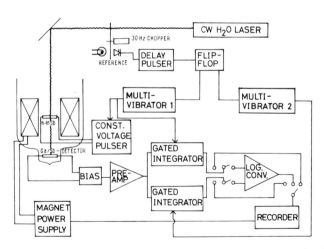

FIG. 5   The experimental setup for observing pulsed electric-field-modulated cyclotron resonance (Kobayashi, 1973).

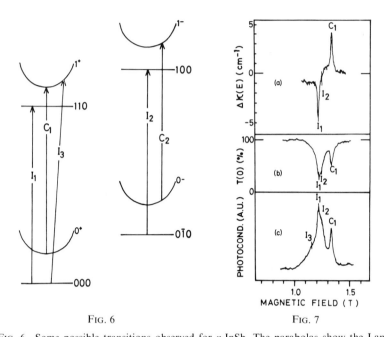

Fig. 6   Fig. 7

Fig. 6   Some possible transitions observed for n-InSb. The parabolas show the Landau subbands, while the horizontal bars indicate the impurity states, with the state designation after Hasegawa–Howard. The transitions I_2 and I_3 are not allowed at extremely high magnetic fields.

Fig. 7   Some features of different type "cyclotron resonance" signals obtained at 119 μm. (a) Pulsed electric-field-modulated cyclotron resonance signals differentiated at relatively low electric field. (b) Ordinary transmission with no electric field. (c) Photoconductance corresponding to cyclotron resonance (Kobayashi and Otsuka, 1974).

I_2 is scarcely recognizable in the $\Delta\kappa(E)$ trace. If the transition assignment in Fig. 6 is right, the marginal appearance of I_2 in the $\Delta\kappa(E)$ trace is understandable, since the rate of electron transfer from $(0\bar{1}0)$ to 0^- would nearly be cancelled by the electron supply from the lower states (000) and 0^+. Existence of the I_2 transition becomes more evident with the help of a photoconductance measurement, as given at the bottom of Fig. 7. Here even the transition I_3, which can be termed as a kind of photoionization, is recognizable. Appearance of the transitions I_2 and I_3, responsible for a "structure" in the overall ICR signal, is rather sample dependent, reflecting the situation in which the impurity states are embedded.

Perhaps more convincing will be the identification of the harmonic resonances. The photoconductance trace can detect the existence of harmonic resonances as high as the sixth order. The origin of such harmonics can be considered in two ways; one is the multiple transition of CCR, and the other a similar thing connected with ICR. The $\Delta\kappa(E)$ observation presented in

Fig. 8 indicates that the harmonics are due mainly to the impurity states, since the signals deflect downward at the magnetic field corresponding to the harmonic resonances shown below. The plotting of the order number against $1/H$ at the peak position of the harmonic resonance also seems to favor the committment of ICR (Kobayashi and Otsuka, 1974). One should, however, note that the photoconductance signals for the second harmonic show a shoulder on the low magnetic-field side. That may indicate existence of two mechanisms contributing to the appearance of the harmonics. Accordingly, one cannot entirely rule out the possibility of the CCR involvement. Be that as it may, one can see that even at the magnetic field corresponding to the fourth harmonic, the impurity states are still definable from the downward deflection of the signal there. The excess donor concentration of the sample employed for this measurement is $1.4 \times 10^{14}\,\mathrm{cm}^{-3}$. Thus, the present technique guarantees the existence of the well-defined impurity states at this degree of carrier degeneracy and at the magnetic field of $\lesssim 0.3$ T.

One of the most important results of the pulsed electric-field-modulated cyclotron resonance measurement for InSb is the elimination of the partition in interpreting for the drop in resistivity observed at $\sim 1\,\mathrm{V\,cm}^{-1}$. In Fig. 9, increase of electron population at each Landau subband $\Delta n_{N\pm}$, as well as the total electron population n_c in the entire conduction band, is presented. The change in population is counted from the equilibrium level. It is the lowest Landau subband 0^+ which is most affected in population. In other words Δn_{0^+} undergoes the most drastic change with increase of the electric field. The only source of electron supply for this increase is the donor level. The electron transfer model suggested by Miyazawa (1969) is clearly justified here. On closer examination of the data, one finds further that Δn_{0^+} eventually decreases corresponding to the growth of Δn_{0^-}, Δn_{1^+}, and Δn_{1^-}. The arrows indicating shoulders also correspond to the appearance of Δn_{0^-}, which influences the 0^+ population more than the population change in the donor level at higher electric fields. Our result, however, does not entirely dismiss the Hitachi model, which puts emphasis on the change in electron mobility. Evidence for change in electron mobility within the conduction band can be seen in Fig. 10, where the "normalized resistivity" $\rho_\perp^*(E)$ is given against electron temperature T_e. The quantity $\rho_\perp^*(E)$ is defined through the relation

$$\rho_\perp^*(E) = \rho_\perp(E)n_c/n_0, \tag{3}$$

with

$$n_0 = N_D - N_A. \tag{4}$$

This normalization is necessary to compensate for the effect due to change in

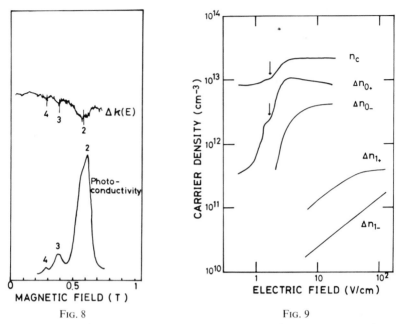

FIG. 8

FIG. 9

FIG. 8   A part of cyclotron resonance harmonics observed through photoconductance (bottom) and that through pulsed electric-field modulation (top). The numbers indicate the orders of harmonics. Note the small but definite downward deflection of the fourth harmonic in the $\Delta\kappa(E)$ signal. The lower-field shoulder of the second harmonic in the photoconductance signal is also obvious (Kobayashi and Otsuka, 1974).

FIG. 9   Change in carrier population at each Landau subband is obtained through pulsed electric-field-modulated cyclotron resonance (119 μm) as a function of electric field. The total carrier density n_c in the conduction band is also shown (Kobayashi, 1973).

carrier density within the conduction band. As a result, the ordinate essentially gives the inverse mobility. The electron temperature giving the abscissa in Fig. 10 is determined from the relative intensity of the first two CCR resonances, namely, $0^+ \rightarrow 1^+$ and $0^- \rightarrow 1^-$ transitions, under the assumption of a Boltzmann-like distribution. More complete description of the electron temperature treatment will be given in Section II.E. What is to be noted here is the deviation of experimental points from the straight line, giving the $T_e^{-3/2}$ dependence of $\rho_\perp^*(E)$. The -3 half-power dependence of resistivity on temperature indicates elastic scatterings by charged centers, or ionized impurities. Our data show that this scattering mechanism prevails between 4.2 and 8 K of electron temperature and that another stronger mechanism starts above 8 K. An electron temperature of 8 K corresponds to ~ 1.8 V cm^{-1} of the applied electric field, under which the drastic change in resistivity is almost at the end. Hence the conclusion is drawn that transfer of electrons

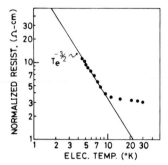

FIG. 10   Resisitivity per carrier (inverse mobility) is given as a function of derived electron temperature. Below $T_e = 8$ K, the experimental points are well aligned with the $T_e^{-3/2}$ line; i.e., the well-known electron scattering by ionized impurities (Kobayashi and Otsuka, 1974).

from donor states to the conduction band certainly holds the main responsibility for the drop in resistivity, but there is something in what the Hitachi group insists. After the fashion of Kotera, Yamada, and Komatsubara, (1972), one may suggest that the first rapid drop of resistivity with increase of the applied electric field is partially due to the electron scattering by ionized impurities, and the subsequent slow decrease of resistivity evidently reflects the onset of a new scattering mechanism, seen in the form of deviation from the $T_e^{-3/2}$ line in Fig. 10. It is ironical that the more the experimental accuracy improves, the more ambiguous the decision on mechanism becomes. Perhaps the most erroneous thing would be to try to interpret physical phenomenon by a single mechanism.

D.   CYCLOTRON RESONANCE IN DIFFERENT FIELD GEOMETRIES

The cyclotron resonance observation described so far is always carried out in the Faraday configuration, and in the hot-electron experiment the applied electric field is set perpendicular to the static magnetic field. This geometry, $E \perp H$, is indeed more direct to compare with theoretical prediction. But the electrons also can be made hot at the geometry of $E \| H$. One should expect a considerable difference in the mechanism of accelerating electrons between these two geometries and, hence, a difference in resultant distribution. The driving force exerted on electrons at the geometry of $E \| H$ is rather easy to understand, since the cyclotron-orbiting electron is simply pulled along by the electric field in the direction parallel to the axis of the spiral motion. The essential feature is quite similar to linear acceleration. On the other hand, the story of $E \perp H$ geometry is rather peculiar. Classical mechanics tells us that electrons under this geometry will be carried away in the direction perpendicular both to E and to H at a constant velocity E/H. Such a difference in feature is illustrated in Fig. 11. The most essential thing here is

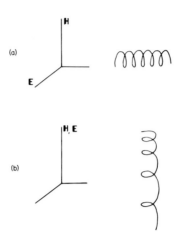

FIG. 11   Electron motion in the presence of electric and magnetic fields at two typical geometries: (a) $E \perp H$ and (b) $E \parallel H$.

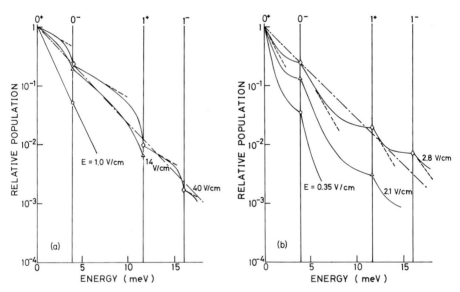

FIG. 12   Expected difference in electron distribution is schematically shown by solid lines
for two geometries: (a) $E \perp H$ and (b) $E \parallel H$, and for several applied electric fields. Relative
electron population is determined from absorption intensity and indicated by symbols at the
bottom of each subband. The chained line connecting the populations at 0^{+} and 0^{-} leads to the
intersubband electron temperature, and the broken lines drawn tangential to the solid line
at each subband bottom are related to the intrasubband electron temperature. See text (Matsuda
and Otsuka, 1978).

that acceleration, and hence the increase of the flow in the energy space, occurs on the occasions of scattering, quite in contrast to the case of $E \| H$, in which acceleration is checked at each scattering. The rate of scattering is proportional to the density of states for electrons. Efficiency of electron heating will then be large if the density of states is large for $E \perp H$, and the contrary will hold for $E \| H$. The form of the density of state function is readily obtainable in the presence of a magnetic field, so one can predict the qualitative shape of the distribution function. At energy slightly over the bottom of each subband, the distribution function for the transverse field geometry will undergo a change milder than that for the longitudinal field geometry. The situation is crudely sketched in Fig. 12a, b.

Our main interest in this section is to see how the difference in electron distribution will be reflected in the behavior of cyclotron resonance. As already mentioned, the rise of electron temperature should cause an effect similar to that due to the rise of lattice temperature. In Section B we saw the variation of cyclotron resonance trace at 84 μm with lattice temperature, and in Section C, the effect of electron transfer at the influence of an electric field on cyclotron resonance is observed. A direct comparison between lattice and electron temperature effects can be made in terms of an absorption experiment rather than a differential method, since the difference in the form of a distribution function will more clearly be seen by watching the absorption curves.

In order to facilitate the direct comparison between two geometries, use is made of the Voigt configuration, in which the selection of geometry can be made by simple switching. The far-infrared beam had to be bent by means of a metallic reflector inside the bore of the superconducting magnet. No essential difference of signal properties is seen in this configuration from those in the Faraday configuration, at least for the wavelengths employed. Typical resonance curves are given in Fig. 13a, b for the two geometries. On close examination of the line shape, linewidth, and relative intensities of resonance peaks, one can definitely tell differences for the same applied field strength: (i) Asymmetry is larger for $E \perp H$, (ii) linewidth is also larger for $E \perp H$, and (iii) at the same electric field of 15 V cm^{-1}, the ICR signal still persists for $E \| H$, even at the emergence of the third quantum line of CCR, although it is already missing for the geometry $E \perp H$, when the third quantum line is not coming yet.

The first two points reflect differences in the distribution function within each Landau subband, while the third, at least partially, shows the difference in the rate of impact ionization at the donor impurities. Existence of those differences means that in either geometry the form of the electron distribution function is deviating from the Maxwellian. Hence we must be careful in using the terminology "electron temperature." In other words, we have to

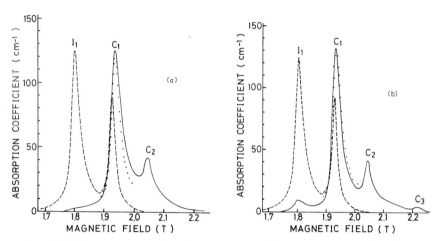

FIG. 13   Cyclotron resonance line shapes are compared between two geometries: (a) $E \perp H$ and (b) $E \parallel H$. For reference, the thermal equilibrium C_1 line is calculated corresponding to the lattice temperature common to the intersubband electron temperature at 15 V cm^{-1} and is shown by dots (normalized at the top to the 15 V cm^{-1} line) (Matsuda and Otsuka, 1978).

introduce a definition of electron temperature convenient for describing the experimental results. Before entering into a discussion of electron temperature, we summarize the results of line shape and signal intensity analyses. In Fig. 14a, b, we give the half-widths of the $0^+ \rightarrow 1^+$ line for geometries $E \perp H$ and $E \parallel H$, respectively. The quantity is measured from the peak position both toward higher and toward lower magnetic fields. One finds such an asymmetry that the lower field half-width is always smaller than the higher field side. It is further shown in Fig. 15 that the peak position is shifted toward the higher field side with increasing applied electric field. The amount of shift is larger for $E \perp H$. Features presented in Figs. 14 and 15 are related to the internal distribution of electrons within the 0^+ subband. The overall electron distribution extending over many subbands can be obtained from the relative intensity of each quantum line. Assuming that the whole population at a subband is concentrated at its bottom, we plot in Fig. 12a, b the population at several given electric fields against energy measured from the bottom of the 0^+ subband for the two geometries. If we are concerned only with the populations at 0^+ and 0^-, all we can do is to assume a Maxwellian distribution and derive the electron temperature from the relative signal intensities of the transitions $0^+ \rightarrow 1^+$ and $0^- \rightarrow 1^-$. The extension of the straight line connecting the population at 0^+ and 0^-, however, misses the populations at 1^+ and 1^-, which are deviating downward from the extended straight line, while those for $E \parallel H$ are deviating upward.

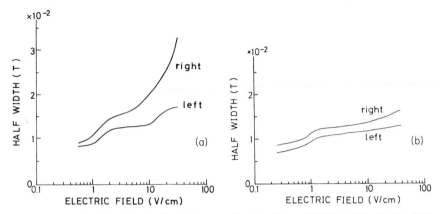

FIG. 14   Half-width of the C_1 line is given as a function of electric field both for the high-field side (indicated "right") and for the low-field side (indicated "left"): (a) $E \perp H$ and (b) $E \parallel H$ (Matsuda, 1979).

The deviation from the straight line means deviation from the Maxwellian. It is naturally smaller for lower electric field.

The simplifying assumption made in presenting Fig. 12a, b should be reconsidered now. Actually, the electrons are not entirely populated at the bottom of each subband but are spread over a certain range within the subband. In other words, one has to think over the intrasubband distribution of electrons. This is certainly entangling the electron temperature argument considerably, since we already have a complexity arising from the difference of geometry. The situation, however, cannot be avoided as we shall discuss in the next section.

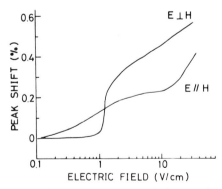

FIG. 15   Peak shift of the C_1 line observed at 119 μm is given as a function of electric field for two geometries (Matsuda, 1979).

E. ELECTRON-TEMPERATURE DETERMINATION

In the preceding three sections we have seen how cyclotron resonance signals can reflect the electron distribution in the conduction band as well as in the impurity states. In this section we shall be concerned with the electron-temperature concept within the conduction band.

We have seen that there are two ways of thinking about electron temperature. The first is to define it from the relative intensity of resonance signals, and the second is to define it from the line shape of each resonance. Since the former deals with more than one Landau subband, we shall call the electron temperature thus determined "intersubband electron temperature" and denote it T_e^{inter}, while the latter will be called "intrasubband electron temperature" being denoted T_e^{intra}. In the analytic form one can write the population at the subband $N^\pm$ as

$$n(N^\pm) = C_1 \exp[-\varepsilon(N^\pm, k_H = 0)/k_B T_e^{\text{inter}}], \tag{5}$$

which in turn defines T_e^{inter}. The expression $\varepsilon(N^\pm, k_H = 0)$ is the energy value at the bottom of the subband $N^\pm$, k_H being the wave number of electrons along the magnetic field. The normalization constant C_1 is so defined as to satisfy the relation

$$n_c = \sum_{N^\pm} n(N^\pm), \tag{6}$$

where n_c is the total density of conduction electrons as before. The form of Eq. (5) is an approximation that can be justified only at the extreme quantum limit in which practically all the states of electrons will be concentrated at $k_H = 0$.

Electron temperature T_e^{intra}, on the other hand, would be written in terms of the distribution functions inherent to particular subbands $N^\pm$ as follows:

$$f_{N^\pm}(k_z) = C_2 \exp\{-[\varepsilon(N^\pm, k_z) - \varepsilon(N^\pm, k_z = 0)]/k_B T_e^{\text{intra}}\}. \tag{7}$$

The constant of normalization C_2 here is so determined as to meet the condition

$$n(N^\pm) = \sum_{k_H} f_{N^\pm}(k_H). \tag{8}$$

In a strict sense, determination of electron temperature is meaningless once the deviation from the Maxwellian becomes obvious. All we can do then is to "define" it in an approximate manner within the reach of analysis techniques. The intersubband electron temperature will be so defined as the same as in the past experience; namely, only from the signal intersities of the $0^+ \rightarrow 1^+$ and $0^- \rightarrow 1^-$ transitions. We forget the rest of the quantum lines; the temperature thus defined is derived just from the slope of the straight line

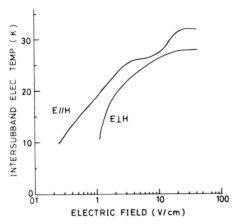

FIG. 16 Intersubband electron temperature determined from the 119 μm measurement is given against electric field for two geometries (Matsuda, 1979).

connecting the populations at the bottoms of 0^+ and 0^-. We take this procedure for both geometries. Such an approximation naturally has a limitation of validity. In reality, the electron temperature cannot be raised indefinitely because of the optical phonon emission. Perhaps the upper limit of 30 K will be reasonable for both geometries in the experiment performed at 4.2 K. Dependence of T_e^{inter}, determined upon the applied electric field, is given in Fig. 16 for two geometries.

Deriving T_e^{intra} is a rather intricate kind of parameter fitting. The absorption line shape can be analytically derived by assuming proper scattering and temperature parameters. It is assumed that the electron scattering is dominated by the ionized impurities. The expression for the inverse relaxation time, by Kawamura *et al.* (1964), is adopted, since its derivation has been most intuitive and a few other more elaborate ones are found to give no better fit with experiment. The expression is

$$1/\tau = 0.915(e^2/\kappa)m^{*-1/4}\omega_c^{-1/4}\hbar^{-3/4}N_I^{1/2} \tag{9}$$

for $k_B T < \hbar\omega_c$ and

$$1/\tau = 0.545(e^2/\kappa)m^{*-1/4}\omega_c^{1/2}(k_B T)^{-3/4}N_I^{1/2} \tag{10}$$

for $k_B T > \hbar\omega_c$. As a matter of fact, using an elaborate theory is of little value in view of distortion of the absorption line shape due to the nonparabolicity of the conduction band. The relevant conductivity expression will be given by

$$\text{Re}\{\sigma_\pm(\omega)\} = (1/Vh\omega_c)\,\text{Re}\left\{\sum_\alpha (f_\alpha - f_{\alpha+1})|J_+^\alpha|^2/[i(\omega - \omega_c - \Delta_\alpha) + \Gamma_\alpha(\omega)]\right\} \tag{11}$$

where f_α and J_+ are the distribution function and the current matrix element, respectively, with α being a set of quantum numbers at the quantum limit. Δ_α and Γ_α are the quantities giving the peak shift and the half-width, respectively. The scattering relaxation time τ is incorporated in Γ_α. The energy dispersion is assumed after the approximation of the three-band model. The effect of k_H then appears in the difference $\omega = \omega_c$ in the denominator, since we have

$$\omega/\omega_c = 1 - 8[(N + \tfrac{1}{2})\hbar\omega_c \mp g_e^*\mu_B H_0 + \hbar^2 k_H^2/2m_e^*]/\varepsilon_g; \tag{12}$$

where g_e^*, μ_B, and ε_g are the electronic g factor, Bohr magneton, and the energy gap, respectively. Perhaps the most essential approximation is that f_α obeys the classical statistics, with the intersubband electron temperature T_e^{inter} derived in the manner already described replacing the lattice temperature. The current matrix element is calculated in terms of the wave functions by Kacman and Zawadzki (1971). The absorption coefficient as a function of magnetic field is written

$$\alpha(H) = (\omega/cn)\,\text{Im}\{\varepsilon\} \tag{13}$$

with

$$\varepsilon = \varepsilon_l + 4\pi i\sigma/\omega, \tag{14}$$

where ε_1, c, and n are the lattice dielectric function, the light velocity, and the real part of the refractive index, respectively. Suitably choosing the line-width parameter Γ_α, one can calculate an absorption line shape. The intra-subband temperature T_e^{intra} will be so determined as the observed higher-field half-width fits the calculated one. This T_e^{intra}, now different from the starting value T_e^{inter}, will then be employed to determine the second value of Γ_α, and the subsequent procedures will be similar. After repetition one obtains the final convergent value of T_e^{intra}, which we eventually assign as the intra-subband electron temperature. The final T_e^{intra} is given against the applied electric field in Fig. 17. One may note, in comparison with the case of T_e^{inter}, a contrast in relative magnitude for $E \parallel H$ and $E \perp H$. Except for very low fields, T_e^{intra} for $E \parallel H$ is lower than that for $E \perp H$. This is obvious, as seen from the linewidth of resonance.

The concept of T_e^{intra} is rather hard to grasp in comparison with T_e^{inter}. In Figs 12a, b, we schematically indicate the distribution of electrons within a subband. It is found from theoretical analysis that our T_e^{intra} is proportional to the inverse of the gradient of the distribution function at the bottom of the subband. In the geometry $E \perp H$, one finds that T_e^{intra} very close to the well-defined T_e^{inter} at fields less than 8 V cm^{-1}. At higher electric fields, on the other hand, T_e^{intra} becomes larger than T_e^{inter}. In the other geometry, $E \parallel H$, T_e^{inter} seems to be smaller than the tentatively determined T_e^{intra}. One has to

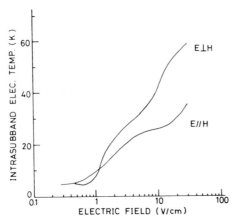

FIG. 17   Intrasubband electron temperature determined from the 119 μm measurement is given against electric field for two geometries (Matsuda, 1979).

admit that defining electron temperature in this geometry is very difficult both theoretically and experimentally.

F. IMPURITY IONIZATION AS A THERMOMETER

So far we have dealt with electron temperature defined through the distribution of conduction electrons. The concept of electron temperature, however, is frequently discussed with an involvement of the impurity states. A most typical case would be the work by Miyazawa and Ikoma (1967) on dc transport, since these authors treated the nonlinear behavior of resistivity in terms of electron transfer between the donor states and the conduction band. In the work of cyclotron resonance by Kobayashi and Otsuka (1974), comparison was made between the electron temperature determined from the relative intensities of CCR quantum lines, i.e., T_e^{inter} in the last section, and that determined after the manner introduced by Miyazawa (1969). Miyazawa defined the electron temperature in terms of the lattice temperature at which the thermal equilibrium resistivity becomes equal to the non-equilibrium resistivity. Agreement between the two electron temperatures is pretty good for lower electric fields ($E < 7\,\mathrm{V\,cm^{-1}}$) at 4.2 K, but discrepancy starts to arise as the electric field is raised further.

The purpose of this section is to show that the role of the impurity states can also be exploited for determining the electron temperature in cyclotron resonance. In Section B we saw the decay of the impurity cyclotron resonance with increasing lattice temperature, and in subsequent sections we further saw the exhaustion of donor electrons on application of an electric field. These observations suggest the possibility of defining electron temperature

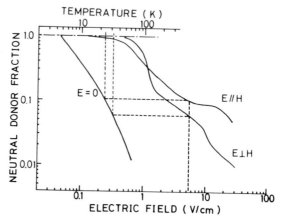

Fig. 18   Fraction of neutral-donor concentration is given as a function of electric field for two geometries. The electron temperature at a given electric field is so derived as to correspond to the lattice temperature, where the thermal equilibrium ($E = 0$) data give the same amount of neutral donors. The correspondence is guided by broken lines (Matsuda and Otsuka, 1978).

in terms of the signal intensity of ICR. Writing the ionization energy of the donor state ε_d, we have, for thermal equilibrium,

$$(N_A + n_c)n_c/(N_D - N_A - n_c) = N_c \exp(-\varepsilon_d/k_B T); \qquad (15)$$

where N_c is the density of states at the conduction band in the presence of a magnetic field. What we shall do is to replace T with an electron temperature at a corresponding value of $N_D - N_A - n_c$ obtained from the ICR intensity under the application of an electric field. In fact, this type of analysis is carried out for the two geometries $E \perp H$ and $E \parallel H$, with reference to the signal intensity of ICR under thermal equilibrium (Matsuda and Otsuka, 1978). The result is shown in Fig. 18.

The electron population at the donor $N_D - N_A - n_c$ is expressed by the ICR intensity relative to the thermal equilibrium value at 4.2 K. Correspondence between the ICR signal intensity under an electric field and that under thermal equilibrium but at elevated lattice temperature is indicated by broken lines. One finds that there is a discrepancy in electron temperature between the two geometries; in other words, directionality of temperature is again seen here.

The electron temperature defined here is certainly derived from the impurity states. But for creating a nonequilibrium distribution, conduction electrons are responsible, i.e., the difference in the impact ionization rate of the colliding electrons is causing the observed directionality of temperature. The electron temperature defined here is very close to T_e^{intra} as discussed in the preceding section, since the impact ionization rate should no doubt be

**cross
section**

E ⊥ H

E ∥ H

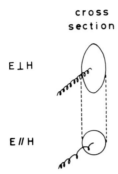

FIG. 19   Possible difference in electron scattering cross section by neutral impurities between two geometries is intuitively described. It depends upon from what angle the incident electron sees the oval-shaped electron cloud at the impurity.

related very strongly to the distribution of electrons within the first Landau subband.

One may note that for a given electric field, the electron temperature corresponding to the geometry $E \perp H$ is higher than the other above 1 V cm^{-1}. In other words, the impact ionization rate is larger for $E \perp H$. The ionization rate, of course, should be proportional to the collision cross section. Under the application of a high magnetic field, the envelope of the wave function of the donor electron becomes ovally shaped. The orbiting radius of the incident electron is smaller than the effective Bohr radius of the donor electron. Under such a circumstance, the geometrical cross section of collision is smaller for $E \parallel H$ than for $E \perp H$, since the electron hits the target as a circle in the former case and as an elongated ellipse in the latter (Fig. 19). Of course, this oversimplified picture does not tell much truth about the actual electronic scattering process in the quantum limit. In fact, the argument loses its power for the electric field below 1 V cm^{-1}. It offers us a natural guide to intuitive thinking, however.

G.   CONNECTION WITH CYCLOTRON EMISSION

During the course of the pulsed electric-field-modulated cyclotron resonance experiment (Kobayashi, 1973; Kobayashi and Otsuka, 1974), it was observed that a certain amount of the laser beam absorption was compensated by the reverse process, i.e., cyclotron emission due to the electrons raised by the electric field up to the higher Landau subbands. The phenomenon was observed also by Gornik (1972) independently of the cyclotron resonance experiment and caused such an interest that a strong far-infrared monochromatic source tunable by magnetic field may be developed. Achieving this kind of source would be all the more attractive from the point of view of a solid-state device if the emission could be made coherent. Unfortunately, attempts to make the emission lase have failed so far, and the prospect for the future is rather discouraging. The system is not very likely even as a simple, incoherent monochromatic source. The drawback here is the co-existence very close to each other of the CCR and ICR emissions. Moreover, one has to take the existence of the harmonic resonance into account. Elimination of undesired wavelengths with the help of a suitable filter or by housing the emitter within a resonant cavity might be a partial solution. But one then loses most of the emitting power, as well as the most important characteristic of the system; namely, tunability through a magnetic field. The idea of cyclotron emission has been extended to GaAs (Waldman *et al.*, 1974) and to HgCdTe (Dornhaus *et al.*, 1974), but the quality of the emission is more or less the same for GaAs and much worse for HgCdTe. Recently, Gornik and Tsui (1978) tried to obtain a far-infrared source, from the hot electrons in two-dimensional system, tunable not by magnetic field but by the gate voltage that controls the carrier density within the

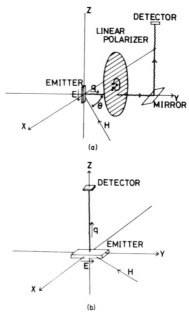

FIG. 20 The experimental setup for seeing the polarization character of the cyclotron emission is given in the upper part. Both Faraday- and Voigt-like configurations can be achieved by rotating the magnetic field in the *xy* plane, through the angle ϕ. Rotation of the linear polarizer around its normal axis is characterized by the angle θ, giving the ordinary and extraordinary modes of polarization. In the lower part, a simplified arrangement is shown examining the difference in field geometries (Matsuda, 1979).

inversion layer of Si–MOS. Describing that work in detail is beyond the scope of this chapter. We shall state here only some comments on the physical quantities obtained from cyclotron emission with reference to the corresponding ones obtained from laser cyclotron resonance. The concept of electron temperature is readily accessible also in the emission experiment. The emission bandwidth is rather broad, however. One further has a problem of the detector system. The bandwidth of the detector response causes the apparent emission to be broader and hence the resolution is poorer.

In order to study the polarization character of the cyclotron emission, an experimental arrangement shown in the upper part of Fig. 20 is set up with the help of a linear polarizer. In this geometry, both Faraday- and Voigt-like configurations can be achieved. The polarization behavior is thus observed in Fig. 21 with the use of an n-GaAs detector. In the Voigt configuration ($H \perp q$), the emission signal varies as the rotating angle ϕ of the polarizer is changed. In the Faraday configuration ($H \parallel q$), on the other hand, it does not vary. Thus the emission is found linearly polarized in the Voigt and circularly polarized in the Faraday configurations, respectively. Once this character is confirmed, and when one is content to see only the result in the Voigt configuration, the simplified argument shown in the lower part of Fig. 20 is more useful, since the merit of this arrangement is clear as described in Section D.

Some cyclotron emission signals observed in the form of photoresponse against magnetic field at various kinds of detector are given in Fig. 22 in the Faraday configuration. For each trace a peak appears when the cyclotron energy meets the maximum photoresponse of the detector. Connecting the

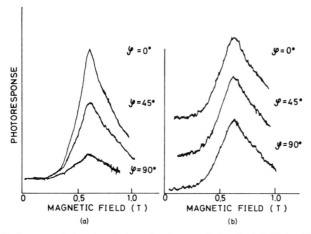

FIG. 21   Cyclotron emission signals through a linear polarizer both in Voigt ($H \perp q$) and in Faraday ($H \parallel q$) configurations (Matsuda, 1979).

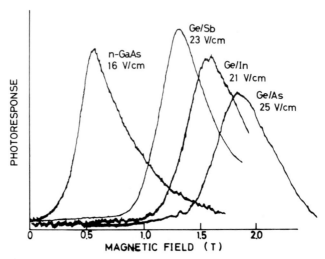

FIG. 22   Photoresponse corresponding to cyclotron emission is observed by several detectors (Matsuda, 1979).

peak positions against the photoionization energies of the detectors, one obtains the magnetic-field dependence of the emission line. The observed peak positions are given in Fig. 23, in comparison with the corresponding calculated cyclotron energies. It is found that the main emission peak consists of two branches: one originating from the reverse process of ICR and the other from that of CCR. Even at the magnetic field that is high enough to allow the complete resolution between CCR and ICR in the absorption

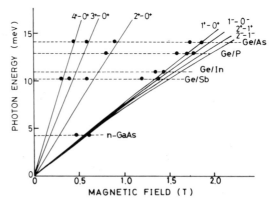

FIG. 23   Peak positions of the cyclotron emission obtained from a typical n-InSb sample are compared with the calculated Landau subband spacings (solid lines). The horizontal lines show the energy of maximum photoresponse at each detector (Matsuda, 1979).

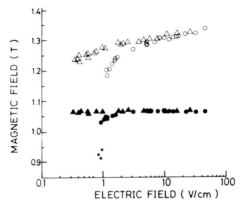

FIG. 24 Peak positions of the cyclotron emission obtained from an n-InSb sample are plotted against electric field for the case when the antimony-doped germanium is used as the detector. Open and solid triangles are from the CCR and ICR emissions in the case of $E \perp H$, respectively, while circles are from those in the case of $E \parallel H$. Crosses are from secondary peaks observed in the case of $E \perp H$ associated with ICR (Matsuda, 1979).

experiment, apparent overlapping is quite appreciable. The ICR emission is more dominant at lower electric fields, but it is overwhelmed by the growing power of the CCR emission at higher electric fields. The CCR emission, on the other hand, is composed of more than one cyclotron transition; namely, $1^+ \rightarrow 0^+$, $1^- \rightarrow 0^-$, and so on, depending on the strength of the applied electric field. Such a situation causes the overall peak position to shift toward higher magnetic fields as the applied electric field is increased. The same is not true for the peak position of ICR. Features are presented in Fig. 24 for two geometries in the Voigt configuration.

Despite the appreciable overlapping, it has still been possible to identify the peak position of the ICR emission in separation from that of the CCR. We find that tunability of CCR as well as ICR is quite good and is consistent with that reported in earlier works. The harmonic-like transitions are also observed, but their peak positions are not affected by the applied electric field. The systematic deviation of the observed harmonic resonance positions from the expected harmonic positions for CCR indicates that the harmonics are associated somehow with the impurity states. In fact, the relative intensity of the harmonics is quite sensitive to the impurity concentration.

It is of interest to see a definite correlation between the resistivity and the emission intensity. In Fig. 25, we typically show the corresponding quantities at a certain sample as a function of the applied electric field. The magnetic field is fixed at 1.3 T and the temperature at 4.2 K. The configuration is Voigt, so that both geometries of $E \perp H$ and $E \parallel H$ are taken. One can see that the rise of emission occurs at the drop of resistivity. The effect of the

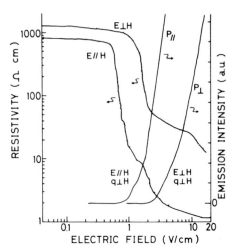

Fig. 25 It is shown for two geometries that the cyclotron emission starts to rise when the resistivity drop is nearly at the end. Lines indicated $P_\perp$ and $P_\parallel$ indicate the emission intensities (Matsuda, 1979).

electric field is recognizable earlier for $E \parallel H$ than for $E \perp H$. This is somewhat similar to the result of the absorption experiment. If one writes the emitting power P as a function of electric field, the electron temperature can be defined through the relation

$$P(E) = P_0 \exp[-(\varepsilon_1 - \varepsilon_0)/k_B T_e], \qquad (16)$$

where P_0 is a constant, and ε_0 and ε_1 are the energies at 0^+ and 1^+ subbands, respectively. This is, of course, an approximation and corresponds to the approximate intersubband electron temperature in the absorption experiment, which is defined in terms of the electron populations only at the 0^+ and 0^- subbands. The electric field dependence of T_e defined in Eq. (16) is very close to T_e^{inter} defined in the absorption experiment for both geometries.

One thus finds that correspondence between absorption and emission experiments is quite good. The information on the electron distribution obtained from the emission experiment, however, is rather limited in comparison with that obtained from the absorption work. For example, the internal electron distribution within a subband, or a concept of T_e^{intra}, is not accessible from the emission approach. The role of the emission work, accordingly, is merely to supplement the absorption work and confirm the consistency in certain aspects. However, the possibility of constructing a new, useful device compensates for the weakness in physical clarity. It is frequently true that more practical things have to remain less physical in our lives.

III. Far-Infrared Study of the Exciton System

A. LANDAU LEVELS AND EXCITONIC TRANSITION

The binding energy of an exciton in germanium is thought to be ~ 3.8 meV while that in silicon is ~ 14.7 meV (Hensel et al., 1977). These magnitudes of energy lie in the far-infrared region. The exciton system in a semiconductor can be regarded, at least in the framework of the effective mass approximation, as a quasi-hydrogen atom. One can indeed observe electronic transitions corresponding to those in the hydrogen atom. Situation in a semiconductor, however, is somewhat complicated by the anisotopic nature as well as by the degeneracy of the energy bands. If the system is exactly hydrogenic, one can find its binding energy very accurately from the observed energy of a certain definite transition, say 1s → 2p. Even with some dissimilarity which exists in reality, perhaps this procedure is still the most accurate for deriving the binding energy.

In this section we focus our attention on the behavior of excitonic transitions in the presence of a magnetic field. We mentioned in Section II that the discrete levels of the impurity states in InSb are distributed over the Landau levels of conduction electrons. The association model of the impurity states with the Landau levels can be launched only in the case of a very strong magnetic field; in other words, the magnetic field should satisfy the condition $\hbar\omega_c/2Ry^* \gg 1$, where Ry^* is the effective Rydberg energy. It has often been remarked, however, that the association between the Landau levels and the impurity states takes place under intermediate or even weaker magnetic fields. In Section II we called a particular impurity transition in InSb at high magnetic field "impurity cyclotron resonance." In the same sense one might as well classify the relevant excitonic transition as a kind of cyclotron resonance. Indeed, one can see a parallel rise in excitonic transition energy as a function of magnetic field with the cyclotron energy of the light hole (Fujii and Otsuka, 1976). As will be seen later, the same is true also for the impurity system not satisfying the high magnetic field condition.

Differing from the case of InSb at high magnetic field, the excitonic transitions observed at intermediate or weaker magnetic fields are quite manifold. In other words, there are many allowed transitions between the discrete levels of the excitonic state if the magnetic field is not too strong. Hence the use of the terminology "cyclotron resonance" is rather misleading. Perhaps simply using the term "magneto-optical transitions" would be more appropriate. Of all the possible magneto-optical excitonic transitions, the most prominent and readily accessible one will be the lowest field 1s → 2p_-type transition, which is associated with the Landau levels of the light hole. The absorption peak corresponding to this transition is frequently utilized as a guide for monitoring the existence of excitons, in particular, when one deals

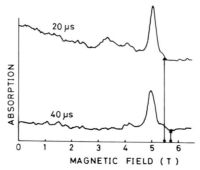

FIG. 26   Time-resolved magnetoabsorption traces of EHD and excitons obtained by D₂O laser. Results for two delay times, 20 and 40 μs, are shown. The EHD signal for 84 μm appears as a broad background plateau, while the excitonic signal is a series of lines. The most outstanding and readily accessible excitonic signal appears at 5 T. The nomial intensity of the EHD signal is indicated by arrows (Otsuka et al., 1977).

with the coexisting system of the electron–hole drops and its surrounding excitons. A typical absorption trace for that transition is shown in Fig. 26 for germanium at the wavelength of 84 μm (Otsuka et al., 1977). The excitonic transition is standing on the broad, background absorption that arises from absorption by electron–hole drops, as will be discussed in Section B. The decay of the exciton signal with time gives an "apparent lifetime" of the exciton system. The actual average life of an exciton is usually expected to be shorter than that, since the exciton system as observed in the signal is coexisting with electron–hole drops and will be continually recruited from the drops through the evaporation of electron–hole pairs. This recruitment is seen most remarkably in the case of a strain-confined, large, electron–hole drop as will be discussed in Sections D and E. In addition to the prominent peak by the $1s \rightarrow 2p_-$ transition, one can observe in Fig. 26 several less-conspicuous peaks that also arise from the interdiscrete-level transitions of the exciton. The identification of all these transitions in the presence of a magnetic field is tedious but, in principle, can be done. In this chapter, however, we shall rather avoid stepping deeply into spectroscopy of the exciton and concentrate our effort on clarifying the kinetics governing the exciton and the electron–hole drop.

B.   TIME-RESOLVED STUDY OF EXCITONS AND ELECTRON–HOLE DROPS

It is of more interest to watch the formation as well as the decay process of an electron–hole drop in the exciton gas than to observe its stationary spectrum. The dynamic process of the electron–hole drop, which we shall henceforth abbreviate as EHD, cannot be discussed without paying attention to the behavior of the exciton gas in which EHD is embedded. There is more than one way of excitation for studying the transient behavior of EHD, as well as its surrounding exciton system through time-resolution techniques. Excitation by pulsed light is followed by a standard rise and fall of the combined system. Repetition of the pulsed excitation leads to the most common time-resolution analysis of the process. Depending on the purpose of investigation, one can change the time-width of the excitation. In some cases one may prolong the pulse so much that it can practically be regarded as a stationary illumination. A sudden switch on or off of excitation with the help of a suitable modulator enables us to observe the growth or decay process of the system in or after the steady-state illumination, respectively. All these kinds of excitation cause results that can be traced through time resolution.

The far-infrared (FIR) laser now comes into play. It can be used either cw or pulsed. A typical use of the FIR study of the strongly excited system is illustrated in Fig. 27. The photoillumination is made either by a Q-switched YAG laser, a xenon flash lamp, or an Ar^+ laser. The FIR laser is operated at 20–30 Hz in synchronized combination with the frequency-divided photo-excitation (10–15 Hz). The FIR absorption, with and without application of the photopulse, is led to the two-channel boxcar system and differentiated.

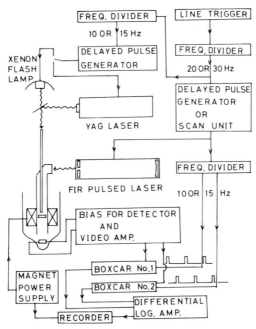

FIG. 27   Block diagram of the far-infrared study setup for the EHD and exciton system in germanium (Otsuka et al., 1977).

Wavelength of the FIR laser, of course, is changed at convenience. Use of 84 or 119 μm is suitable in a sense for decay analysis, since the absorption signal arising from EHD looks like a nonresonant plateau at low magnetic fields and can easily be distinguished from that arising from the free-exciton system. The feature is already seen in Fig. 26. The relative strength of absorption by EHD will be measured by the height of the plateau at the foot of the $1s \to 2p_-$ exciton signal for each delay time after the end of the photo-excitation, as indicated by arrows in Fig. 26, while that of the exciton signal is measured by the height of the $1s \to 2p_-$ signal as stated before. A typical absorption versus delay-time plot is given in Fig. 28 for both the exciton and EHD. A very striking thing in this case is that the exciton signal does not change very much for the first 40 μs, though the decay of EHD is quite evident during the same time interval. The commonly accepted rate equations for electron–hole pairs, as well as for excitons can be written as

$$\dot{n}_d = -n_d/\tau_0 - aT^2 n_d^{2/3} \exp(-\phi/k_B T) + b n_d^{2/3} n_{ex} \tag{17}$$

and

$$\dot{n}_{ex} = aN T^2 n_d^{2/3} \exp(-\phi/k_B T) - (n_{ex}/\tau_{ex}) - b n_d^{2/3} n_{ex} N. \tag{18}$$

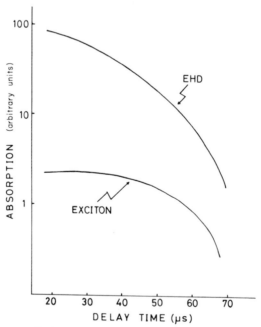

FIG. 28   A decay profile for excitons and EHD observed by far infrared. Apparently, the exciton signal intensity remains almost constant for the first 40 μs, while the EHD signal decays steadily (from data taken by T. Nakata).

Here n_d is the number of pairs in an EHD; n_{ex} the density of excitons; τ_0 the recombination time of electron–hole pairs in EHD; τ_{ex} the lifetime of excitons; ϕ the binding energy of EHD, commonly known as the work function for evaporation of the electron–hole pair from the surface of EHD; a and b the constants associated with thermionic emission and back-flow of excitons from and to EHD, respectively; and finally N the density of EHD, or the number of drops per unit volume. The observed constancy of n_{ex} (or $\dot{n}_{ex} = 0$) during the first stage after excitation assures us the relation

$$n_{ex} = aNT^2 n_d^{2/3} \exp(-\phi/k_B T)/[(1/\tau_{ex}) + bN n_d^{2/3}]. \tag{19}$$

The right-hand side of Eq. (19) contains a time-dependent quantity n_d. In order for n_{ex} to be independent of time passage, it is necessary to have the condition

$$1/\tau_{ex} \ll bN_d^{2/3}. \tag{20}$$

In that case, one obtains

$$n_{ex} = (a/b)T^2 \exp(-\phi/k_B T). \tag{21}$$

Meanwhile if one combines the condition $\dot{n}_{ex} = 0$ with Eq. (17), the relation

$$\dot{n}_d = (-n_d/\tau_0) - (n_{ex}/\tau_{ex} N) \tag{22}$$

is obtained. Assuming here, as is frequently done, that the density of EHD N does not vary with time, the integration of Eq. (22) yields

$$n_d = A \exp(-t/\tau_0) + C, \tag{23}$$

where C is a new constant of integration. Differentiating with respect to t, we have

$$\dot{n}_d = -(A/\tau_0) \exp(-t/\tau_0) \tag{24}$$
$$= -(A/\tau_0) \exp(-t/\tau_0) - (C/\tau_0) - (n_{ex}/\tau_{ex} N) \tag{25}$$

according to Eq. (22). The constant C is thus found to be

$$C = -n_{ex}\tau_0/\tau_{ex} N, \tag{26}$$

giving

$$n_d = A \exp(-t/\tau_0) - (n_{ex}\tau_0/\tau_{ex} N). \tag{27}$$

If we write $n_d = n_{d0}$ at $t = 0$, Eq. (27) can be written

$$n_d = n_{d0} \exp(-t/\tau_0) + n_{ex}\tau_0[\exp(-t/\tau_0) - 1]/\tau_{ex} N. \tag{28}$$

The above kinetics leading to Eq. (28) are, of course, justified only during the time interval in which the condition $\dot{n}_{ex} = 0$ is guaranteed.

One can repeat the time-resolution experiment at various temperatures. Since the back-flow coefficient b is proportional to $T^{1/2}$, one can expect

$$n_{ex} \propto T^{3/2} \exp(-\phi/k_B T). \qquad (29)$$

From the collection of the values of n_{ex} at different temperatures and at time-independent stages, one finds that $\phi = 14.7$ K (1.27 meV). This value is somewhat smaller than the average of optically determined values, but fairly in agreement with those derived after thermodynamic analysis. The analytic procedure taken here is certainly full of crude assumptions, and the absolute value of ϕ thus derived cannot be relied upon too much. In fact, the time interval in which n_{ex} remains practically constant depends on temperature to some extent and, of course, the condition cannot be regarded as being universal. But the resultant close fitting of ϕ with values derived by other people seems to favor the general consistency of the analytic steps.

What has been described so far in this section is merely a partial aspect of the time-resolution study. More features leading to new physical quantities will be introduced in due order.

C. RESPONSE OF THE ELECTRON–HOLE DROP TO THE FAR INFRARED

When one sweeps the magnetic field at a fixed far-infrared wavelength for magnetoabsorption by photoexcited germanium, a number of peaks are obtained. Some of them are ascribed to absorption by various excitonic transitions and others are due to EHD. A typical feature obtained with a laser wavelength of 337 μm is illustrated in Fig. 29 with temperature as a parameter (Fujii and Otsuka, 1975). When temperature is high, or around 4.2 K, a number of well-resolved peaks are observed. With lowering temperatures, many of the dominant peaks observed at 4.2 K diminish or become obscure. Some of the peaks more recognizable at higher temperatures are due to excitonic transitions; others are associated with EHD. Since the spectra are obtained 10 μs after the photoexcitation, most of the originally born excitons are considered dying and the observed excitonic transition signals are mainly provided by the excitons evaporated from the drop. As temperature is lowered, the evaporation rate becomes smaller and, hence, the excitonic peaks get harder to observe. Accordingly, examination of temperature dependence of the peaks is the first step in distinguishing between excitonic and drop-associated spectra.

Once the distinction is made, one can study the response of EHD to far infrared quite systematically. The general characteristics of the EHD behavior in far-infrared spectroscopy can be classified into two categories:

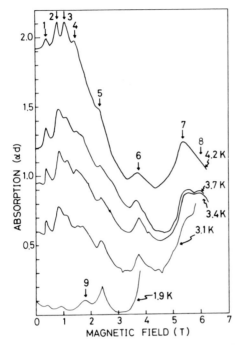

FIG. 29 Magnetoabsorption spectra of photoexcited germanium obtained by HCN laser with temperature as a parameter. As temperature is lowered, excitonic signals (indicated 1 through 4) become less pronounced (Fujii and Otsuka, 1975).

one is the case in which $kR \ll 1$ and the other $kR \gg 1$, where k is the wave number of the radiation within the drop and R the radius of the drop. On rare occasion, the case of $kR \sim 1$ is treated.

The condition $kR \ll 1$ corresponds to the ordinary drop, sometimes called the α drop, which is produced by an appropriate photoillumination; the opposite condition $kR \gg 1$ corresponds to the case of a large drop, which can be produced only with the help of an inhomogeneous uniaxial stress on top of the photoillumination. In the case of a large drop having a size of several hundred microns one can come across the dimensional resonance for millimeter waves corresponding to occurrence of the condition $kR = m\pi$, where m is an integer. In this section, however, we shall restrict ourselves to the ordinary drop, which has an average size of a few microns.

For dealing with ordinary drops, the external wave number k_0, i.e., the wave number in the crystal dielectric, plays an explicit role. The interaction of incident radiations with small drops can be described in terms of the limiting-case Mie theory, in which both $kR \ll 1$ and $k_0 R \ll 1$ are guaranteed. Then the extinction cross section is practically governed by the absorption cross section and one obtains to the fifth order of $k_0 R$:

$$Q_{ext} \sim Q_{abs} \sim \left(\frac{2\pi}{k_0^2}\right)\left[2(k_0 R)^3 \operatorname{Im} \frac{\tilde{\varepsilon} - 1}{\tilde{\varepsilon} + 2}\right.$$

$$\left. + (k_0 R)^5 \left(\tfrac{1}{15} \operatorname{Im} \tilde{\varepsilon} + \tfrac{1}{6} \operatorname{Im} \frac{\tilde{\varepsilon} - 1}{\tilde{\varepsilon} + \frac{3}{2}}\right)\right]. \tag{30}$$

Here $\tilde{\varepsilon}$ is the dielectric function inside the drop divided by that outside. The scattering terms appear only at the sixth power of $k_0 R$, which we have neglected. In the Faraday configuration the ratio of the dielectric functions for two circular polarizations can be expressed as

$$\tilde{\varepsilon}_\pm = 1 - \sum_j \frac{\omega_{pj}^2}{\omega(\omega \mp \omega_{cj} + i/\tau_j)}, \tag{31}$$

where ω is the angular frequency of the incident radiation, ω_{cj} the cyclotron frequency, ω_{pj} the plasma frequency, τ_j the relaxation time for the carriers of the jth kind.

An absorption peak by EHD will appear when the dielectric function $\tilde{\varepsilon}$ satisfies such condition that makes Q_{abs} infinite. To the crudest approximation, Q_{abs} is proportional to $\operatorname{Im}[(\tilde{\varepsilon} - 1)/(\tilde{\varepsilon} + 2)]$. The pole originating from this term is due to the electric dipole and we may call it electric dipole resonance. The nextcomer, which is proportional to $\operatorname{Im} \tilde{\varepsilon}$, is due to the magnetic dipole. The pole corresponding to this term appears at the same magnetic field as the cyclotron resonance field of each carrier kind. We may call this pole magnetic dipole resonance, or sometimes induction cyclotron resonance. The prediction of resonances made in such a manner is not readily compared with experimental results because of several complications. First, the laser beam is linearly polarized and, hence, both electronlike and holelike resonances contribute to absorption. In addition to such overlapping of resonances, the line broadening due to finite relaxation times considerably obscures the peak positions. Second, use of the ordinary cyclotron mass from each kind of carrier raises some problem, since one may have to take account of the many-body effect. Third, the shape of a drop may be distorted as one goes to high magnetic fields. Fourth, the quantum oscillations due to crossing of the Landau level with the Fermi level cause dips in absorption. Fifth, some geometric effects can cause an extra-large peak. Sixth, different laser wavelengths often yield completely different spectra. And so on. Thus one has to face much complexity in analyzing the magnetoplasma spectra by EHD. Such complexity can be utilized, however, to detect some subtle nuances of EHD that cannot readily be predicted from the luminescence experiment.

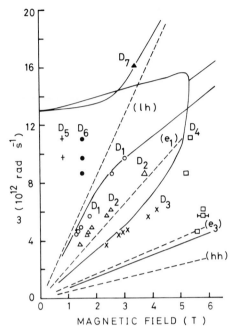

FIG. 30   Frequency versus magnetic-field diagram of the drop-associated signals indicated D_1 through D_7. Cyclotron resonances of free carriers are indicated by broken lines (Nakata *et al.*, 1979).

Typical peak positions of magnetoplasma resonance due to EHD are indicated $D_i(i = 1, 2, 3, \ldots)$ in Fig. 30 against magnetic field and laser wavelengths (Nakata *et al.*, 1979). The solid lines are theoretical predictions for occurrence of the electric dipole resonance with the simplest approximations. The direction of the applied magnetic field is parallel with $\langle 111 \rangle$ and hence there are two kinds of electron valleys having degeneracy 1 and 3. The cyclotron resonance positions for these two kinds of valleys are given by dashed lines, with indications of (e_1) and (e_3), respectively. Those for holes are also given by indicators (lh) and (hh) for light and heavy holes, respectively. We assign D_1, D_3, and D_7 to be resonances by electric dipole. For D_2, we tentatively assign magnetic dipole resonance, since it is close to the (e_1) line. However, we have to admit another possibility (Gavrilenko *et al.*, 1976). D_4, D_5, and D_6 are unaffected by wavelength. The occurrence of D_4 corresponds to exhaustion of the (e_1) electrons. D_5 and D_6 occur at the fields of the Fermi energy minima, indicating increase of the work function.

The linewidth analysis of D_3 has been most convenient for examining the behavior of the relaxation time. Temperature dependence of linewidth is

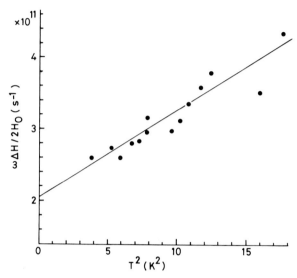

FIG. 31 Linewidth of the D_3 line, a representative electric dipole resonance, is plotted against temperature. A quadratic temperature dependence is obvious (Nakata *et al.*, 1978).

investigated by the use of 337 μm, and the result is given in Fig. 31. The straight line drawn through the data points satisfies the equation

$$\omega \Delta H / 2H_0 = \alpha + \beta T^2, \tag{32}$$

with

$$\alpha = (2.1 \pm 0.1) \times 10^{11} \, \text{s}^{-1} \tag{33a}$$

and

$$\beta = (1.2 \pm 0.1) \times 10^{10} \quad \text{deg}^{-2} \, \text{s}^{-1}. \tag{33b}$$

The T^2 dependence of the linewidth strongly indicates that the collisional relaxation time within EHD is dominated by the carrier–carrier scattering (Landau, 1957) and not by phonons or impurities. As will be described later, the same conclusion is drawn for the large drop produced by application of an inhomogeneous uniaxial stress by analyzing the line shape in the Alfven-wave resonance. Such analysis is rather unique to magnetoplasma spectroscopy and furnishes a good example of merit that is not available in the luminescence experiment.

Another feature of the magnetoplasma approach is seen in the quantum oscillation. So far as our experience is concerned, oscillations in resonance absorption with magnetic field is most clearly seen in the $H \| \langle 111 \rangle$ case at the laser wavelength of 119 μm (Fujii and Otsuka, 1975, Nakata *et al.*, 1979).

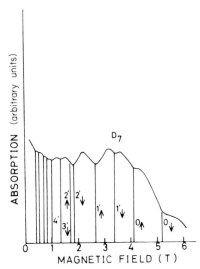

FIG. 32   A kind of quantum oscillations is observed at 119 μm for the EHD signal D_7 (see Fig. 26) when $H\|\langle 111\rangle$. Each dip corresponds to a crossing of the electron Landau level (unprimed for the one-valley and primed for the three-valley) by the electron Fermi level. Arrows up and down indicate the spin states (Nakata *et al.*, 1978).

The optimum condition for observing the oscillation is rather delicate. It depends strongly on delay time after photoillumination and somewhat on intensity of the far-infrared radiation. Some of the best data is shown in Fig. 32. It is seen that the oscillations, or the dips in absorption, are primarily determined by the behavior of electrons. That there is less contribution by holes can also be seen in the Alfven-wave resonance of the large drop. One may further note the appearance of the splitting due to the g factor. It should be remarked here, however, that the magneto-oscillation can also be observed in luminescence (Martin *et al.*, 1977).

From the dips of oscillations one can derive the electron–hole pair density within EHD as a function of the external magnetic field. The one derived from our observation in the geometry of $H\|\langle 111\rangle$ is given in Fig. 33. The density is 2.2×10^{17} cm^{-3} at zero field and it remains the same up to 2 T. Then one comes across a decrease to 2.0×10^{17} cm^{-3} at 2.5 T. After that the density starts to rise again and reaches the value 2.7×10^{17} cm^{-3} at 5 T. An eventual increase of the electron–hole pair density at high magnetic field is confirmed also in the luminescence experiment at the geometry of $H \| \langle 100\rangle$ (Martin *et al.*, 1977). In this geometry no such a decrease of density as that observed in our geometry of $H \| \langle 111\rangle$ is recognized. A possible reason for the dip in density observed in our measurement is considered to be due to a geometrical effect. As mentioned before, if the

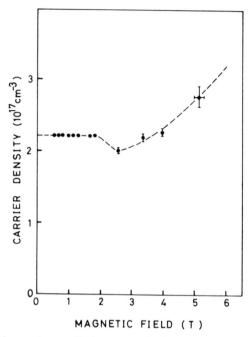

FIG. 33   From the quantum oscillations as seen in Fig. 28, one can derive the approximate electron–hole pair density within EHD as a function of magnetic field. The dip near 2.5 T is considered to arise from a geometrical reason inherent to the case of $H \parallel \langle 111 \rangle$ (Nakata *et al.*, 1978).

magnetic field is applied along $\langle 111 \rangle$, two kinds of valley systems, (e_1) and (e_3), are produced. The Landau levels of the former system are raised by the magnetic field faster than for the latter system. Hence the crossing of the Landau level with the electron Fermi level occurs earlier for the (e_1) system. Then the electrons in the (e_1) system start spilling out of the valley. In other words, the effective density of states in the conduction band to accommodate electrons will be decreased due to the runaway of this valley. Thus the density of available electrons has to be affected. Once all the (e_1) electrons are spilled out, the remaining (e_3) valleys are all equivalent and will no longer see any geometrical cut-down of the density of states. On the contrary, there will be an increase of the density of states because of the decreasing cyclotron radius. In the geometry of $H \parallel \langle 100 \rangle$, the electron valleys are all equivalent and no partial spilling out is expected, so a monotonic increase of the pair density with magnetic field is expected. Unfortunately, the quantum oscillation is not so easy to observe in the $H \parallel \langle 100 \rangle$ geometry as in the $H \parallel \langle 111 \rangle$ case.

Another possible geometric effect is seen in the appearance of the D_4 peak. It is nearly independent of laser wavelength and, as already mentioned, appears at the field that corresponds to the exhaustion of the (e_1) electrons.

As to the possible shift of the resonance peaks due to the renormalization effect of the carrier masses, we shall avoid entering into speculative arguments.

D. INVESTIGATION OF THE LARGE ELECTRON–HOLE DROP

So far we have restricted ourselves to topics of ordinary electron–hole drops that are produced within an unstrained crystal by photoexcitation. One of the most exciting findings in the recent study of EHD has been the creation and detection of the so-called large electron–hole drop (LEHD), which has a macroscopic size of several hundred microns and can be produced only under the application of an inhomogeneous uniaxial stress (Markiewicz et al., 1974). The observation of dimensional resonance in microwaves, or formation of the Alfven-type standing waves within a drop, furnishes direct evidence that the reported large drop is really a continuous plasma and not a mere aggregate of droplets. In addition to the Alfven-wave resonance observation, the ultrasonic experiment has further brought forth the Fermi liquid nature of LEHD most clearly (Ohyama et al., 1976). One finds, from the quantum oscillations, the electron–hole pair density within LEHD to be 6.2×10^{16} cm^{-3} and the intercarrier collision time to be 6.0×10^{-11} s at 1.8 K and for the stress along $\langle 011 \rangle$. These values are in qualitative analyses that will be mentioned here.

It is not our intention to describe the method of LEHD formation. We will rather concentrate on the physical properties obtained through the use of microwaves and far infrared.

The most striking aspect of the magnetoplasma effect observable through microwaves is the Alfven-wave dimensional resonance. The electromagnetic waves penetrating into plasma are classified as Alfven and helicon waves, both of which can be defined only for $\omega_c/\omega > 1$, where ω is the frequency of the applied ac field (Alfven, 1950). The former is waves propagating when $n_e = n_h$, the latter when $n_e \neq n_h$. Accordingly, in EHD only the Alfven waves can be built up. If the size of EHD and the penetrating waves satisfy certain geometric conditions, one can expect appearance of a kind of standing wave, resulting in strong absorption of electromagnetic energy. This is called "dimensional resonance." For a spherical plasma, as in the case of EHD, the condition for resonance is rather intricate. One may generally put

$$kR = \gamma_{ij} \tag{34}$$

for postulating a resonance. Here γ_{ij} is the jth zero of the ith spherical Bessel function, and k the wave number. If one combines the resonance condition,

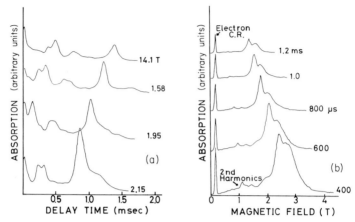

FIG. 34   Time-resolved Alfen-wave dimensional resonance patterns: (a) Time variation with fixed magnetic field, (b) magnetic field variation with fixed delay time (Honbori, 1979).

Eq. (34), with Eq. (32) and takes some approximations, one obtains the relation

$$H = [4\pi n(m_e + m_h)]^{1/2}(R\omega/\gamma_{ij}) \tag{35}$$

$$\propto R\omega n^{1/2}; \tag{35a}$$

where H is the resonance magnetic field and R the corresponding radius of LEHD. Thus we find that the resonance field is proportional to the drop radius. There are two ways of tracing the Alfven-wave resonance. One is to see absorption as a function of magnetic field at a fixed delay time after the photoexcitation. The other is to see it as a function of delay time at a fixed magnetic field. Typical recorder traces for both cases are given in Fig. 34a, b. The apparent twin structure, more obvious in the case of the magnetic field sweep (Fig. 34b), is due to the electronlike and holelike polarizations. With an infinitely large magnetic field, the structure is expected to disappear. For LEHD, absorption by magnetic dipole is more dominant than that by electric dipole on account of the large radius. If one writes the magnetization M, the absorbed power is given by

$$P \propto (\omega/2) \operatorname{Im}(\mathbf{M} \cdot \mathbf{H}) \tag{36}$$

$$= (\omega/2)H^2 V\chi'' \tag{37}$$

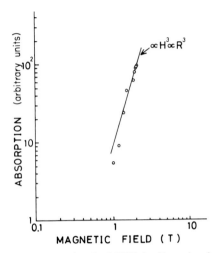

FIG. 35   Millimeter wave absorption by LEHD in dimensional resonance is given as a function of magnetic field. The straight line shows absorption proportional to the third power of magnetic field (Honbori, 1979).

where χ'' is the imaginary part of the susceptibility and V the volume of LEHD, which is proportional to R^3. Since χ'' is proportional to R^2 for small R and becomes independent of R for large R, we have

$$P \propto R^5 \qquad \text{for small } R \tag{38}$$

and

$$P \propto R^3 \qquad \text{for large } R. \tag{39}$$

Figure 35 shows the intensity of the main absorption as a function of resonance magnetic field H, which is proportional to the drop radius. For higher magnetic fields, or larger R, the R^3 dependence of absorption is quite good.

It has been shown that the half-width of the resonance line ΔH gives the carrier relaxation time τ through the relation

$$\Delta H / H = 1/\omega\tau. \tag{40}$$

One can readily confirm the proportionality between ΔH and H. The relaxation time given here is actually the average over electrons and holes, or the quantity given by

$$\tau_{\mathrm{av}} = \tau_e \tau_h (m_e + m_n)/(m_e \tau_h + m_h \tau_e). \tag{41}$$

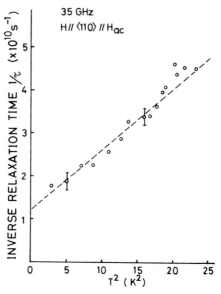

FIG. 36 Temperature dependence of the inverse relaxation time for LEHD obtained from microwave dimensional resonance. The quadratic dependence on temperature is again evident as in the far-infrared case for EHD (see Fig. 27) (Honbori, 1979).

By measuring the half-width ΔH as a function of temperature, one can find the temperature dependence of the carrier relaxation time. The result is presented in Fig. 36. One finds again here, as in Section C, that the inverse relaxation time has a quadratic dependence on T in accordance with Landau's prediction (1957) on the carrier–carrier scattering relaxation, which is written in his original form:

$$1/\tau = \gamma_0[(k_B T)^2 + (\hbar\omega/2\pi)^2] \quad \text{for} \quad k_B T \lesssim \hbar\omega. \tag{42}$$

Here γ_0 is a constant. Since we are holding ω as fixed, the observed behavior shows the law of the type given by Eq. (42) quite well. In other words, the most dominant scattering relaxation process within LEHD is through carrier–carrier interaction and not through carrier–phonon or carrier–impurity scattering. To our knowledge, such a clear T^2 dependence as presented in Fig. 36 has never been observed for electrons in metals. This observation, together with that for ordinary drops, as given in Fig. 31, constitutes an experimental verification of the Landau theory. Difference in absolute magnitude of $1/\tau$ between Fig. 31 and Fig. 36 is attributed to the frequency dependence and also to the different electron–hole pair densities between EHD and LEHD. The most reliable value of the carrier-pair density in LEHD is 6.2×10^{16} cm^{-3}, as reported in the work by Ohyama et al., for the acoustic attenuation. It should be mentioned here that a crude T^2 dependence of $1/\tau$ is also noted in the acoustic attenuation work.

From Fig. 34a, one can plot R as a function of delay time, since the resonance field is proportional to radius, as seen in Eq. (35a). The time constant for radial decay is found as long as 1500 μs. Such an incredible lifetime of LEHD has been confirmed long since (Wolfe et al., 1975), and it is attributed to the smaller electron–hole pair density in comparison with that in an ordinary drop. It will be seen later that there is another reason to make the life of LEHD longer than that of an ordinary drop. Detailed features in microwave dimensional resonance cannot be described here but will be left to references. We shall rather turn our eyes to aspects obtained by far infrared.

At the use of far infrared, the condition $kR \gg 1$ can always be taken for granted for LEHD. If one takes the wavelength 119 μm, for example, one can observe magneto-optical absorption by excitonic transitions on top of the broad background signal of LEHD. At first sight, the feature is quite similar to the the case of ordinary drops as treated in Section B. There is a difference, however, between the two cases. Comparison is made in Fig. 37. First, the exciton signal is somewhat broader in linewidth for the LEHD case. A more striking feature for the LEHD case is persistence of the exciton signal with delay time. In fact, the apparent decay-time constant of the exciton system is found to be almost exactly as long as the radial decay-time constant. This feature will be discussed further in the next section.

E. EXISTENCE OF EXCITONIC ATMOSPHERE AROUND THE LARGE DROP

It was stated in Section D that the apparent life of the exciton system becomes very long when one observes the excitonic transition in the presence of LEHD. As generally accepted, the average lifetime of an isolated exciton is of the order of several microseconds. The observed exciton signal, accordingly, may not necessarily be reflecting persistence of an identical group of excitons. As a matter of fact, if one measures the intensity of the excitonic signal as a function of delay time, a time constant of 1500 μs is obtained (Fig. 38). The decay character of the exciton system and that of LEHD are thus found quite parallel. It is believed that what is actually happening is that a region in which excitons can coexist with LEHD is maintained during the entire decay process of LEHD. Coexistence of excitons with a drop is not surprising so long as one admits the evaporation of electron–hole pairs out of the drop. In fact, we made a kinetic analysis in Section B based on the coexistence assumption for ordinary drops. However, the decay behavior of the exciton signal in the present case is different from that shown in Fig. 28. There is no time-independent period in the decay process, but a perfect parallelism with the radial decay of LEHD. In order to excavate the real situation, the

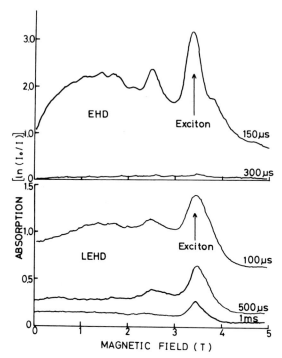

FIG. 37 Time-resolved recorder traces of magneto-optical transitions for EHD and LEHD cases obtained through 119 μm. In the latter case, both excitonic and LEHD signals remain for a much longer time. See the delay times indicated on the right (from data taken by T. Ohyama).

far-infrared attenuation by the LEHD signal (the broad background in Fig. 37) is plotted against delay time. The decay-time constant thus derived is 750 μs, or exactly half the radial decay-time constant which is 1500 μs. This is indeed an outstanding fact worth noting. It thus seems necessary to build new kinetics for LEHD and its associated exciton system.

As essential difference of the LEHD case from the ordinary EHD case is the confinement of the system within a strain-potential system which can be written as, with radius R in front:

$$\dot{n}_d = \pi R^2 v_T \bar{n}_{ex} - (4\pi n_0 R^3/3\tau_0) - 4\pi R^2 A T^2 \exp(-\phi/k_B T) \qquad (43)$$

and

$$\dot{N}_{ex} = 4\pi R^2 A T^2 \exp(-\phi/k_B T) - \pi R^2 v_T \bar{n}_{ex} - (N_{ex}/\tau_{ex}). \qquad (44)$$

Here n_0 is the carrier pair density in LEHD; τ_0 the total lifetime of LEHD; $\bar{n}_{ex}$ the average of the exciton density defined within the potential well; v_T the thermal velocity of excitons; τ_{ex} the lifetime of excitons; and ϕ the

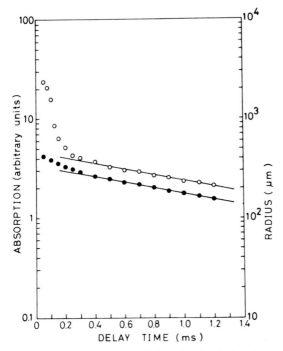

FIG. 38 Long-life decay features of the far-infrared excitonic signal (open circles) and the drop radius derived from the microwave dimensional resonance (solid circles). The almost perfect parallelism of the two decays ensures $N_{ex} \propto R$ (from data taken by T. Ohyama).

work function, A a constant connected with a and defined in Eqs. (17) and (18) through relation $4\pi(3/4\pi n_0)^{2/3}A = a$. We have further written $N_{ex} = V_{eff}\bar{n}_{ex}$, where V_{eff} is the effective volume of the region around LEHD where one can find excitons coexisting with LEHD. Those excitons in the above region were born through evaporation and are now constrained by the strain force and pulled back to the surface of LEHD. In other words, each exciton evaporating from LEHD experiences a return trip to its surface. The time needed for the trip can be calculated from classical mechanics to be of the order of 10^{-8} s. This is much shorter than the average recombination time of an exciton, which is estimated to be $\lesssim 10^{-5}$ s. In other words, excitons have little chance to recombine during their traveling time; or, there exists a layer shell around the LEHD body, filled with excitons popping up and absorbed down. The entire system will be regarded to be in quasi-steady state and the rate equations are solved to yield

$$R = R_0 \exp(-t/3\tau_0) \tag{45}$$

and

$$N_{ex} = 4A V_{eff} T^2 \exp(-\phi/k_B T)/v_T, \qquad (46)$$

with

$$V_{eff} = 4\pi R^2 \overline{\Delta R}. \qquad (47)$$

Here R_0 is the initial radius of LEHD, and $\overline{\Delta R}$ appearing in Eq. (47) is the radial extent of the exciton system jumping up and down.

The physical meaning of $\overline{\Delta R}$ is established through the existence of a strain potential. The center of LEHD is expected to lie on the minimum of the potential. We shall assume a spherically symmetric potential $V(\mathbf{r})$ with $\mathbf{r} = 0$ at the minimum. The simplest approximation for the form of $V(\mathbf{r})$ is a parabola, or putting $V(\mathbf{r}) = \alpha r^2$. If one takes this approximation, and treats the exciton system subject to this potential like an ideal gas, some elementary calculation yields

$$\overline{\Delta R} \sim k_B T/2\alpha R. \qquad (48)$$

This expression has been derived under the assumption $R \gg \Delta R$ and is justifiable only by confirming the consistency. The estimated value of α so far is 130 K mm^{-2} (Wolfe et al., 1978). For a fairly large drop of $R = 300$ μm at $T = 4$ K, for example, we obtain $k_B T/2\alpha R = 50$ μm. Since $\overline{\Delta R}$ in Eq. (48) is inversely proportional to R, the condition $R \gg \Delta R$ can be guaranteed only for a very large drop. Lowering temperature can offer some remedy against breakup of the validity of Eq. (48). We shall, accordingly, restrict our argument only to LEHD having a size of the order of ~ 200–300 μm. With the help of Eq. (48) one can rewrite Eq. (46):

$$N_{ex} = (8\pi A R k_B T^3/\alpha v_T)\exp(-\phi/k_B T). \qquad (49)$$

This expression shows proportionality between N_{ex} and R. The observed parallel decay character between the drop radius and the excitonic signal is thus understood.

The existence of the excitonic "atmosphere" around the LEHD body is worth noting. It is the parabolic potential that causes the birth of such atmosphere and is somewhat similar to the aerial atmosphere surrounding a planet that is lying at the center of a gravitational potential.

One has to be careful in such a case about the physical meaning of "evaporation" of an electron–hole pair from the LEHD surface. In the case of an ordinary drop, overcoming the work function ϕ guarantees the pair for an escape from the drop. A mere release from the drop surface, however, does not mean an eventual escape in the LEHD case. Unless the released pair has a large excess kinetic energy, it is impossible to reach the region where the strain potential no longer has an influence. Accordingly, it is more practical to define an effective work function ϕ_{eff} which is dependent on R and larger than ϕ. Only through this ϕ_{eff}, can the electron–hole pairs in LEHD escape. Aided by the internal decay, the total rate of change in electron–hole pair number can be written, in place of Eq. (43), as

$$\dot{n}_d = (-4\pi n_0 R^3/3\tau_0) - 4\pi R^2 A T^2 \exp(-\phi_{\text{eff}}/k_B T). \qquad (43a)$$

In this expression, we neglect the back-flow of the pairs having kinetic energy greater than ϕ_{eff}. From the new rate in Eq. (43a), one obtains

$$R(t) = (R_0 + 3\gamma\tau_0) \exp(-t/3\tau_0) - 3\gamma\tau_0, \qquad (50)$$

where

$$\gamma = (A T^2/n_0) \exp(-\phi_{\text{eff}}/k_B T). \qquad (51)$$

The observed attenuation of the far-infrared beam due to LEHD ends, naturally, when $R(t)$ becomes zero. Therefore, we introduce a cut-off time t_c, such that $R(t_c) = 0$. An approximation which is valid for $t < 3\tau_0$ yields the relation

$$t_c = (n_0 R_0/A T^2) \exp[\phi_{\text{eff}}(R = 0)/k_B T]. \qquad (52)$$

It is easy to find t_c experimentally. The far-infrared attenuation is measured at zero magnetic field. Plotting $\ln(T^2 t_c)$ against T^{-1} as seen in Fig. 39,

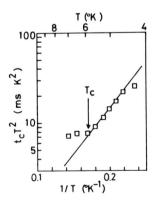

FIG. 39 From the cut-off time analysis of the far-infrared LEHD absorption at various temperatures, one can derive the limiting effective work function for $R = 0$. The solid line through the data is drawn on the parameter fitting $\phi_{\text{eff}}(R = 0) = 23$ K. Deviation of the experimental points from this line gives the critical temperature of 5.9 K for LEHD (from data taken by T. Ohyama).

$\phi_{\text{eff}}(R = 0)$ can be derived by parameter fitting. The best choice gives $\phi_{\text{eff}}(R = 0) = 23$ K. The abrupt deviation of the experimental points from the straight line in Fig. 39 occurs near 5.9 K, and this is practically independent, above a certain level, of excitation power. We presume the temperature at which deviation occurs to be the critical temperature for LEHD. As a matter of fact, it is impossible to observe any LEHD-associated signal above this temperature, including the dimensional resonance experiment through millimeter waves.

The effective work function is a function of R, and the above-derived value of 23 K for $\phi_{\text{eff}}(R = 0)$ is a minimum value of ϕ_{eff}, since the exciton-holding potential $V(\mathbf{r})$ is zero at $\mathbf{r} = 0$. It was found (Ohyama, 1978) that the far-infrared attenuation is nicely explained if one assumes the form

$$\phi_{\text{eff}}(R) = \phi_0 + \bar{\alpha}R^2, \tag{53}$$

with $\bar{\alpha} = 250$ K mm^{-2}, and $\phi_0 = \phi_{\text{eff}}(R = 0) = 23$ K. Though this fitting has been made at $T = 4.8$ K, the general character would be more or less the same at different temperatures. The relation (53) yields, for example, $\phi_{\text{eff}} = 46$ K for $R = 300$ μm. Since the average of the work function for the ordinary drop is of the order of 20 K, one can imagine how hard it is for electron–hole pairs to get away from the surface of LEHD. Incidentally, one may note here that the value of $\phi_0 = \phi_{\text{eff}}(R = 0)$ is very close to the value of the work function for the ordinary drop. One more thing to be mentioned is the similarity of the form for the strain potential and the effective work function [Eq. (53)]. It is no surprise, since the latter should evidently reflect the strain-potential form. A rigorous argument, however, will distinguish $\bar{\alpha}$, the coefficient of the R^2 term in ϕ_{eff}, from α, the coefficient of the parabolic strain potential.

Finally, mention should be made about the far-infrared attenuation character. As already seen, the decay-time constant for the far infrared is almost exactly half the radial decay constant. In other words, the extinction cross section for the far infrared is proportional to R^2, quite in contrast to the case for small droplets, in which the extinction cross section is proportional to the cube of the drop radius at the crudest approximation. This is not surprising. When the wavelength of the incident radiation is much less than the radius of the target body, the wave theory predicts the extinction cross section $2\pi R^2$ for a total reflector. The present situation is somewhat more delicate, and perhaps one should deal with the problem in the category of anomalous diffraction as termed by van de Hulst (1957). The behavior of anomalous diffraction, however, approaches that of a total reflector in the large radius limit, so that the observed proportionality of the cross section to R^2 is not unreasonable.

IV. Related Studies

A. ELECTRON-EXCITON SCATTERING STUDY THROUGH CYCLOTRON RESONANCE

We have seen that the linewidth of cyclotron resonance gives the inverse of the carrier-scattering relaxation time. The linewidth analysis can be made more direct in the case of electron resonance in silicon or germanium, where the conduction band is nearly perfectly parabolic and the simplest assumption

$$2/\omega\tau = \Delta H/H_0 \tag{54}$$

is acceptable with least reluctance. Here H_0 is the resonance field and ΔH the half-width of the Lorentzian absorption curve. The half-width ΔH is broadened when one introduces impurities into the crystal. The electron scattering by neutralized impurities or by ionized impurities has been studied rather extensively, using mostly 35 GHz waves (Otsuka et al., 1966a, 1973). As a rule, the related measurement is carried out in steady-state conditions. Under the steady bandgap light illumination, all the impurities are neutralized and one usually sees the electron scattering by neutralized impurities. The number of scatterers is unchanged during the course of experiment at liquid helium temperatures.

One may look at neutralized impurities, or to be more exact, at neutralized tri- and pentavalent impurities in silicon and germanium, to find model images of hydrogen atoms in vacuum (Pearson and Bardeen, 1949). Scattering of electrons by these impurities can be analyzed in terms of electron–hydrogen atom (Erginsoy, 1950) or positron–hydrogen atom collision (Otsuka et al., 1964, 1966a). In Section III, we saw the transient behavior of excitons through resonance techniques; in other words, one cannot overlook existence of excitons in an experiment of cyclotron resonance so long as carriers are produced by photoillumination. The exciton is certainly a kind of hydrogen atom-like center. One might then think that contribution of electron–exciton collision should be taken into account when analyzing the cyclotron resonance linewidth. As a matter of fact, in the past steady-state measurements, intensity of photoillumination was minimized in order to avoid the contribution of carrier–carrier interaction. This procedure was equally, or even more, effective in avoiding the contribution of electron–exciton collisions. The carrier–carrier scattering comes to light as one raises the intensity of photoillumination (Kawamura et al., 1964). In that case, the possibility of contributions from the electron–exciton interaction should be seriously considered. Since an exciton is only of temporary existence, one has to take a time-resolution measurement. Perhaps the first time-resolved cyclotron resonance measurement was carried out by ourselves for pure

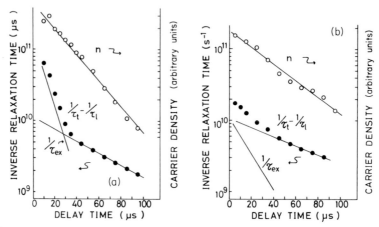

FIG. 40 (a) The inverse relaxation time (linewidth) and carrier density (open circles) are given at 4.2 K against delay time in the strongly photoexcited cyclotron resonance of germanium. $1/\tau_t$ indicates the total linewidth and $1/\tau_1$ the contribution from scatterings by lattice. The difference (solid circles) gives sum of the contributions from electron–exciton and electron–carrier scatterings. The former ($1/\tau_{ex}$) can be separated from the kink the electron–exciton interaction is more dominant, while on the other side the electron–carrier interaction is mainly causing the line broadening. (b) As temperature is lowered to 2.9 K, the kink becomes less pronounced, indicating condensation of excitons into EHD (Yoshihara, 1971).

germanium crystal using 35 GHz waves (Ohyama *et al.*, 1971). The excitons, as before, are produced by means of a xenon flash lamp. In Fig. 40a, two quantities, absolute intensity of the electron resonance line (linewidth × peak height) and electron resonance linewidth (with subtraction of the contribution from electron–phonon scatterings $1/\tau_l$), are given at 4.2 K against delay time after the photopulse having the length of ∼1 μs. The former is proportional to the electron density and the latter to the difference of the inverse relaxation times, or $1/\tau_t - 1/\tau_l$, $1/\tau_t$ giving the total observed linewidth. One finds a kink in the $(1/\tau_t - 1/\tau_l)$ versus delay-time data. Our interpretation is that in the time interval before the kink occurs, we are observing electron–exciton interaction. Since the life of an exciton is short, contribution of electron–exciton collisions to the linewidth becomes unobservable some 40 μs after the photopulse. After the kink, it is considered that the linewidth is mainly governed by the electron–carrier interaction. This idea comes from the observed $n^{1/2}$ dependence of $1/\tau_t - 1/\tau_l$ after the kink, which can readily be found by replotting the data of Fig. 40a in the $(1/\tau_t - 1/\tau_l)$ versus n form. The one-half power dependence on the carrier density obviously reflects the electron–carrier interaction as treated by Kawamura *et al.* (1964). It is further found that a quadratic dependence of $1/\tau_t - 1/\tau_l$

on the carrier density exists before the kink. That is believed to arise from the bimolecular nature of the electron–hole pair, thus strongly supporting the idea of electron–exciton interaction. Accordingly, subtraction of the contribution from the electron–carrier interaction, as indicated by the less-steep gradient, will lead to the genuine contribution from the electron–exciton interaction, which we show by $1/\tau_{ex}$. Other original interpretations by Ohyama et al. (1971) inherent to this observation, such as lifetimes of free carriers as well as that of excitons, should now be considered in inclusion of the drop formation. The lifetimes derived from the observed decay process are evidently overestimated, since the effect of recruitment through evaporation from the drop was not taken into account. Of interest, however is that, the kink appearing in the $(1/\tau_t - 1/\tau_1)$ versus delay-time plot becomes obscure as temperature is lowered (Fig. 40b). In other words, the exciton system becomes harder to detect through linewidth analysis. This is consistent with our experience in far-infrared measurements. The fact is that excitons have been absorbed by drops, or, in what is the same effect, little evaporation is taking place. In the absence of the drop concept, such a phenomenon remained just an enigma, since one naturally expected more excitons at lower temperatures.

In order to add further support to existence of the electron–exciton interaction, electric-field pulses are applied to the sample for 3 μs each in synchronization with photopulses, with intention of causing impact dissociation of excitons through accelerated carriers. If the carrier density is increased after the electric pulse, one may accept the implicit existence of excitons. Typically, the experiment is carried out so that the electric field is switched on 10 μs after the photopulse. The boxcar gate is then opened after another lapse of 8 μs. The resultant signal of electron cyclotron resonance shows a striking increase in magnitude with a considerable plasma shift (Fig. 41). If one applies the same electric pulse after the occurrence of the kink in Fig. 40, neither the enhancement of electron resonance nor the plasma shift is observed. These observations show that right after the photopulse, there are plenty of excitons to be dissociated as a result of carrier impact, while after a certain time interval there are very few excitons left and no appreciable enhancement is expected through carrier impact.

Perhaps the weakest part of the above work will be the absence of estimation of the exciton density. If it were available, one could readily derive the cross section for electron–exciton collision. As a rough estimate, we expect the cross section to be fairly close to that of electron–neutral donor collision. One might argue that an exciton can be regarded as either a donor or acceptor, so that there is an equal possibility that electron–exciton collision should be compared with electron–neutral acceptor collision. The latter is similar to positron–hydrogen atom collision and its cross section is by an

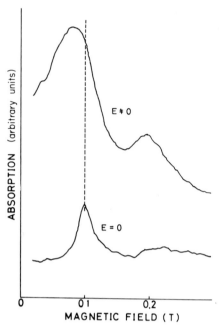

FIG. 41   Electron cyclotron resonance signal is largely enhanced on application of an electric field. Moreover, the peak position is considerably shifted toward lower magnetic field. These effects are thought to arise from decomposition of excitons on impacts by electrons (Yoshihara, 1971).

order of magnitude smaller than the cross section of the electron–neutral donor collision (Otsuka et al., 1966a). Our choice of the electron–neutral donor collision for simulation arises from consideration of the exchange effect. In other words, we have two electrons in the electron–exciton collision process. The exciton, in the meantime, is to be simulated with an acceptor when one deals with hole–exciton collision, since we have to consider exchange between two holes. Thus an exciton becomes either donorlike as it interacts with an electron or acceptorlike as it interacts with a hole. If one assumes that the electron–exciton collision can be treated just like the electron–neutral donor collision, one can derive from Fig. 40a that the density of excitons right after the photopulse is $\sim 4 \times 10^{14}$ cm^{-3}. It is of interest that this value is very close to the density of excitons coexisting with drops at 4 K. That this figure is large enough to cause condensation into droplets is generally accepted today.

A somewhat different way of analyzing to estimate the cross section for electron–exciton collision is taken for electron cyclotron resonance in silicon (Ohyama et al., 1973). The results obtained indicate that although the magnitude of the cross section stands between the two models, namely electron–donor and electron–acceptor collision models, it is closer to the former. Following the experimental result, theoretical approaches dealing with pure three-body collision have been developed (Matsuda et al., 1975; Elkomoss and Munschy, 1977). The calculated results are in good agreement with the experimental after the proper use of effective mass values.

B. ELECTRON SCATTERING BY IONIZED IMPURITIES

We mentioned in Section A that impurities in silicon or germanium crystals are neutralized by the bandgap light illumination at low temperature. Low-temperature transport study by means of cyclotron resonance, accordingly, sees a big difference from that by the dc method, since the latter almost always encounters dominant electron-scattering effects of ionized impurities. Historically, the first report of studies of ionized impurity scattering by cyclotron resonance came from Sekido et al. (1964). These authors endeavored to produce ionized impurities even under the bandgap light illumination by introducing critical amounts of compensating dopants, raising the ambient temperature, and minimizing the intensity of the illuminating light. They are believed to have succeeded in acquiring a certain amount of ionized impurities and in carrying out analyses making use of the Brooks–Herring formula for relevant electron scattering. Since the arrival of FIR lasers, however, one is advised to make free carriers by driving electrons out of the neutral donor sites. The resonance itself is to be observed by microwaves. Indeed, this type of experiment was carried out by the present author's group on antimony-doped germanium with negligible compensation (Otsuka et al., 1973). Pumping of electrons out of the donor sites was made by an ordinary discharge-type pulsed H_2O laser at the wavelength of 119 μm. The available number of free electrons was adjusted by controlling the laser output power through insertion of black polyethylene sheets in the path of the laser beam. In Fig. 42 we show the free-carrier concentration—which is equal to the concentration of ionized impurities—plotted against the number of black polyethylene sheets, which change the laser power output linearly. It is seen that the full power output is not strong enough to eject all the electrons out of the impurity levels. Cyclotron resonance is carried out again at 35 GHz. Illumination of bandgap light is now unnecessary. Contribution of neutral impurities to the cyclotron resonance linewidth can easily be handled away so that one can estimate the genuine contribution of ionized impurities. Naturally, no resonance for holes is observable. A definite advantage of this experiment over the one carried out by Sekido et al., is the uniqueness of the ionized impurities; i.e., ionized antimony. In the work by Sekido et al. one had to admit both ionized donors and ionized acceptors, the

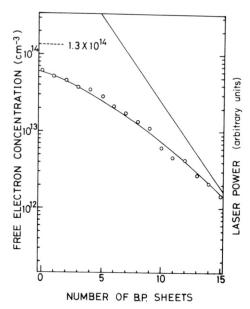

FIG. 42   Free carriers are produced in antimony-doped germanium through far-infrared laser (119 μm) illumination. The intensity of laser beam is controlled by black polyethylene sheets, the number of which is given in the abscissa. The free-carrier concentration (open circles), given in the ordinate, tends to the total donor concentration as the laser beam intensity (the solid straight line above) is increased (Otsuka et al., 1973).

latter being copper, singly, doubly, or triply ionized, depending on circumstance. Moreover, the neutral impurity in the present case is just one kind, in contrast to the case of Sekido et al. where both neutral donors and acceptors were encountered. Another difference is that with FIR laser excitation an experiment can be done even at liquid helium temperatures. An outstanding result drawn from this experiment is that neither the Brooks–Herring (1951) nor the Conwell–Weisskopf (1950) theory accounts for the observed line broadening. In other words, the observed broadening is found to be much smaller than theoretical predictions. The result is shown typically for 3.2 K in Fig. 43. It is not our intention to go into further details in this chapter, but it would be worth pointing out the invalidity of the Born approximation when one deals with electron scattering at low temperature.

Finally, the importance of the electron-scattering study at low temperature should be emphasized, since it is directly connected with efficiency or quality of a photoconductive far-infrared detector. More extensive works dealing with different dopants are awaited, not only for germanium but also for general prospective materials, making use of the many wavelengths in the far infrared now available through the optical pumping technique.

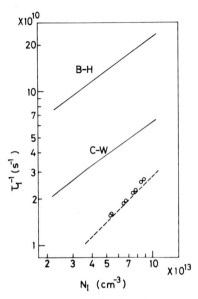

FIG. 43 Experimentally derived inverse relaxation time of electrons by ionized impurity scatterings (open circles) is plotted against impurity concentration at a fixed temperature of 3.2 K. For comparison, theoretical predictions by Brooks–Herring and by Conwell–Weisskopf are also given. Discrepancy between calculations and experiment is quite obvious at this temperature (Otsuka et al., 1973).

C. MAGNETO-OPTICAL ABSORPTION OF IMPURITY STATES

There are millions of possible electronic transitions between discrete levels of impurity states in semiconductors which can be surveyed by far-infrared lasers. The amount of energy associated with each transition is a function of applied magnetic field. Exhaustive study of such transitions may lead to a new type of solid-state laser device, especially if one focuses attention on the transitions between excited states in compound semiconductors. Indeed, this kind of investigation has been initiated by British workers (Skolnick et al., 1977), but it will take more time even to see a partial success. In this section we shall just keep into magneto-optical transitions at a typical impurity center in a typical semiconductor, namely indium impurity in germanium crystal. The system to be surveyed is in thermal equilibrium and in that sense deviates from the proper theme of this chapter. Nevertheless, we take it up here for its possible connection with the nonequilibrium study. In fact, we have seen for InSb the potent contribution of impurity centers to the nonequilibrium study. Not necessarily in the same manner, but in a more or less modified way, the impurity center in silicon or germanium may also have a variety of contributions. The submillimeter detector device by D^- or

A^+ centers is just one example (Norton, 1976). Moreover, the impurity centers may act as a nucleus in EHD formation, as will be discussed in Section D. They may serve as a source of carriers in the off-equilibrium transport study or scattering centers as well. Indium impurities in germanium crystal would make a typical example. Indeed, one finds a certain connection between cyclotron transition of free carriers and inter-discrete-level transition within this impurity. A definite similarity has been observed between conduction electrons and donor-bound elections in InSb; in other words, the term "cyclotron resonance" is justified to some degree even for impurity-bound electrons. But one then needs a criterion for justification; namely, the relation $\gamma \equiv \hbar\omega_c/2\,\mathrm{Ry}^* \gg 1$. The heavy inequality is readily satisfied for InSb even under a magnetic field of the order of 1 T, since the electronic effective mass is extremely small. The same cannot be expected for impurities in germanium. Nevertheless, as we shall see, some association between the discrete impurity levels and free-carrier Landau levels is evident for indium-doped germanium case (Otsuka et al., 1976). There is little doubt of similarity for other dopants.

The sample employed here is a specially grown germanium crystal doped with indium at the concentration of $N_A = 3.7 \times 10^{14}$ cm^{-3}. The compensation is quite negligible, as confirmed by the absence of the far-infrared absorption signal inherent to arsenic or antimony impurities. The main wavelengths employed in a discharge-type laser containing H_2O or D_2O vapor are as follows: H_2O (118.65, 85.564, 78.4, and 48.0 μm), D_2O (84.284 μm). Two kinds of detector are used: arsenic-doped germanium for wavelength shorter than 100 μm and antimony-doped germanium for 118.65 μm. Every measurement has been done under the Faraday configuration.

Absorption peaks appear to be quite numerous as one sweeps the magnetic field. In Fig. 44, we take photon energy as the ordinate and magnetic field as the abscissa for $H \parallel \langle 111 \rangle$. The absorption peaks observed, corresponding to several laser wavelengths, are shown by circles. Probable connections are made by solid lines A through E for different wavelengths. Theoretical calculation for a kind of 1s → 2p type transition is indicated by broken lines; they are associated with the bound hole states at the indium acceptor site. We assume that in order to account for the A series, one has to associate the 1s-like state with the hole Landau quantum state 1_0, and the $2p_-$-like state with the Landau state 2_1, where the numbering 1_0 and 2_1 are after the manner introduced by Suzuki and Hensel (1974). Similarly the B series is to be compared with the $1s(2_0) \to 2p_-(1_1)$ transition. Calculation has been carried out in a way similar to but modified from Yafet et al., the details being described in the Appendix. The relevant cyclotron transitions in the valance band are given in Fig. 45 for $H \parallel \langle 111 \rangle$ under the restriction that $k_H = 0$, k_H being the wave number component along the direction of the magnetic

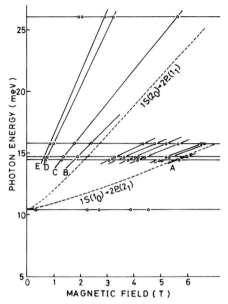

FIG. 44   Magneto-optical transitions between discrete levels of indium impurities in germanium are plotted against magnetic field. Several series, indicated A through E, are easy to find. Two numerical calculations based on the association model of impurity states with Landau levels are given by broken lines and can be compared with A and B series. See text, Appendix, and Fig. 45 (Otsuka *et al.*, 1977).

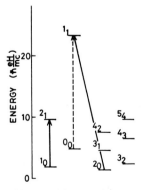

FIG. 45   Some cyclotron transitions within the valence band of germanium are indicated after the manner of Suzuki and Hensel. The impurity transition associated with the cyclotron transition arising here from the level 0_0 is expected to be seen only at elevated temperature.

field. The high field condition $\gamma \gg 1$ is not satisfied in the present case. On the contrary, we are mostly dealing with the case of $\gamma \lesssim 1$. The association model of the discrete impurity states with the free-carrier Landau levels then does not seem acceptable. Nevertheless, one finds indication of association by the closeness of the A and B series to the calculated lines based on the association model. On inspection of Figs. 44 and 45, one may wonder about the whereabouts of the possible transition $1s(0_0) \rightarrow 2p_-(1_1)$. The energy difference between 0_0 and 2_0 is about 0.5 meV at 1.5 T. Separation between the associated levels $1s(0_0)$ and $1s(2_0)$ will be of the same order. It is expected then that population of holes in $1s(0_0)$ is considerably smaller than that in $1s(2_0)$ at 4.2 K. In other words, much less absorption corresponding to the transition $1s(0_0) \rightarrow 2p_-(1_1)$ is expected at this temperature. Even if any appreciable absorption is observed at 4.2 K, its intensity should come down very rapidly as one goes to lower temperatures. No indication of such a drastic change in absorption intensity has been detected for the peaks showing up in the absorption trace at 4.2 K. So one may conclude that the wanted transition is missing in our spectra observed below 4.2 K. On the contrary, our tentative exploration with somewhat elevated temperature finds appearance of a new peak. More exhaustive investigation with many different wavelengths will confirm whether we have been right or wrong in identifying the peak to be the wanted transition. The rather unexpected result of association of discrete bound states with Landau states in the continuum is also seen for free (Fujii and Otsuka, 1976) and bound excitons (Yamanaka et al., 1978). In other words, "cyclotron resonance" is observable for excitons in the sense that impurity cyclotron resonance is observable for neutral donors in InSb. Justification for using the terminology "cyclotron resonance" for excitons or impurity states in silicon or germanium will be more complete if one goes to higher magnetic fields, where the condition $\gamma \gg 1$ is more readily established.

D.   IMPURITIES, EXCITONS, AND ELECTRON–HOLE DROPS

In this section, we will more or less be dealing with impurity states in a semiconductor. Excitons, although they vanish after a lapse of time, may also be regarded as a kind of hydrogenic impurity. The impurity in the style of hydrogen atom is one of the most important things in understanding the physics of semiconductors. Also in the far-infrared investigation, the role of the hydrogenic impurity state has been much emphasized. For future work, the situation will remain the same. The role of an impurity will be looked upon with keen interest, in particular with regard to the nucleation of electron–hole drops. It is well known that, in general, problems of both homogeneous and inhomogeneous types of nucleation are to be considered. For electron–hole drops, the difference between these two types is more delicate. In the homogeneous case, density fluctuation of excitons is responsible for nucleation and an embryo starts from an exciton. The exciton, as stressed in Section A, can be regarded either as a neutral donor or neutral acceptor. If the inhomogeneous nucleation starts at an impurity site, donor or acceptor, the feature must be very similar to the homogeneous case.

The excitonic molecule, if it really exists, can be compared with an impurity catching an exciton, the latter being similar to an asymmetric molecule. The sole, but very important, difference lies in the zero-point energy of constituents. The adiabatic approximation cannot be utilized for the excitonic molecule. No image of vibration or rotation model inherent to a diatomic molecule is justifiable for the excitonic moleculelike quartet, and perhaps an image of a light nucleus would be more appropriate. The homogeneous nucleation model requires the formation of an excitonic molecule $(ex)_2$ first, and then $(ex)_3$, $(ex)_4$, ..., $(ex)_{N_c}$, where N_c is the number of electron–hole pairs needed to form the critical size of electron–hole drop, which is now similar to a heavy nucleus. What has been reported so far, however, is no more than formation of trions, (eeh) or (ehh) (Kawabata et al., 1977), and $(ex)_2$ in silicon (Kulakovskii and Timofeev, 1977; Gourley and Wolfe, 1978). In other words, detection of a free multiexciton complex before reaching the critical size has not yet been heard of. Thus, the actual process of homogeneous nucleation is still unsolved. On the other hand, a symptom of inhomogeneous nucleation is found in the controversial "bound multiple exciton complex" (BMEC) (Pokrovskii, 1972; Sauer, 1973; Kosai and Gershenzon, 1974). Sauer et al. withdrew the model afterward (Sauer and Weber, 1976, Sauer et al., 1977), but it is still highly possible that there exists something more complicated than a single bound exciton (Thewalt, 1977).

Presuming that BMEC exists, next comes the question of whether it grows into an embryo for EHD. So long as one sticks to the luminescence technique alone, it will be difficult to draw a final conclusion, since the appearances of luminescence spectra are more or less the same and no drastic change in interpretation can be expected beyond the present stage. One possible breakthrough would be reinforcement by far-infrared magneto-spectroscopy. It is stated in Section III.C that a very minute change in attribute of EHD can be reflected on the far-infrared spectra. The same will be true for the embryo stage of EHD. There may be some difficulty, however, in separating this kind of approach from the problem of interaction between impurities and EHD, the latter not necessarily being in the embryo state. Controls in time resolution, excitation intensity, and so on will be important factors in laying open the process of nucleation.

As a preliminary stage of investigation, we mention here some results of EHD study associated with impurities. The first one is the effect of impurity

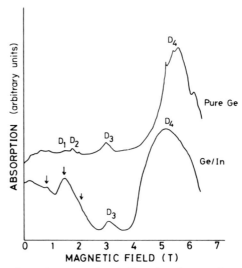

FIG. 46   Magnetoabsorption trace of an indium-doped germanium obtained at 420 μm. The arrows indicate presence of peaks affected by impurities. For comparison, the same is also shown for pure germanium (Nakata *et al.*, 1979).

on the carrier-pair density within EHD (Nakata *et al.*, 1979). In Fig. 46, we show some difference in absorption spectra at 420 μm between pure and indium-doped germanium. One can see a few more peaks for indium-doped germanium ($N_{In} \sim 8 \times 10^{15}$ cm^{-3}) at low magnetic fields. On the other hand, the D_1 and D_2 peaks observed for pure germanium are missing, while the D_3 and D_4 peaks remain. What one may note here is enhanced linewidths of D_3 and D_4. The three new peaks at low fields, positioning at 0.79, 1.39, and 2.10 T, are not affected by laser wavelength. We recall that for pure germanium one obtains two frequency-independent peaks at 0.83 and 1.52 T. The first two peaks above for the doped sample are very close to these, possibly suggesting the same origin. Doping of indium, accordingly, seems to cause a shift of these peaks. Now, presuming that the Fermi level minimum causes these peaks, one comes to the conclusion that the carrier-pair density of EHD for our indium-doped sample is decreased to $1.9 \sim 2.0 \times 10^{17}$ cm^{-3}. Similar results have been experienced for antimony-doped samples with different antimony concentrations. The overall impurity-concentration dependence of the EHD carrier-pair density is plotted in Fig. 47 for more than one impurity, together with the theory by Smith (1976) for the antimony-doped case. Results obtained through a different method by Benoit a la Guillaume and Voos (1972) are also included.

The effect of impurity on EHD is very interesting but still full of uncertainty. Nominating the impurity as a candidate of the nucleation center for EHD

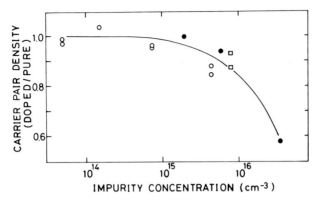

FIG. 47   Variation of carrier-pair density within EHD is given against impurity concentration. Both antimony and indium impurity results are included. The solid circles are after the luminescence experiment by Benoit a la Guillaume and Voos. Theoretical prediction by Smith (solid line) gives a good fitting to experimental points (Nakata *et al.*, 1979).

raises the idea that introduction of impurities encourages formation of EHD and hence its stability. Indeed, an indication supporting this idea is found in luminescence work (Alekseev *et al.*, 1970). But we also have other experimental evidence showing that the total lifetime of EHD decreases upon introduction of impurities. In other words, impurity centers may behave as sinks of EHD as well. In view of encouraging the no-phonon recombination channel between electrons and holes through impurities, enhancement of annihilation may certainly lead to a shorter life of EHD. Such double dealing of impurity centers makes things quite complicated. Perhaps one important factor would be the amount of impurities. A small amount of impurity is certainly effective in raising EHD, but an excessive amount is not welcomed. Existence of good and evil impurities for EHD may also be a possibility.

As a concluding remark it may be said that there is still a vast amount of room left for speculation in the set of impurities, excitons, and EHD. A considerable amount of work by luminescence has been reported for this field, but very little by far infrared. Advent of a tunable far-infrared laser will no doubt accelerate new developments in this still uncultivated field, including the attractive topic of EHD nucleation.

Appendix: The Functional Forms in the Variational Calculation

The variational calculation we carried out in Section IV.C is essentially the same as that made in the past by Wallis and Bowlden (1958), Yafet *et al.* (1956), or by Kohn and Luttinger (1955). In the previous calculations, however, isotropic mass and the variational parameter ε, which measures the ratio of the transverse to longitudinal spread of the wave function with respect to the magnetic field direction, are assumed always to be less than unity. Owing to the introduction of mass anisotropy, ε may become more than unity and it is inappropriate to rely on the old results. Accordingly, we have derived the form that can be used most generally for $\varepsilon > 0$. Instead of the functional form $L(x)$ defined by Wallis and Bowlden, i.e.,

$$L(x) = \ln \frac{1 + \sqrt{1 + x^2}}{1 - \sqrt{1 - x^2}}, \qquad (x < 1),$$

we introduce the new functional form

$$K(x) = \begin{cases} \dfrac{1}{2} \displaystyle\int_{-1}^{1} \dfrac{dy}{(x^2 - 1)y^2 + 1} = \dfrac{1}{\sqrt{x^2 - 1}} \arctan \sqrt{x^2 - 1}, & x > 1, \\[2mm] 1, & x = 1, \\[2mm] \dfrac{1}{2\sqrt{1 - x^2}} \ln \dfrac{1 + \sqrt{1 - x^2}}{1 - \sqrt{1 - x^2}} & x < 1. \end{cases}$$

$$\tag{A1}$$

This function has the property that

$$\frac{d}{dx} K(x) = \frac{1}{x(x^2 - 1)} - \frac{x}{x^2 - 1} K(x). \tag{A2}$$

We show below the trial wave functions as well as the energy forms in terms of $K(x)$: the variational parameters are a and b, with use of $\varepsilon \equiv a/b$.

(i)   In case of the Yafet–Keyes–Adams-type approach

$$\psi_{1s} = \exp[-(\rho^2/4a^2) - (z^2/4b^2)], \tag{A3}$$

$$E_{1s} = \frac{1}{2a^2} + \frac{\alpha}{4b^2} + \frac{\gamma^2}{2} a^2 - \sqrt{\frac{2}{\pi} \frac{2}{b}} K(\varepsilon);$$

where

$$a = \frac{\alpha}{4} \sqrt{\frac{\pi}{2}} \, \varepsilon(\varepsilon^2 - 1)[1 - K(\varepsilon)]^{-1}, \tag{A4}$$

$$\gamma^2 a^4 - 1 - \frac{\alpha}{2} \varepsilon^2 + 2\sqrt{\frac{2}{\pi}} \varepsilon a K(\varepsilon) = 0,$$

$$\alpha \equiv m_c^*/m_\parallel^*;$$

and

$$\psi_{2p-} = \rho \exp[-(\rho^2/4a^2) - (z^2/4b^2)]e^{i\phi}, \tag{A5}$$

$$E_{2p-} = \frac{1}{a^2} + \frac{\alpha}{4b^2} + \gamma^2 a^2 + \gamma$$

$$- \sqrt{\frac{2}{\pi}} \frac{1}{b(\varepsilon^2 - 1)} [1 + (\varepsilon^2 - 2)K(\varepsilon)]; \tag{A6}$$

where

$$a = \frac{\alpha}{4} \sqrt{\frac{\pi}{2}} \varepsilon(\varepsilon^2 - 1)^2 [(\varepsilon^2 + 2)K(\varepsilon) - 3]^{-1},$$

$$2\gamma^2\alpha^4 - 2 - \frac{\alpha}{2}\varepsilon^2 + \sqrt{\frac{2}{\pi}} a \frac{\varepsilon}{\varepsilon^2 - 1} [1 + (\varepsilon^2 - 2)K(\varepsilon)] = 0.$$

(ii)   In case of the Kohn–Luttinger-type approach

$$\psi_{1s} = \exp[-(\rho^2/4a^2) - (z^2/4b^2)^{1/2}], \tag{A7}$$

$$E_{1s} = \frac{1}{6a^2} + \frac{\alpha}{12}\frac{1}{b^2} + 2\gamma^2 a^2 - \frac{1}{b}K(\varepsilon); \tag{A8}$$

where

$$a = \frac{\alpha}{6}\varepsilon(\varepsilon^2 - 1)[1 - K(\varepsilon)]^{-1},$$

$$4\gamma^2 a^4 - \frac{1}{3} - \frac{\alpha}{6}\varepsilon^2 + a\varepsilon K(\varepsilon) = 0;$$

and

$$\psi_{2p-} = \rho \exp[-(\rho^2/4a^2) - (z^2/4b^2)^{1/2}]e^{i\phi}, \tag{A9}$$

$$E_{2p-} = 6\gamma^2 a^2 + \frac{1}{5a^2} + \frac{\alpha}{20}\frac{1}{b^2} + \gamma$$

$$- \frac{3}{8}\frac{1}{b(\varepsilon^2 - 1)} [1 + (\varepsilon^2 - 2)K(\varepsilon)]; \tag{A10}$$

where

$$a = \frac{4}{15}\alpha\varepsilon(\varepsilon^2 - 1)^2 [(\varepsilon^2 + 2)K(\varepsilon) - 3]^{-1},$$

$$12\gamma^2 a^4 - \frac{2}{5} - \frac{\alpha}{10}\varepsilon^2 + \frac{3}{8} a \frac{\varepsilon}{\varepsilon^2 - 1} [1 + (\varepsilon^2 - 2)K(\varepsilon)] = 0.$$

ACKNOWLEDGMENTS

The author is indebted to T. Ohyama, K. L. I. Kobayashi, K. Fujii, T. Yoshihara, O. Matsuda, H. Nakata, Y. Okada, I. Honbori and many other collaborators in his laboratory for cooperating in making this article.

REFERENCES

Alekseev, A. S., Bagaev, V. S., Galkina, T. I., Gogolin, O. V., and Penin, N. A. (1970. *Fiz. Tverd. Tela* **12**, 3516–3520 [*Engl. transl.: Sov. Phys.–Solid State* **12**, 2855–2861].
Alfven, H. (1950). "Cosmical Electrodynamics." Oxford Univ. Press, London and New York.
Benoit à la Guillaume, C., and Voos, M. (1972). *Solid State Commun.* **11**, 1585–1588.
Brooks, H. (1951). *Phys. Rev.* **83**, 879.
Burstein, E., Picus, G. S., and Gebbie, H. A. (1956). *Phys. Rev.* **103**, 825–826.
Button, K. J., Gebbie, H. A., and Lax, B. (1966). *IEEE J. Quantum Electron.* **9E2**, 206–207.
Conwell, E., and Weisskopf, V. F. (1950). *Phys. Rev.* **77**, 388–390.
Dornhaus, E., Happ, H., Müller, H., Nimtz, G., Schablitz, N., Zorplinski, P., and Bauer, G. (1974). Proc. Int. Conf. Phys. Semicond., *12th Stuttgart* (M. H. Pilkuhn, ed.), pp. 1157–1161. Teubner, Leipzig.
Dresselhaus, G., Kip, A. F., and Kittel, C. (1955). *Phys. Rev.* **98**, 368–384.
Elkomoss, S. G., and Munschy, G. (1977). *J. Phys. Chem. Solids* **38**, 557–563.
Erginsoy, C. (1950). *Phys. Rev.* **79**, 1013–1014.
Fujii, K., and Otsuka, E. (1975). *J. Phys. Soc. Jpn.* **38**, 742–749.
Fujii, K., and Otsuka, E. (1976). *In* "Molecular Spectroscopy of Dense Phases" (M. Grosmann, S. G. Elkomoss, and J. Ringeissen, eds.), pp. 143–146. Elsevier, Amsterdam.
Fukai, M., Kawamura, H., Sekido, K., and Imai, I. (1964). *J. Phys. Soc. Jpn.* **19**, 30–39.
Gavrilenko, V. I., Kononenko, V. L., Mandel'shtam, T. S., and Murzin, V. N. (1976). *Pis'ma Zh. Eksp. Teor. Fiz.* **23**, 701–704 [*English transl.: Sov. Phys.–JETP Lett.* **23**, 645–648].
Gornik, E. (1972). *Phys. Rev. Lett.* **29**, 595–597.
Gornik, E., and Tsui, D. C. (1978). *Solid State Commun.* **21**, 139–142.
Gourley, P. L., and Wolfe, J. P. (1978). *Phys. Rev. Lett.* **40**, 526–530.
Hasegawa, H., and Howard, R. E. (1961). *J. Phys. Chem. Solids* **21**, 179–198.
Hensel, J. C., Phillips, T. G., and Thomas, G. A. (1977). *In Solid State Phys.* **32**, 87–314.
Honbori, I. (1979). Master Thesis, Graduate School of Science, Osaka Univ., unpublished.
Kacman, P., and Zawadzki, W. (1971). *Phys. Status Solidi (b)* **47**, 629–642.
Kaplan, R. (1969). *Phys. Rev.* **181**, 1154–1162.
Kawabata, T., Muro, K., and Narita, S. (1977). *Solid State Commun.* **23**, 267–270.
Kawamura, H., Hayashi, Y., and Fukai, M. (1961). *J. Phys. Soc. Jpn.* **16**, 2352.
Kawamura, H., Fukai, M., and Hayashi, Y. (1962). *J. Phys. Soc. Jpn.* **17**, 970–974.
Kawamura, H., Saji, H., Fukai, M., Sekido, K., and Imai, I. (1964). *J. Phys. Soc. Jpn.* **19**, 288–296.
Kazarinov, R. F., and Skobov, V. G. (1962). *Zh. Eksp. Teor. Fiz.* **42**, 1047–1053 [*English transl.: Sov. Phys.–JETP* **15**, 726–730].
Kobayashi, K. L. I. (1973). Ph.D. Thesis, Graduate School of Science, Osaka Univ., unpublished.
Kobayashi, K. L. I., and Otsuka, E. (1974). *J. Phys. Chem. Solids* **35**, 839–849.
Kobayashi, K. L. I., Otsuka, E., Takeuchi, N., and Yajima, T. (1971). *Jpn. J. Appl. Phys.* **10**, 1704–1709.
Kobayashi, K. L. I., Komatsubara, K. F., and Otsuka, E. (1973). *Phys. Rev. Lett.* **30**, 702–705.
Kohn, W., and Lutinger, J. M. (1955). *Phys. Rev.* **98**, 915–922.

Komatsubara, K. F., and Yamada, E. (1966). *Phys. Rev.* **144**, 702–707.

Kosai, K., and Gershenzon, M. (1974). *Phys. Rev.* **139**, 723–736.

Kotera, N., Yamada, E., and Komatsubara, K. F. (1972). *J. Phys. Chem. Solids* **33**, 1311–1324.

Kulakovsky, V. D., and Timofeev, V. B. (1977). *Pis'ma Zh. Eksp. Teor. Fiz.* **25**, 487–491 [*English transl.: Sov. Phys.–JETP Lett.* **25**, 458–461].

Kurosawa, T. (1965). *J. Phys. Soc. Jpn.* **20**, 937–942.

Kurosawa, T., and Yamada, E. (1973). *J. Phys. Soc. Jpn.* **24**, 603–612.

Landau, L. D. (1957). *Zh. Eksp. Teor. Fiz.* **32**, 59–66 [*English transl.: Sov. Phys.–JETP* **5**, 101–111].

Lax, B., Zeiger, H. J., and Dexter, R. N. (1954). *Physica* **20**, 818–828.

Markiewicz, R. S., Wolfe, J. P., and Jeffries, C. D. (1974). *Phys. Rev. Lett.* **32**, 1357–1360.

Martin, R. W., Störmer, H. L., Rühle, W., and Bimberg, D. (1977). *J. Luminescence* **12/13**, 645–649.

Matsuda, O. (1979). Ph.D. Thesis, Graduate School of Osaka Univ., unpublished.

Matsuda, O., and Otsuka, E. (1978). *Solid State Commun.* **26**, 925–928.

Matsuda, K., Hirooka, M., and Sunakawa, S. (1975). *Prog. Theoret. Phys.* **54**, 79–92.

Miyazawa, H. (1969). *J. Phys. Soc. Jpn.* **26**, 700–709.

Miyazawa, H., and Ikoma, H. (1967). *J. Phys. Soc. Jpn.* **23**, 290–305.

Murase, K., and Otsuka, E. (1968). *J. Phys. Soc. Jpn.* **25**, 436–442.

Murase, K., and Otsuka, E., (1969). *J. Phys. Soc. Jpn.* **26**, 413–420.

Murotani, T., and Nishida, Y. (1972). *J. Phys. Soc. Jpn.* **32**, 986–998.

Nakata, H., Fujii, K., and Otsuka, E. (1978). *J. Phys. Soc. Jpn.* **45**, 537–544.

Nakata, H., Fujii, K., Ohyama, T., and Otsuka, E. (1979). *J. Magn. Magn. Mat.* **11**, 127–130.

Norton, P. (1976). *J. Appl. Phys.* **47**, 308–320.

Ohyama, T. (1978). *Inst. Phys. Conf. Ser.* **43**, 375–378.

Ohyama, T., Murase, K., and Otsuka, E. (1970). *J. Phys. Soc. Jpn.* **29**, 912–924.

Ohyama, T., Yoshihara, T., Sanada, T., Murase, K., and Otsuka, E. (1971). *Phys. Rev. Lett.* **27**, 33–34.

Ohyama, T., Sanada, T., and Otsuka, E. (1973). *J. Phys. Soc. Jpn.* **35**, 822–825.

Ohyama, T., Hansen, A. D. A., and Turney, J. L. (1976). *Solid State Commun.* **19**, 1083–1086.

Oka, Y., and Narita, S. (1970). *J. Phys. Soc. Jpn.* **28**, 674–683.

Otsuka, E., Murase, K., Iseki, J., and Ishida, S. (1964). *Phys. Rev. Lett.* **13**, 232–233.

Otsuka, E., Murase, K., and Iseki, J. (1966a). *J. Phys. Soc. Jpn.* **21**, 1104–1111.

Otsuka, E., Murase, K., and Yamaguchi, K. (1966b). *J. Phys. Soc. Jpn.* **21**, 1249–1254.

Otsuka, E., and Yamaguchi, K. (1967). *J. Phys. Soc. Jpn.* **22**, 1183–1190.

Otsuka, E., Ohyama, T., and Murase, K. (1968). *J. Phys. Soc. Jpn.* **25**, 729–739.

Otsuka, E., Fujii, K., and Kobayashi, K. L. I. (1973). *Jpn. J. Appl. Phys.* **12**, 1600–1605.

Otsuka, E., Okada, Y., Fujii, K., and Sakakima, H. (1976). *Sci. Rep. Osaka Univ.* **25/2**, 1–10.

Otsuka, E., Ohyama, T., Nakata, H. and Okada, Y. (1977). *J. Opt. Soc. Am.* **67**, 931–935.

Partl, H., Müller, W., Kohl, F., and Gornik, E. (1978). *J. Phys. C Solid State Phys.* **11**, 1091–1103.

Pearson, G. L., and Bardeen, J. (1949). *Phys. Rev.* **75**, 865–883.

Pokrovskii, Ya. (1972). *Phys. Status Solidi (a)* **11**, 385–410.

Sauer, R. (1973). *Phys. Rev. Lett.* **31**, 376–379.

Sauer, R., and Weber, J. (1976). *Phys. Lett.* **36**, 48–51.

Sauer, R., Schmidt, W., and Weber, J. (1977). *Solid State Commun.* **24**, 507–509.

Sekido, K., Fukai, M., and Kawamura, H. (1964). *J. Phys. Soc. Jpn.* **19**, 1579–1586.

Skolnick, M. S., Carter, A. C., Couder, Y., and Stradling, R. A. (1977). *J. Opt. Soc. Am.* **67**, 947–951.

Smith, D. L. (1976). *Solid State Commun.* **18**, 637–639.

Suzuki, K., and Hensel, J. C. (1974). *Phys. Rev. B* **9**, 4184–4218.

Thewalt, M. L. W. (1977). *Can. J. Phys.* **55**, 1463–1479.

Van de Hulst, H. C. (1957). "Light Scattering by Small Particles." Wiley, New York.

Waldman, S., Chang, T. S., Fetterman, H. R., Stillman, G. E., and Wolfe, C. M. (1974). *Solid State Commun.* **15**, 1309–1312.

Wallis, R. F., and Bowlden, H. J. (1958). *J. Phys. Chem. Solids* **7**, 78–89.

Wolfe, J. P., Markiewicz, R. S., Kittel, C., and Jeffries, C. D. (1975). *Phys. Rev. Lett.* **34**, 275–277.

Wolfe, J. P., Markiewicz, R. S., Kelso, S. M., Furneaux, J. E., and Jeffries, C. D. (1978). *Phys. Rev. B* **18**, 1479–1503.

Yafet, Y., Keyes, R. W., and Adams, E. N. (1956). *J. Phys. Chem. Solids* **1**, 137–142.

Yamada, E., and Kurosawa, T. (1973). *J. Phys. Soc. Jpn.* **24**, 603–612.

Yamanaka, M., Muro, K., and Narita, S. (1978). *J. Phys. Soc. Jpn.* **44**, 1222–1230.

Yoshihara, T. (1971). Master Thesis, Graduate School of Science, Osaka Univ., unpublished.

Index